Wi-Fi 7 In Depth

Wi-Fi 7 In Depth

Your guide to mastering Wi-Fi 7, the 802.11be protocol, and their deployment

Jerome Henry
Brian Hart
Binita Gupta
Malcolm Smith

♦♦Addison-Wesley

Library of Congress Control Number: 2024945657

ISBN-13: 978-0-13-532361-8
ISBN-10: 0-13-532361-4

1 2024

GM K12, Early Career and Professional Learning
Soo Kang

Director, ITP Product Management
Brett Bartow

Executive Editor
Nancy Davis

Development Editor
Christopher Cleveland

Managing Editor
Sandra Schroeder

Production Editor
Mary Roth

Copy Editor
Jill Hobbs

Indexer
Timothy Wright

Proofreader
Donna E. Mulder

Technical Reviewer
Samuel Clements

Interior Designer
codeMantra

Cover Designer
Chuti Prasertsith

To my amazing wife Michelle and son Richard,
for their love and support.
—Brian Hart

To my husband Shashank for his unwavering support, love,
and encouragement every step of this journey. To my son Vayun,
for being the joy and sunshine in my life. And to my parents,
who taught me to dream big and persevere.
—Binita Gupta

Contents at a Glance

Contents

Chapter 3: Building on the Wi-Fi 6 Revolution   58

Chapter 4: The Main Ideas in 802.11be and Wi-Fi 7   98

Chapter 6: EHT MAC Enhancements for Multi-Link Operation   **176**

Chapter 7: EHT MAC Operation and Key Features   222

Chapter 8: Wi-Fi 7 Network Planning   268

Chapter 9: Future Directions

Index

Foreword

As I sat on a plane waiting for the rest of the passengers to board, a mother and her young son of about 4 years came down the aisle. The little boy had some gaming device in his hand, and as they passed my seat, I heard the boy say, "Mom, does the plane have Wi-Fi?" He didn't know how the plane connected to satellite and/or terrestrial systems, and he didn't understand the cloud that supported his device. He just knew that if there was Wi-Fi his device worked, and without it … well, it didn't.

In the ever-evolving landscape of wireless technology, Wi-Fi has stood as a beacon of innovation and a facilitator of the unceasing demand for connectivity. Predictably, the thirst for wireless connectivity is driving the need for expanded bandwidth. Yet, the requirements of next-generation connectivity extend beyond this, calling for a broader set of enhancements. Can wireless be predictable? Can it be "deterministic"? We are on a journey to ever strive toward these goals. Wi-Fi 7 is the next step. Of course, Wi-Fi 7 offers more capacity, but it is more: It offers new ways to protect against interference, new ways to improve reliability, and new capabilities to bound worse-case latency. This book is a comprehensive guide to understanding, implementing, and optimizing Wi-Fi 7, the latest iteration in a lineage of wireless protocols that have shaped the modern world.

The importance of Wi-Fi 7 cannot be overstated. Studies from the Wi-Fi Alliance and the Wireless Broadband Alliance have underscored its significance, highlighting how this new standard will address the burgeoning need for a more robust, efficient, and versatile wireless experience. Wi-Fi 7 is poised to redefine the boundaries of what is possible, offering unprecedented speeds, reduced latency, and greater capacity. It is a response to the insatiable appetite for data and the increasing complexity of wireless environments, from smart homes and offices to entire smart cities.

The authors of this book have been at the forefront of Wi-Fi's evolution. Their leadership in innovative development and contributions to standards has been nothing short of amazing. They have not only witnessed but also shaped the transformation of Wi-Fi, infusing the industry with their vision and expertise. Through their work, they have ensured that Wi-Fi remains a vital, dynamic force in the technological landscape.

As you embark on this journey through the pages of this book, you are invited to explore the intricacies of Wi-Fi 7, to understand its significance, and to appreciate the dedication of those who have brought it to life. Whether you are a network engineer, a product manager, a researcher, or simply a technology enthusiast, this book will serve as an invaluable resource, illuminating the path to a more connected and capable world.

Welcome to the era of Wi-Fi 7. Let us begin.

—Matt MacPherson
CTO of Wireless, Cisco Systems

Preface

In many ways, this century resembles Wi-Fi. Every few years, new developments fundamentally change the way we work and communicate. Each time we look back a few years, we realize that today we have more information to absorb and more new technologies at our disposal. What was once concluded to be impossible is now experimented with or achieved sooner and faster than we thought. The development of Wi-Fi seems to follow the same trends. Six years separated Wi-Fi 4 (802.11n) from Wi-Fi 5 (802.11ac), but then 802.11ac Wave 2, Wi-Fi 6 (802.11ax), Wi-Fi 6E, and now Wi-Fi 7 followed each other at intervals of just a few years. Each time, it seems that the next generation solves major problems that the previous generations surprisingly ignored. This fast pace means that some features that were heralded as fundamental changes end up being barely adopted by the vast majority of the market actors.

Wi-Fi 7, and its underlying standard IEEE 802.1be, have been caught up in this trend. Wi-Fi 6E made news by bringing 802.11 to a new, legacy-free segment of the spectrum. This was the largest spectrum addition in the history of Wi-Fi, and the absence of legacy devices in this new band allowed Wi-Fi to implement a multiplicity of efficient features. Reading the press reports, it might seem as if, by contrast, Wi-Fi 7 is just about adding Multi-Link Operations (MLO). This ability to establish multiple active links to an AP device is certainly nice, but limited to a single AP device. It also resembles the "dual connection" feature that many vendors had been enabling for several years.

This reading of the merits of each generation misses the forest for a few (headline-worthy) trees. Wi-Fi has now established itself as *the* access technology of choice. All year round, new trends, new challenges, and new possibilities are presented to the IEEE Wireless Next Generation (WNG) Standing Committee. Each issue that seems to become mainstream is selected to be solved as part of the next 802.11 amendment. The 802.11be project was started at a time when myriad new applications were appearing on the market, such as wireless Time-Sensitive Networking (TSN) in the industrial Internet of Things (IoT), cloud gaming, and the trio of augmented reality (AR), virtual reality (VR), and extended reality (XR). These applications not only require high bandwidth, but also are very sensitive to delay and jitter (the variation in delays between two subsequent frames or packets). No Wi-Fi system could reasonably claim to support such applications, especially at the high-density scale of 30 or more headsets connected to the same access point in a classroom. At the same time, video-conferencing had become common, and new IoT applications—self-driving robots and the like—promised better efficiency, albeit at the cost of a near-zero tolerance for packet losses. The 802.11 standard, designed during the years when the main philosophy was "best effort," could readily accept retry rates of 10%, but was widely incompatible with any industrial requirements.

The 802.11be amendment and its associated Wi-Fi Alliance certification (Wi-Fi 7 release 1 launched in January 2024, and release 2 is targeted to launch by the end of 2026) were designed to solve these challenges and bring efficiency at the scale that these new applications demand. However, the solutions in the 802.11 standard are many and must be implemented in combination with each other. Some of the solutions are also complex, even when their purpose can be described in just a few words. For example, MLO, including the potential to send and receive at the same time, supposes a receive circuit located near a transmit circuit that operates at a power level several billion times higher. This engineering challenge results in multiple types of MLO devices, ranging from those that can truly send and receive at the same time to those that really have only one radio, with multiple hybrid forms in between. They will all be labeled MLO, with some additional acronyms in small print revealing precisely what they can and cannot do. Similarly, some features solve a given problem, but only in some scenarios; thus, they will act as stepping stones until the next generation (likely Wi-Fi 8) adds the elements needed to solve the issue in all cases.

In other words, 802.11be and Wi-Fi 7 can be complex, and their successful implementation requires a deep dive into the details. As we were finalizing the last elements of Wi-Fi 7, we concluded that a book was necessary—one that would explain the main features of 802.11be; detail what Wi-Fi 7 certification includes and what it leaves aside; and offer sufficient technical background and information to help you master each aspect of 802.11be and Wi-Fi 7, while recognizing both their potential and their limitations, and providing insights into why those limitations exist.

Audience and Background

This book is about 802.11be and Wi-Fi 7. Although the early chapters provide the fundamental information necessary to understand the 802.11 background upon which 802.11be is built, you still need to be familiar with 802.11 basics and the core ideas behind the 802.11be precursor amendments (802.11n, 802.11ac, and 802.11ax). You should also be familiar with Wi-Fi deployments, or at least the process of deploying access points and achieving satisfactory coverage.

The primary audience for this book is networking professionals who work with Wi-Fi and need to understand 802.11be and Wi-Fi 7 in detail, so as to integrate Wi-Fi 7 devices into an existing wireless network or to manage a Wi-Fi 7 network. If your task is to design or deploy a Wi-Fi 7 network, either as an overlay to an existing network or as a brand-new (greenfield) deployment, this book will provide the information you need to be successful. If you are a technical person and just curious about the inner workings of 802.11 (and, in particular, the 802.11be generation), this book will give you ample information on the key processes that make Wi-Fi work the way it works, from a small home to a large enterprise or industrial network. If you are working with 802.11 and unsatisfied by the prospect of reading countless resources that paraphrase the standard without truly explaining it, this book should bring your smile back: We took care, while explaining the main 802.11 features, to detail the context in which they were designed, so you can see the intentions and constraints as you read about the chosen mechanism.

Structure

We strongly believe that understanding 802.11be and Wi-Fi 7 is not merely about dissecting the frames added with the 802.11be amendment (although this dissection is definitely important). That is, we think it is also critical to integrate these frames and associated exchanges in the landscape where they were found. This second component is often difficult to accomplish if you have not participated directly in the IEEE 802.11 work streams, because the frames are just a vehicle to achieve a goal that was discussed over multiple years, in countless video-conferencing calls and face-to-face meetings in small or large committees. All the exchanges and the context that led to the choice of a particular frame are clear in the minds of the few hundreds (sometimes few dozens) of people who contributed to a given section of the amendment, but invisible to all others. If you open 802.11be for the first time, and if you have not participated in IEEE work for the previous generations of 802.11 amendments, the context may be missing. The vocabulary you use may not have the exact same meaning as it has in an IEEE context.

Chapter 1, "Wi-Fi Fundamentals," is intended to be a catch-up chapter. The bedrock of the 802.11be amendment is the assumptions and mechanisms found in the more than 6000 pages of the 802.11 baseline standard. Chapter 1 outlines the foundations upon which 802.11be is built, along with the vocabulary you need to understand the 802.11be building blocks. As the responses of the first readers of our manuscript revealed, 802.11 and networking professionals may also find useful information in this chapter, as several mechanisms and terms are only partially defined in the text of the 802.11 standard.

The 802.11 standard is built to support backward compatibility. The difficult task of inventing improvements while ensuring that older devices still perform as intended means that the construction of a new amendment is a delicate exercise, where any imbalance can risk toppling the entire edifice. Chapter 2, "Reaching the Limits of Traditional Wi-Fi," examines performance improvements that were made over several generations of 802.11 advances. The chapter also details their limitations and challenges, which serve as the starting point for the 802.11be work. This chapter will help you better understand 802.11 roaming, quality of service (QoS), and security, but also appreciate where these enhancements break.

The 802.11be amendment expanded on features developed with 802.11ax and the associated Wi-Fi 6 and Wi-Fi 6E programs. Chapter 3, "Building on the Wi-Fi 6 Revolution," examines these programs in detail, because Wi-Fi 6 fundamentally changed the way clients could be scheduled. In turn, porting 802.11 into the 6 GHz spectrum opened the door to new enhancements that were simply not possible in other bands.

The goal of Chapter 4, "The Main Ideas in 802.11be and Wi-Fi 7," looks into the main ideas of 802.11be and Wi-Fi 7 is not to provide details of the frames and the exchanges, but rather to give you a clear view of the context, the challenges that 802.11be attempted to solve, and the solution that became part of the Wi-Fi 7 program.

Chapter 5, "EHT Physical Layer Enhancements," then dives into the protocol, focusing on the Physical layer (PHY) aspects of 802.11be and Wi-Fi 7. Most features have a PHY and a MAC component, but their implications are different. This chapter details the changes in transmission and reception techniques and in the way the channel is used.

Chapter 6, "EHT MAC Enhancements for Multi-Link Operation," then goes up one layer and focuses on the MAC layer. This chapter is about MLO and the related protocols and procedures. Although MLO is a major component of the 802.11be/Wi-Fi 7 programs, all too often this core element is presented in a more narrow or incomplete way in the literature. Chapter 6 aims to remedy this problem.

Chapter 7, "EHT MAC Operation and Key Features," then examines the other MAC features in 802.11be and Wi-Fi 7, including the critical QoS and priority scheduling mechanisms. Each of them could have been the subject of a separate chapter, but in combination the elements in Chapter 4 (context and goals), Chapter 5 (PHY aspects), and this chapter should give you all the information you need to understand them in depth.

The many new elements and mechanisms brought forward by 802.11be raise questions about the coverage model. Can you design a Wi-Fi 7 network the way you designed Wi-Fi 6 and earlier networks? How do you measure the performance of a Wi-Fi 7 device? Chapter 8, "Wi-Fi 7 Network Planning," dives into these questions, helping you become comfortable with Wi-Fi network planning, surveys, deployment, and performance assessment.

Although 802.11be brings forward many features, its designers were well aware that some questions would be left unanswered. Chapter 9, "Future Directions," examines where the standard stopped and provides a glimpse into where the conversations are going for the next generation after 802.11be. This chapter will give you an idea of what Wi-Fi 8 might look like.

Acknowledgments

This book would not have been possible without the whole Pearson team supporting us, the authors. We are grateful to Nancy Davis, who gave us the opportunity to work on this project. We know that a big challenge for such a book is in the choice of authors. The publisher needs to trust that, although they probably have the technical expertise needed, those authors will also be able to convey the intent and the complexities of the technology and protocols in a way that can be used by readers in need of precise yet practical information.

Many of our colleagues and industry friends accepted the challenge of reading and reviewing chapters and even entire parts of the book. Their feedback helped us turn our collection of nuggets of knowledge into a usable flow. Chris Cleveland then made sure that our thoughts would be structured in a logical and usable way, and that the message we conveyed would be crisp and clear.

Sam Clements, whose 802.11 expertise and attention to details are immense, accepted our invitation to join us. As our first external reader, he made countless technical edits to make the content more accurate and easier to read.

Jill Hobbs then took the time to scrutinize each sentence, each expression, and each table. She made sure that their meaning would be accurate and understandable without ambiguity, irrespective of the cultural and language background of the reader.

Behind these people, many others invested time and effort to make this book possible and accessible. Pearson team, thank you for giving us the tools to share what we learned!

About the Authors

Jerome Henry is a Distinguished Engineer in the Office of the Wireless CTO group at Cisco Systems. He is a Certified Wireless Networking Expert (CWNE No. 45) and Wireless CCIE (No. 24750), and has authored multiple books and video courses focusing on wireless technologies, ranging from introductory to expert levels. Jerome has been a member of the IEEE since 2006, and was elevated to Senior Member in 2013. He has contributed to the development of multiple amendments, has been in leadership positions of several Task Groups, and received an award for his contributions to the 802.11-2020 revision of the standard. Jerome has authored multiple research papers in the wireless technology field, and participates in multiple Wi-Fi Alliance working groups, with a strong focus on wireless efficiency, IoT, and low power applications. As an educator, he has more than 10,000 hours of experience in the classroom, and was awarded the IT Training Award Best Instructor silver medal. He is based in Research Triangle Park, North Carolina.

Brian Hart is a Principal Engineer in the Cisco Wireless group. Brian has been working in wireless since 1996, after graduating with a PhD in wireless physical layer signal processing from the University of Canterbury, New Zealand. He authored many peer-reviewed research papers on receiver design for time-varying channels while working as a Research Fellow and Fellow at the Institute of Advanced Studies at the Australian National University. In 2000, he joined the pioneering Wi-Fi startup Radiata Communications, which was acquired by Cisco in 2001. At Cisco, Brian was the PHY systems engineer for the pioneering 802.11a/b/g baseband processing ASIC used in Cisco's AP1131 and AP1252 Wi-Fi APs. Brian has been an IEEE 802.11 Working Group voting member since 2006 and has made hundreds of contributions across tens of groups, including 802.11ac, 802.11ax, and 802.11be. He was elected co-chair of the VHT PHY and HE MAC ad hoc groups. He is also active at the Wi-Fi Alliance on related feature selection and certification activities. Brian has performed detailed simulation modeling of most 802.11 PHY generations, with selected prototyping, and helped productize advanced location algorithms. He is based in the San Francisco Bay Area in California.

Binita Gupta is a Senior Technical Leader in Wireless Engineering group at Cisco Systems. She has more than 20 years of experience leading technology development of various wireless systems, products, and services across Wi-Fi, 5G, private networks, and IoT. Before joining Cisco in 2023, Binita served as a wireless systems architect at Meta, Intel, and Qualcomm, driving wireless system design for products, and leading standards development in IEEE 802.11, Wi-Fi Alliance (WFA), and Wireless Broadband Alliance (WBA). In her role as a voting member at IEEE 802.11, Binita has been deeply involved in the development of the 802.11be (Wi-Fi 7) and 802.11bn ("Wi-Fi 8") amendments and has made numerous contributions in those Task Groups. She also participates in WFA Working Groups and has made contributions to the Wi-Fi 7, QoS Management, and EasyMesh certification programs. Additionally, she has led Wi-Fi and 5G convergence projects at the WBA as Chair, Co-Chair, and Chief Editor. Binita holds a master's degree in computer science from the University of Illinois at Urbana–Champaign, and a bachelor of technology degree in computer science from IIT, Kharagpur, India. She is based in San Diego, California.

Malcolm Smith is a Principal Engineer at Cisco Systems and serves as a wireless advisor in Cisco's Networking CTO office. Malcolm leads Wi-Fi university research and related strategy, innovation, and partner ecosystem development. Malcolm has 30 years of wireless experience, including leading technology development, marketing, and productization of various generations of both 802.11-based Wi-Fi and 3GPP-based cellular radio systems. Malcolm has received several awards, including Cisco's Pioneer award for what is now WBA Open Roaming. He is the Chair and Co-Chair of the Industrial IOT and Wi-Fi 7 Working Groups, respectively, within the WBA and Co-Chair of the XR MSTG within the WFA. He is based in Dallas, Texas.

Chapter **1**

Wi-Fi Fundamentals

basic service set: [BSS] A set of stations (STAs) that have successfully synchronized using the MLME-JOIN.request service primitive and one STA that has used the MLME-START.request primitive.

802.11-2020, Clause 3

From its very beginning, the IEEE 802.11be project toyed with concepts that questioned the very foundations of Wi-Fi. For example, how can a single client device associate to two Access Point (AP) devices' radios? How do the AP and client devices manage the security of the association in that case? How would frames be reassembled if they are split over several, unsynchronized radio links? The answers to these questions are not obvious, because they have to be built upon six previous generations of a Wi-Fi standard that takes backward compatibility as a serious requirement.

To help you understand the design decisions that will be exposed further in Chapter 4, "The Main Ideas in 802.11be and Wi-Fi 7," and later (and in some way, in Chapter 3, "Building on the Wi-Fi 6 Revolution," as well), this chapter introduces the fundamental concepts of 802.11 and Wi-Fi—at least those that are relevant to a clear understanding of 802.11be and Wi-Fi 7. Some of the ideas summarized in the sections that follow will undoubtedly sound very familiar to you, but you might find yourself coming back to these paragraphs as you read the chapters on Wi-Fi 7, as these fundamental concepts create strict constraints that limited the freedom of the 802.11be designers.

Evolution of 802.11 Standards

Before jumping into the details of the 802.11 architecture, it might be useful to look at a map of the 802.11 amendments. The original 802.11 standard was published in 1997. Since then, multiple initiatives have contributed to deeply modifying and improving the standard. In most cases, these efforts have taken the initial form of a Topic Interest Group (TIG) that studies new use cases or developments that may affect the standard. The TIG can then turn into a Study Group, and subsequently into a Task Group (TG), whose goal is to define the modifications to the standard needed to fulfill the use cases defined by the TIG. Each TG is assigned a code letter, in lowercase and alphabetical successive order (e.g., 802.11a, 802.11b, . . . , 802.11aa, 802.11ab, . . .). Some letters or combinations of letters are skipped if they are associated with other well-known acronyms or could cause confusion.

Historically, some TGs have targeted improvements to the Physical layer (PHY) and the MAC layer (MAC) to increase throughput, bandwidth, and/or speed. They have often brought other features along the way, but are most known for these improvements. Table 1-1 lists these task groups.

TABLE 1-1 802.11 Task Groups Known for Speed or Bandwidth Improvements

TG Code Name	Project Start Year	Amendment Publication Year	Some Key Improvements
802.11	1991	1997	Up to 2 Mbps in 2.4 GHz, DSSS modulation over 20 MHz channels
802.11b	1997	1999	Up to 11 Mbps in 2.4 GHz, CCK modulation over 20 MHz channels
802.11a	1997	1999	Up to 54 Mbps in 5 GHz, OFDM modulation over 20 MHz channels
802.11g	2000	2003	Up to 54 Mbps in 2.4 GHz, OFDM modulation over 20 MHz channels
802.11n	2003	2009	Up to 600 Mbps over up to 40 MHz channels, in 2.4 GHz and 5 GHz, with up to 4 spatial streams, MIMO
802.11ac	2008	2013	Up to 6.9 Gbps in 5 GHz over up to 160 MHz channels, 8 spatial streams, downlink multiuser MIMO
802.11ax	2014	2021	Up to 9.6 Gbps over up to 160 MHz channels, in 2.4 GHz, 5 GHz then 6GHz, OFDMA modulation, downlink and uplink multiuser MIMO

Other TGs have focused on improvements related to specific 802.11 functions, such as security or quality of service (QoS), or to specific environments, such as mesh or Japan operations for the 4.9 GHz band. Many of these TGs produced features that are useful in their own right, but not directly related to the topics of this book. Table 1-2 lists the main TGs of interest.

TABLE 1-2 Other 802.11 Task Groups of Interest for This Book

TG Code Name	Project Start Year	Amendment Publication Year	Group Focus
802.11i	2001	2004	Security improvements
802.11e	2000	2005	QoS improvements
802.11k	2002	2008	Radio resources measurements
802.11r	2006	2008	Fast roaming
802.11v	2004	2011	Wireless network management
802.11aa	2008	2012	Video transport streams
802.11ah	2010	2016	Sub 1 GHz operations
802.11ai	2010	2016	Fast initial link setup

Once a task group completes its work, it publishes a document called an amendment (e.g., "the 802.11ax amendment"). At regular intervals, a maintenance group integrates these amendments into the main standard text. The maintenance group is called TGm, followed by a letter. For example, in 2007, TGma published a revision of the standard incorporating all amendments published since the original 802.11-1997 standard (and its 802.11-1999 revision). TGmb integrated more amendments and published 802.11-2012; then TGmc concluded with the publication of 802.11-2016; and TGmd ended with the publication of 802.11-2020. TGme should complete its work in early 2025. In addition to integrating amendments, the maintenance group can incorporate minor fixes or additions to the main body of the 802.11 standard.

The 802.11 Architecture

A key element to understanding the state of the IEEE 802.11 standard, as it leads to 802.11be, is to remember that its inception dates from the 1990s. Over the nearly three decades of its existence, the standard has deprecated only a few features. However, each time this has been a difficult decision to make. The world is a large place and 802.11 is widely adopted. Without the certainty that a given feature is not critical to anyone, or without the clear conclusion that a feature is unwanted (e.g., Wired Equivalent Privacy [WEP], which presented major security risks), deprecating a feature might have unwanted consequences on existing deployments. Therefore, in most cases, new features add to older ones. Concepts that were once adopted are maintained, even if newer technologies suggest that they should be modified or redefined.

Components of 802.11 Architecture

From its very inception, 802.11 defined the station (STA) as the core entity behind 802.11 exchanges. A STA is not a physical device, but rather a logical entity that can be addressed uniquely (i.e., it has a single and unique MAC address) and that provides 802.11 physical (PHY) and Medium Access Control (MAC) functions. The STA can be positioned inside a regular client device (e.g., laptop, smartphone), but the STA (in the 802.11 sense) is not the device itself.

STAs can communicate with each other directly. This direct communication is useful in multiple scenarios, in particular those in which the STAs have a special relationship (when your smartwatch needs to communicate with your smartphone, for example). The relationship is, of course, not limited to a pair: A single STA can communicate with two or more other STAs. This example supposes that the STAs are in range and can communicate with each other—in other words, that each STA sends 802.11 signals that can be successfully received (demodulated and decoded) by the intended other STA(s). The set of STAs that can communicate with each other form a Basic Service Set (BSS), and the area their signals cover forms a Basic Service Area (BSA). When more than two STAs are in the BSS, it is not necessary that all STAs should be able to communicate with all other STAs of the BSS. For example, some STAs may be out of range of other STAs.

These exchanges among STAs are limited to the 802.11 wireless space and medium. To allow STAs to communicate beyond that space, a common practice is to implement in the BSS a STA that incorporates a connection to the Local Area Network (LAN). The system and interface allowing such connection is called a Distribution System (DS). The LAN is defined by 802 standards as a network of devices covering a limited geographic area. In most implementers' minds, the LAN is a wired network (e.g., Ethernet), but there is no such requirement in the 802.11 definition of the DS. As long as the DS allows for the extension of the communication beyond the BSS, it fulfills its purpose. Practically, the DS is a logical concept, connecting the STA to a portal behind which a wider network operates. A STA that implements an interface to the DS and allows other STAs in the BSS to access the DS is called an Access Point (AP). Figure 1-1 illustrates the BSS, non-AP STA, and AP concepts.

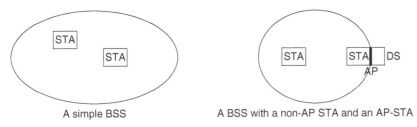

A simple BSS   A BSS with a non-AP STA and an AP-STA

FIGURE 1-1   STA, AP, BSS, and DS

This terminology is not without problems and can cause a lot of confusion. For most Wi-Fi practitioners, the STA is the client in your hand and the AP is that thing on the ceiling that connects to the Ethernet network and also provides 802.11 services. But for the 802.11 standard, an AP is a functional unit that includes a STA and a DS Access Function (DSAF) and enables access to the DS for other STAs over the wireless medium. To distinguish the client STA from the AP STA function, the 802.11 standard sometimes refers to the client STA as a non-AP STA. This book sometimes uses STA to refer to non-AP STAs when the context is clear, and AP for AP-STAs, and non-AP STAs whenever ambiguity is possible. Throughout the 802.11 standard, many additional functions are exclusive to APs, but keep in mind that they are implemented on APs because of the particular position that APs have in the BSS. These features do not define what an AP is.

In enterprise networks (and some homes), it is common to see two or more AP devices connected to the same LAN. When these AP devices present the same configuration (in particular, the same SSID and the same security) so that a client with a given Wi-Fi profile could connect to either of them, they become indistinguishable from each other for the purposes of the 802.11 Logical Link Control (LLC) layer (the upper part of the Data Link layer in 802.11—see the "Network Layers and the 802.11 Frame" section later in this chapter) and the upper layers (including the casual user) on the client (non-AP STA). The term "indistinguishable" is associated to the ESS definition in the standard, and may be confusing for the 802.11 practitioner. The AP devices are different physical objects; their MAC addresses are different. However, they would appear similar to the non-AP STA upper Data Link layer, the LLC. The LLC layer is in charge of the logical link, so it is indifferent to the details of the AP (STA side) link MAC address and other practical details differentiating the connection to AP A from the connection to AP B (if both have the same general configuration with same SSID). In that case, the two or more BSSs having the same SSID and connected by a single DS form an Extended Service Set (ESS), as illustrated in Figure 1-2.

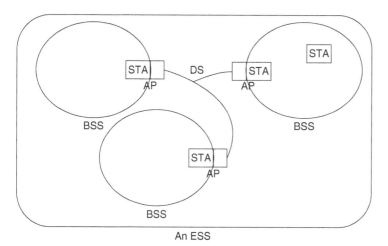

FIGURE 1-2   ESS

The 802.11 standard is sometimes ambiguous on the notion of ESS. As you will see in Chapter 2, "Reaching the Limits of Traditional Wi-Fi," the 802.11 practitioner often has an expectation of coverage continuity between the APs of an ESS. In other words, if you reach the edge of a BSS, lose your connection completely, stay unconnected for a while as you keep moving, and then later regain connectivity in the next BSS, does the BSS pair still form an ESS? Yes, 802.11 does not impose a continuity requirement.[1] Two BSSs can be disjointed (they could be on different continents), but if they are connected by the DS (and have the same SSID and so on), they would still form an ESS. Also note that the ESS is only the 802.11 part, which is the set of all these BSSs connected by a single DS. The DS (and the LAN, perhaps, implementing the DS) is not part of the ESS.

802.11 PHY Schemes

802.11be, like several prior amendments to the 802.11 standard, focuses on improving transmission efficiency (the quantity of informational bits sent per unit of time and channel bandwidth). To understand the enhancements that this amendment brings, you need to understand the basic 802.11 bands and modulations.

802.11 Main Bands

Over its 20-plus years, the 802.11 standard has been defined to operate in many bands: 54/790 MHz,[2] 860/900 MHz, 2.4 GHz, 3.6 GHz, 4.9 GHz, 5 GHz, 5.9 GHz, 6 GHz, 45 GHz, and 60 GHz Some of them are widely used, and will be the primary focus of this book.

---

1 In this chapter and beyond, these references point to the clause of the 802.11-2020 standard 4.3.5.2, where you can find more details, or the exact text, of the concept discussed here.

2 In fact, subsets of this range, depending on the regulatory domains, were defined in 802.11af-2013 (TV whitespace).

The 2.4 GHz band, described in 802.11-1997,[3] is allowed for 802.11 operations over 83.5 MHz worldwide,[4] ranging from 2.400 GHz to 2.4835 GHz. The band was initially allowed for Frequency-Hopping Spread Spectrum (FHSS, obsoleted in 802.11-2012) and Direct-Sequence Spread Spectrum (DSSS) transmissions (1 and 2 Mbps in 802.11-1997 and 802.11-1999, then 5.5 and 11 Mbps in 802.11b-1999), and then was expanded with 802.11g-2003 to Orthogonal Frequency-Division Multiplexing (OFDM) transmissions. The standard describes 11 to 14 channels in this band (depending on the regulatory domain). As the channels are spaced every 5 MHz, but DSSS and OFDM operations require 22 MHz-wide and 20 MHz-wide channels, respectively, only three non-overlapping channels (1, 6, and 11) are reasonably possible in that band in a given space.

The 5 GHz band, described first in 802.11a-2003, is divided into multiple sub-bands, whose availability depends on the regulatory domain.[5] This book uses the U.S. Federal Communications Commission (FCC) numbering convention for these segments. UNII-1 includes four 20 MHz channels, and ranges from 5150 to 5250 MHz. UNII-2A includes four 20 MHz channels, and ranges from 5250 to 5330 MHz. UNII-2C includes eleven 20 MHz channels, and ranges from 5490 to 5710 MHz. UNII-2C/3 includes one 20 MHz channel, and spans the 5710 to 5730 MHz range. UNII-3 includes five 20 MHz channels, and ranges from 5735 to 5735 MHz. Other channels were allowed in the 2010 decade, including UNII-3/4, including one 20 MHz channel and ranging from 5835 to 5855 MHz, and UNII-4, including two 20 MHz channels and ranging from 5855 MHz to 5895 MHz.

The 6 GHz band was added for Wi-Fi 6E operations and is covered in Chapter 3, "Building on the Wi-Fi 6 Revolution."

DSSS Transmissions

DSSS transmissions are allowed in the 2.4 GHz band, with data rates of 1, 2, 5.5, and 11 Mbps. This mode is of limited interest for efficient Wi-Fi operations in general (it provides only slow data rates, compared to more recent transmission modes and techniques) and for Wi-Fi 7. However, backward-compatibility requirements mean that newer transmission techniques in 2.4 GHz must account for the possible presence of DSSS devices, and the simplicity and long range of DSSS transmissions make them attractive even for recent devices (in particular, Internet of Things [IoT] devices) whose primary target is not throughput, but rather transmission robustness (with limited battery consumption). For this reason, DSSS is not obsoleted, despite its 20-plus years of existence.

At its core, DSSS works by spreading the signal across a wider bandwidth than what is minimally required for transmitting the data. After modulating each 1 or 2 bits of data into Binary Phase Shift Keying (BPSK) or Quadrature Phase Shift Keying (QPSK) constellation points at a 1 MHz rate, each constellation point (BPSK or QPSK) is replicated 11 times. Each copy is then multiplied by "chipping code" or "spreading code"—and specifically the Barker sequence (+1, −1, +1, +1, -1, +1, +1, +1, −1,

3 The standard also described infrared operations, in the 805–950 nm range; these operations were obsoleted by the 2012 version of the standard.
4 This frequency range is defined worldwide as suitable for industrial, scientific, and medical (ISM) unlicensed transmissions.
5 802.11-2020 annex E describes the various allowed channels for the different recorded regulatory domains, as of 2020.

$-1, -1^6$)—to produce 11 chips. These chips are sent rapidly in sequence, 11 times faster than the stream of constellation points, or 11 Mchip/s (spreading the signal 11 times wider in the frequency domain). As no or minimal pulse shaping is employed, the resulting signal consumes 22 MHz of bandwidth, but is very resistant to interference given the 11-fold repetition.

Complementary Code Keying (CCK, in 802.11b) is an advancement over this simple scheme. Instead of 1 or 2 bits in and 11 BSPK/QPSK-spread chips out, the CCK function defines how 4 or 8 bits at 1.375 MHz are transformed into 8 QPSK chips at 11 Mchip/s. The CCK function has some parallels to an Inverse Fast Fourier Transform (IFFT) algorithm, and it can be leveraged in the receiver to simplify the demodulation processing so as to recover the data.

In the time domain, these different modulation techniques allow the transmission of 1, 2, and 5.5 or 11 Mbps (via spread BPSK, spread QPSK and two variations of CCK, respectively). The possible presence of DSSS devices in the 2.4 GHz band requires that any more-complex transmission technique needs to implement principles that will allow a DSSS device to know that other transmissions are in progress (and thus refrain from transmitting at the same time).

OFDM Transmissions

Spreading the signal is attractive for robustness, but is soon limited in terms of data rate increases. Given the finer silicon chip feature size and remarkable Radio Frequency (RF) innovations that made it possible to leverage the conventional silicon chip fabrication process, first 802.11a-1999 (in the 5 GHz band) and then 802.11g-2003 (same procedure as 802.11a, but ported to the 2.4 GHz band) introduced a new code, the OFDM transmission technique, that became the baseline for later 802.11 amendments.

At its core, OFDM aims to split a high-data-rate transmission into several lower-data-rate streams that are transmitted simultaneously over multiple frequencies, thereby dividing the signal into smaller signals, sent in parallel ("multiplexed"). The 802.11a/g amendment divided the 20 MHz-wide channel into 64 "subcarriers" (or "tones"), of which 52 are used for active transmission. The subcarrier in the center frequency is left unused (null) to help the receiver center the channel—specifically, if using a direct-conversion architecture, to ignore its wandering DC offset, since that frequency range does not overlap any data tone in the signal. The subcarriers at the top and bottom of the channel are left unused to limit the risk of overlap between neighboring channels active at the same time. This model is similar in concept to the 802.11n model shown at the top of Figure 1-4, except that 802.11 slightly changes the usage of the subcarriers.

Of the 52 active tones, 4 (spread across the channel) send a known sequence; they are used to help the receiver synchronize with the transmitter. Any residual carrier frequency offset or phase noise arising between uncompensated differences between the transmitter and receiver oscillators and phase-locked loops can be measured and removed. The same information can be used to measure and mitigate differences between the rate of the transmitter's Digital-to-Analog Converters (DACs) and the receiver's Analog-to-Digital Converters (ADCs). Finally, the pilots can compensate for subtle amplitude variations during the transmission, such as when the Power Amplifiers (PAs) at the transmitter incrementally

6 See 802.11-2020, clause 15.4.4.4

warm up during a transmission. These 4 tones are called pilot tones. The other 48 tones carry the transmitted data (after encoding and modulation).

The tones can be viewed as spirals along the time axis, which, when viewed side-on or top-down with the time axis running from left to right, look like a sinusoid or co-sinusoid versus time. The frequency of rotation of each spiral tone is 20 MHz/64 = 312.5 kHz apart from its neighbor, so 1/312.5 kHz = 3.2 μs always contains an integer number of cycles of the spiral. The transmitter modulates each tone by a constellation point (or symbol) for 3.2 μs, plus a little extra time, so that the receiver can soak up both the direct signal from the transmitter and the delayed copies (echoes), as some rays of the transmitted data take an indirect (longer) path by first bouncing off walls and other scatterers before reaching the receiver. This extra time is called the *cyclic extension* or *guard interval*, and in 802.11a/g it was 0.8 μs in length. This combination of elements (an integer number of cycles per 3.2 μs, the 0.8 μs cyclic extension, and the receiver processing just 3.2 μs out of the received signal every 4 μs) enables each subcarrier to be extracted at the receiver as a distinct (orthogonal) signal and be processed (to a large degree) without regard to interference from the other subcarriers.

The constellation point (or symbol) incorporates multiple coded bits, using either BPSK, QPSK, or Quadrature Amplitude Modulation (QAM). With BPSK, the spiral tone is rotated 0 or 180 degrees around the time axis to represent the carried data (1 or 0, respectively). With QPSK, blocks of 2 bits, $[B_0B_1]$, are encoded together, and the spiral tone is rotated –135, +135, –45, or +45 degrees to represent the dibits 00, 01, 10, or 11, respectively. With QAM, the two non-time dimensions (called in-phase and quadrature) of the spiral are separately scaled by a discrete signed number. For example, for 16-QAM, the in-phase axis of the spiral tone is scaled by one of 4 values {±1, ±3}. Similarly, the quadrature axis of the spiral tone is scaled by one of these 4 values, too; that gives $4 \times 4 = 16$ different constellation points in total to select from. A common representation of that effect is the idea that the tone peak seems to hit a virtual target, as shown in Figure 1-3, whose position represents a 4-bit value (0000, 0001, …, 1111), for a total of 16 possible values. 64-QAM involves 8 scale factors for each of the in-phase and quadrature axes, or 64 different constellation points in total to pick from.

To prevent a small amount of interference in the time domain from destroying the receiver's ability to demodulate an entire symbol, the 11a/g PHY design incorporates a forward error-correcting code with a selectable amount of redundancy (called the coding rate). The options are labeled ½, ⅔, and ¾, which signifies that redundant information makes up ½ ⅓, and ¼ of the coded data stream, respectively.[7]

The subsequent amendments have all leveraged the OFDM technology. Each of them adds improvements to the way 802.11 operations are performed, but also, as silicon technology improves, increases the width of the channel and the complexity of the allowed modulation. For example, 802.11n-2009 allows 40 MHz bandwidth operations up to 64-QAM and with a new code rate ⅚ [8] (recall that 802.11a/g was limited to 20 MHz and a 64-QAM rate of ¾); 802.11ac allows 160 MHz bandwidth operations and up to a 256-QAM rate of ⅚.[9]

7 See 802.11-2020, clause 17.3.10.2
8 See 802.11-2020, clause 19.3.19.1
9 See 802.11-2020, clause 21.3.18.1

64-QAM

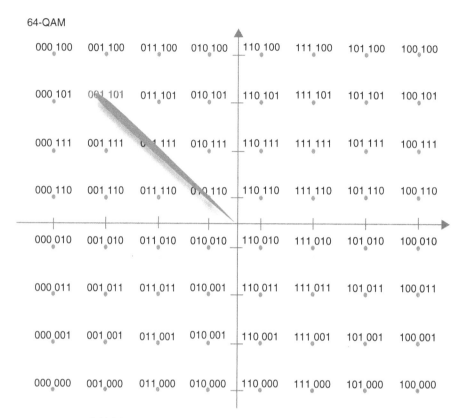

FIGURE 1-3   QAM Transmission

The 802.11n-2009 amendment also improves the subcarrier structure in 20 MHz by reducing the number of null tones, from 12 with 802.11a/g, to 8; it still keeps 4 pilots, but increases the number of tones that carry data from 48 to 52. The 802.11n and 802.11ac amendments allow for larger channel operations with greater efficiency. The channel width increases because 2 (or 4) adjacent channels are bundled together into a single larger channel.[10] When such bundling is performed, the tones at the upper edge of the lower channel and at the lower edge of the upper channel fall into the middle of the transmission channel, and no longer need to be null tones. Additionally, some of the pilots can be removed. Thus, a 40 MHz-wide channel comprises two 20 MHz channels or 128 (64 × 2) tones. But instead of using 112 of them (56 × 2), with 104 tones (52 × 2) for data and 8 (4 × 2) for pilots, the 40 MHz transmission (with 802.11n/ac) uses 114 of the tones: 108 tones for data and 6 for pilots. The 80 MHz transmission in 802.11ac uses a channel with 256 tones (64 × 4), of which 8 are pilots and 234 are used for data transmission (thus leaving 14 zero subcarriers—6 and 5 tones on the sides as a guard band, and 3 in the center). Figure 1-4 illustrates these principles.

10 802.11ac-2013 allows 160 MHz bandwidth operations, but with the idea that two 80 MHz channels are used. They do not need to be contiguous.

The ability of the receiver to correctly demodulate each tone depends on the received signal level (which, in turn, depends on the transmitted power and its attenuation from transmitter to receiver), the number of tones used (i.e., bandwidth), the constellation size, the coding rate, and the implementation quality of the receiver's RF and baseband processing. Thus, the standard provides guidance on the minimum receive (Rx) sensitivity expected from a proper 802.11 receiver for each channel width, constellation size, and code rate. For example, a receiver is expected to receive correctly 90% of the time a 20 MHz signal of length 1000 octets (or 4096 octets), with a BPSK modulation and coding rate of ½, when the signal is received at –82 dBm or better.[11] However, the receiver is expected to be successful at 80 MHz only if the signal is received at –76 dBm or better. And if the modulation is increased to a 256-QAM rate of ⅚ instead of a BPSK rate of ½, success is expected only for a signal of –51 dBm or better.

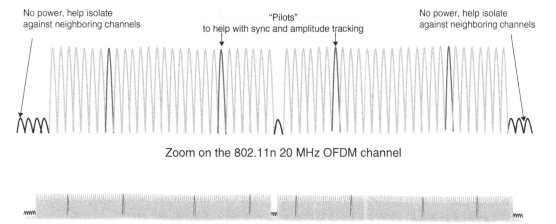

Zoom on the 802.11n 20 MHz OFDM channel

Null, guides, and tones in the 802.11ac 80 MHz channel

FIGURE 1-4   OFDM Transmission

Network Layers and the 802.11 Frame

The terms used to describe the 802.11 frame leverage and augment the Open System Interconnection (OSI) model. In the OSI model, as the application sends data to the network stack, that data may be segmented by the Transport layer (Layer 4). The segment (data with L4 header) is then sent to the Network layer (Layer 3), which applies the matching header, making the segment a packet. The packet is then sent to the Data Link layer (Layer 2, where 802.2 and subsequently 802.11 operate), where a local header is applied, making the packet a frame. The frame is then transmitted as a waveform by the Physical layer (Layer 1, also addressed by 802.11).

The Data Link and Physical layers are further subdivided by functions. The top of the Data Link layer includes the Logical Link Control (LLC) sublayer, which is in charge of interfacing the 802.11 stack with the upper layers. The LLC for 802.11 is defined by the 802.2 set of standards. It takes the packet

11 See 802.11-2020, clause 17.3.10.2

received from the Network layer, and validates whether the packet can be carried by the medium or needs to be fragmented. The result of that decision is the payload that will be transmitted to the lower part of the Data Link layer. That payload is called a MAC Service Data Unit (MSDU). Its name comes from the fact that the lower part of the Data Link layer is called the Medium Access Control (MAC) sublayer. The MAC takes the MSDU and encapsulates into a frame (i.e., adds a MAC header and a trailer to the payload). The result is called the MAC Protocol Data Unit (MPDU). The MPDU is then passed onto the Physical layer (PHY). From the MAC layer viewpoint, the MSDU is the payload, and the MPDU is the final product. However, from the PHY viewpoint, the MPDU is the payload; thus, in the PHY layer, it is called the PHY Service Data Unit (PSDU). The PHY first adds the PHY header, which consists of a preamble and additional information (detailed later in this section). The result is the Physical Protocol Data Unit (PPDU), which is what is actually transmitted over the medium.

The MSDU, and therefore the PPDU, carry a single data payload in legacy 802.11. The 802.11n amendment introduced a key improvement, which all later amendments leverage: the ability to aggregate. Several MSDUs can be aggregated into a single A-MSDU. A common PHY and MAC header starts the frame, and the payload consists of one or more subframes, each with its own LLC, IP headers, and payload (and possibly additional padding so the subframe size is a multiple of 8 octets). The subframe LLC header, using 802.2, can include the Source Service Access Point (SSAP) and the Destination Service Access Point (DSAP), which can represent the logical addresses of the Network layer entities that created or are intended to receive the message. "Access point" in this context is unrelated to the 802.11 AP, and the LSAP and DSAP are usually protocol names. Therefore, a constraint of the A-MSDU (with a single common PHY and MAC header) is that all payloads need to be intended for the same destination MAC address (which would respond with a single ACK frame for the entire A-MSDU; see the "DCF-Based Access" section later in this chapter).

Another method aggregates MPDUs instead of MSDUs. In that case, the PHY header is shared in common, and one or more MPDUs are aggregated into a single frame; they may potentially have different intended target MAC address receivers. Conceptually, the transmitter gains access to the medium, then sends a train of MPDUs during a particular Transmission Opportunity (TXOP; see the "HCF and EDCA" section at the end of this chapter). The train of MPDUs is acknowledged with a block acknowledgment, which provides a bitmap indicating which frame was received properly. This represents another advantage over A-MSDU, where a single subframe error causes the entire A-MSDU acknowledgment to fail.

Structure of the Basic 802.11 Frame

The MAC sublayer and the PHY layer perform functions that are useful to better understand the operations introduced by Wi-Fi 7 and are described in detail in Chapter 6, "EHT MAC Enhancements for Multi-Link Operations," and Chapter 7, "MAC Improvements of Wi-Fi 7 and 802.11be." The PHY layer inserts a preamble. With OFDM, the preamble starts with a (Legacy) Short Training Field (L-STF),[12] which is a known signal sent on a small selection of the available subcarriers. The receiver can use this

---

12 This section uses 802.11n/ac terminology for simplicity. The preamble in 802.11a/g has a similar structure, albeit simpler, and uses terminology that was deemed less clear when subsequent amendments were developed.

field to gain an awareness that a PPDU is arriving, and also align its receive function with the trans-mitted signal (i.e., find the channel center frequency). The L-STF lasts for 8 µs, and is followed by a second field, the Legacy Long Training Field (L-LTF). The L-LTF also lasts 8 µs, but is a known signal sent over more subcarriers than the L-STF, thus allowing the receiver to perform finer synchronization and timing alignment with the transmitted signal.

The Legacy Signal Field (L-SIG), which follows the L-LTF, contains crucial information about the PPDU. In particular, the SIG field mentions the data rate at which the payload (Data field) is modu-lated as well as its length. This field has evolved considerably over the Wi-Fi generations: First, it was the explicit octet count (for 11a/g), then an "FYI" of the PPDU duration (for 11n) and finally an essential signal of the PPDU duration (for 11ac onward). These fields, illustrated in Figure 1-5, are encoded and modulated with the strongest technique available—in most cases, 6 Mbps, with a BPSK rate of ½. As new amendments were developed, new PHY header fields were inserted (either as part-replacement of the legacy preamble described earlier or in a greenfield scenario [i.e., the little-used 11n HT-GF PHY format]) or, more commonly, added behind the legacy header and before the payload field (when backward compatibility was required).

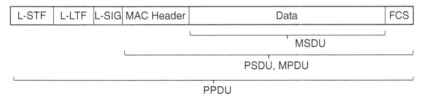

FIGURE 1-5   PHY Header in an OFDM (11a/g) PPDU

The PHY preamble is followed by the MAC header. The MAC header structure can change depending on the type of frame to be transmitted, and detailing all possibilities is beyond the scope of this chapter. However, some fields are always present, and others are present in most frames.

In particular, all frames begin with a Frame Control field,[13] which reports the frame type (management, control, or data) and subtype.[14] The field also indicates whether the frame is sent toward the DS (ToDS) or is forwarded from the DS (FromDS). The retry bit indicates whether the (unicast) frame was attempted before but its expected acknowledgment was not received by the sender (see the "DCF-Based Access" section in this chapter). The power management bit indicates whether the sending STA will stay available after this frame or will go in dozing mode.

All frames also include a Duration/ID field.[15] In most data frames, this field is used to indicate the expected remaining duration of the exchange.

---

13 See 802.11-2020, clause 9.2.4.1
14 See 802.11-2020, Table 9-1, for a complete list.
15 See 802.11-2020, clause 9.2.4.2

The header can then include three or four MAC addresses. This section is key for Wi-Fi 7, as it is intended to represent the Source Address (SA; the STA that initially sent that MSDU), the Transmitter Address (TA; the STA that is transmitting this frame with the included MSDU), the Receiver Address (RA; the intended target of this transmitted frame), and the Destination Address (DA; the expected final destination of the MSDU). When one of these addresses is the AP MAC address of the BSS, that address is called the Basic Service Set Identifier (BSSID). The fourth address is not always used. More specifically, it is used only when transiting between two APs over the 802.11 medium (e.g., in mesh networks). The fourth address is not present when not used. The three or four addresses have different meanings, depending on which entity is sending the frame to which other entity (in particular, to or from an AP-STA, or to and from a non-AP STA). This direction is represented by the value of the ToDS from FromDS bit, as displayed in Table 1-3.

TABLE 1-3   Values of the Four Addresses in the 802.11 MAC Header

To DS	From DS	Address 1	Address 2	Address 3	Address 4
0	0	RA = DA	TA = SA	BSSID	N/A
0	1	RA = DA	TA = BSSID	SA	N/A
1	0	RA = BSSID	TA = SA	DA	N/A
1	1	RA	TA	DA	SA

Another field of interest follows the third address field: the Sequence Control field.[16] It is a critical field, because all data frames (and also management frames) need to be assigned a sequence number. This requirement is important for duplicate detection and recovery.[17] A unicast frame (or an A-MSDU or A-MPDU) needs to be acknowledged by its receiver as part of the channel access rules (discussed later). But when the acknowledgment (ACK) for a unicast frame is not received by the frame transmitter, that failure may occur because the frame was not received by the intended receiver, or because the ACK was not properly detected by the transmitter of the unicast frame. Retransmission is possible, and can happen many times. It is therefore crucial for the receiver to determine whether the current frame is a duplicate. The retry bit alone is not sufficient to make this determination; instead, the sequence number is used for that purpose. Strict rules apply regarding how the sequence number should be set by the transmitter and used by the receiver. One of the outcomes of a properly set sequence number (in combination with timestamps) is the ability for the receiver to reassemble fragmented MSDUs[18] that are transmitted across individual frames. Naturally, this requirement establishes rigid constraints on the possibility of sending MSDUs across two different links with MLD (see Chapter 6).

As Figure 1-6 illustrates, in data frames, the header can—and most always does nowadays—also include a QoS Control field.[19] The primary purpose of this field (in the context of this book) is to carry the Traffic ID (TID) indicating the QoS category associated with the transmitted payload (see Chapter 2).

---

16 See 802.11-2020, clause 9.2.4.4
17 See 802.11-2020, clause 10.3.2.14 for a detailed discussion.
18 See 802.11-2020, clause 10.4 for fragmentation; 10.5 for defragmentation.
19 See 802.11-2020, clause 9.2.4.5

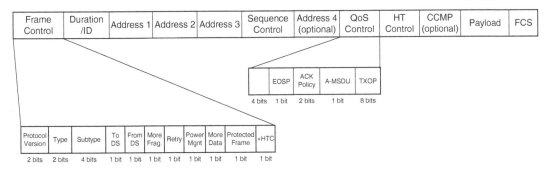

FIGURE 1-6 MAC Header in a Data Frame

The header can also extend into the payload space and include more fields, especially when encryption is used. For example, with the Counter Mode Cipher Block Chaining Message Authentication Code Protocol (Counter Mode CBC-MAC Protocol) or CCM Mode Protocol (CCMP), an 8-octet CCMP header includes a packet number (PN), an Ext IV, and a key ID. The PN is incremented at each subsequent frame. The MAC header is then followed by the carried payload, and a Frame Check Sequence (FCS) field,[20] which is a 32-bit CRC used to validate that the frame did not become corrupted somewhere between its transmission and its reception.

The 802.11 Connection Process

Throughout this section, we consider the most commonly used 802.11 BSS type: the infrastructure BSS where clients connect to an 802.11 network through an AP. Before being able to forward a data frame through the AP, a non-AP STA needs to become associated with that AP.[21] As the standard targets data communication with the DS as the *raison d'être* of the AP, it defines the requirement to associate as a mechanism to inform the DS which AP will be able to forward data frames to a target non-AP STA. However, association is, in fact, also required to transmit data frames, through the AP, to other non-AP STAs in the BSS and beyond. The AP radio, from the transmitting STA's perspective, is then seen as a "beyond the AP" function, although 802.11 clarifies that the Wireless Medium (WM) is distinct from the DS.[22]

Scanning Procedures

A prerequisite for association is the awareness of the presence of an AP on a given channel. As the number and frequencies of channels are known, a STA can set its receiver to each candidate frequency in turn and attempt to discover there the presence of an AP. In the standard, a STA may be configured

---

20 See 802.11-2020, clause 9.2.4.8
21 See 802.11-2020, clause 4.5.3.3
22 See 802.11-2020, clause 4.3.5.1

for an SSID (the SSID parameter in the STA MLME-SCAN.request primitive) and can discover APs with two different methods:[23]

■ **Passive scanning:** With passive scanning,[24] the STA sets its receiver to a particular channel frequency, and listens for AP beacons for a duration of the STA's Station Management Entity (SME) choosing, defined by the STA MinChannelTime and MaxChannelTime parameters (in the same MLME-SCAN.request primitive). The range is implementation-specific, but is expressed in TU (a TU is a unit of 1024 μs), as the STA needs to stay on the channel long enough to hear the next beacon (the interval between two beacons is configurable, but is commonly set to 100 TUs), if it reached the channel just after the last beacon. Quite obviously, this method is inefficient when the number of channels to scan is large. A probe request/response exchange (discussed next) can consume approximately 0.4 ms of medium time (providing the STA with the information it needs), while a beacon interval is by default 102.4 ms (2500 times longer).

■ **Active scanning:** With active scanning,[25] the STA sends a probe request frame, following the channel contention rules detailed in a later section. The request can identify the target SSID, indicating that only the APs servicing that SSID should respond. Alternatively, the SSID field can be null (no value), indicating that all APs are expected to respond, regardless of the individual SSID they serve. The request can be sent to a target MAC address (if a single, known AP is the target) or to the broadcast address (if all APs should respond). The standard sets strict rules for the AP response. In most cases, an AP receiving a probe request must respond with a probe response.[26] The default probe response frame has a structure similar to that of the beacon, minus a few elements, such as the TIM, that are relevant only for associated STAs. The response can be sent to the broadcast address (with 802.11ai and later) or to the MAC address of the requesting STA. The STA can send a single probe request on a given channel, process the responses, if any, and then move to the next channel. In practice, it is common for STA implementations to send one request, process the responses, and repeat the process a second time (after 10 or 20 ms of silence), in case the initial request collided or some of the responses were missed or collided.

The active scanning procedure is faster than the passive method, but is also noisier. A single STA, sending two or more probe requests, can cause each AP device on a channel to send two or more probe responses at short intervals, with no other value than providing the STA with information about the AP—which the STA would have learned about if it had waited for the next beacon or had listened to other STA probe exchanges. The 802.11ai-2016 amendment (detailed in the "FILS" section of Chapter 2) attempted to streamline this procedure, by allowing two enhancements:

■ Before sending a probe request, a STA setting its receiver on the channel should wait a short duration[27] (20 ms). That gives it a chance to hear potential broadcast probe exchanges (in which

23 See 802.11-2020, clause 11.1.4.1
24 See 802.11-2020, clause 11.1.4.2.1
25 See 802.11-2020, clause 11.1.4.3.2
26 See 802.11-2020, clause 11.1.4.3.4
27 See 802.11-2020, clause 11.1.4.3.2

case the STA should not send a probe request, as it already heard the answer it needed), and perhaps the AP's next beacon. In the FILS probe request, the STA also indicates a maximum timer (the ActiveScanningTimer[28]) within which it expects a response.

■ When receiving a FILS probe request, the AP computes the time to the next beacon. If the beacon comes before the maximum timer indicated in the probe request, the AP simply does not answer, and instead sends the beacon when its time comes. The beacon can also include an 802.11k-2008 reduced neighbor report, which indicates the channels, SSIDs, and BSSIDs (MAC address) of neighboring APs known to the local AP. This information allows the STA to expedite its discovery, by focusing on the channels and SSID of interest.

Authentication and Association Procedures

Another prerequisite for association is authentication. Here again, the term is ambiguous, because it is often used in security contexts to express a form of identity validation (for a user or a device), thereby protecting the network from unauthorized access. In 802.11, authentication is expected to be a link-level operation,[29] and without either end-to-end or user-to-user authentication. The original purpose of 802.11 authentication was to confirm that the STA is a valid 802.11 device—that is, capable of performing 802.11 MAC exchanges, and therefore 802.11 PHY exchanges, with the STA within the AP. Nevertheless, the connection with security keeps coming back into the standard, as in the 802.11 security section (discussed in Chapter 2). These schemes are important, because some of them align well with the concept of Multi-Link operations introduced in 802.11be, whereas others do not.

Once authentication is completed successfully, the STA can register to the AP (and obtain an Association Identifier, AID) through the association process. This association allows the AP to inform the DS about which AP (itself) can forward frames to that STA MAC address.[30]

In the early days of 802.11, association would complete the connection setup, and the AP would start forwarding the STA's data frames immediately. In an enterprise environment, that conclusion would come too early. In such a case, access to the network is reserved for corporate devices, so verification of the device's identity (or its user's identity) is needed before traffic is allowed. This is especially critical for wireless access technologies like 802.11, where any device in range of the AP may attempt to send frames and explore (and possibly attack) the network. Likewise, this type of precaution also makes sense in home networks. There, connecting devices may not need to prove their individual identity, but they may need to prove that they are part of the trusted home device group.

In 2004, the 802.11i amendment introduced an additional 802.1X authentication step, derived from parallel work on the IEEE 802.1X standard,[31] that introduces the idea of a Robust Security Network

28 See 802.11-2020, clause 11.1.4.3.2
29 See 802.11-2020, clause 4.5.4.2
30 See 802.11-2020, clause 4.5.3.3 and 12.6.9
31 802.1X was also published in 2004. The standard was later updated in 2010 and 2020. IEEE 802.11, in its 2020 version, refers to the 2010 update.

(RSN). After the association process completes, the AP is ready to forward the STA's frames. However, its radio interface now also includes, above the 802.11 MAC function, an IEEE 802.1X Port Access Entity (PAE) that controls the forwarding of data to and from the MAC function.[32] The STA implements a supplicant PAE role, and the AP an authenticator PAE role. The STA also implements an Extensible Authentication Protocol (EAP[33]) peer function, and the AP an EAP authenticator function. An Authentication Server (AS) is optionally deployed on the DS.

Without RSN, the AP radio (its port, or PAE) is said to be *uncontrolled*. At the end of the association procedure, the AP port starts forwarding the STA's frames without further validation. With RSN[34] using 802.1X authentication, the AP port is *controlled*. At the end of the association procedure, the port is still unauthorized. In that state, it allows only EAP exchanges. The AP and the STA start an EAP dialog, through which the STA establishes a connection to the AS. With 802.11, the AP converts the STA EAP-encapsulated messages into a RADIUS[35] encapsulation, before forwarding them to the AS. The AP performs the reverse operation in the other direction.

Through their dialog, the AS validates the identity of the STA supplicant (and in most cases, the STA also validates the identity of the AS). Then, each on their side, they derive the same Master Session Key (MSK). The AS then passes the MSK to the authenticator[36] (the AP). The AP and the STA select the first *n* bits[37] of the MSK, and use them as the Pairwise Master Key (PMK). The AP and the STA then undergo a 4-frame exchange (called a 4-way handshake) to generate a common Pairwise Transient Key (PTK) used to encrypt unicast exchanges between them. The AP also passes a Group Transient Key (GTK) to the STA, which is used to encrypt multicast and broadcast frames. Both the PTK and the GTK can be changed during the session. Once the 4-way handshake completes, the STA's and the AP's respective controlled ports become authorized, and can start passing data to their MAC function: the STA to send data frames toward the AP, and the AP to forward frames for the STA to and from the DS.

Channel Access Rules

Once a STA is associated, it can start transmitting data frames. Even before the STA associates, it may need to send discovery frames. All of these transmissions are subject to the same channel access rules. Channel access, until 802.11ax, was solely driven by the general principle of collision avoidance (in 802.11, it is named Carrier Sense Multiple Access, with Collision Avoidance [CSMA/CA]). 802.11ax added triggered access to collision avoidance. One legacy mechanism, called the distributed Coordination Function (DCF), treated all traffic the same way. However, it was soon superseded by the Hybrid Coordination Function (HCF), a mechanism that recognized different types of traffic should have different statistical chances of getting onto the medium.

---

32 See 802.11-2020, clause 4.3.8
33 https://www.rfc-editor.org/rfc/rfc3748
34 You will learn in Chapter 2 that 802.11ai-2016 introduces some additional subtlety.
35 802.11 describes only RADIUS, but 802.1X also allows EAPOL and Diameter/RFC 3588.
36 See 802.11-2020, clauses 12.6.10.2 and 12.6.14.
37 256 bits or more, depending on the authentication scheme in use; See 802.11-2020, clause 12.7.1.3.

Clear Channel Assessment Principles

The ideas behind collision avoidance are simple. First, Listen Before Talk (LBT). If another transmission is detected, then wait until that transmission completes.

Second, after that detected transmission completes, wait a little longer in case other transmissions still need to complete (e.g., with an ACK; see the "DCF-Based Access" section later in this chapter). Then, wait a random amount of time before transmitting. This delay accounts for the possibility that two or more STAs might need to transmit and might detect another transmitting STA. If they were to start right after the other transmission completed, they would detect the end of the transmission the same way and at the same time, and would then start transmitting at the same time—and therefore collide. By waiting a random amount of time, it becomes more likely that one of the STAs would start before the other; then the other would detect the new transmission and wait for its completion.

Third, after transmission, confirm that collision did not occur, whenever possible. These principles apply equally to each transmitter in the BSS—in other words, they apply to the AP in the same way as to each individual STA.

This idea of time (random or not) implies that the STAs must operate at about the same pace. Although the clocks implemented in 802.11 devices are not calibrated (and thus may operate at different speeds), each subsequent generation of PHY defined an inaccuracy tolerance to bound the differences within an acceptable range.[38] Then, each PHY defines a few key timers. One of them is an interval value, called aSlotTime, which is the speed at which the STAs are expected to count. Then, many of the time-based operations in 802.11 are expressed in aSlotTime units (in combination with the few other key timers). Most PHYs of interest for this book use a transmission technique built upon OFDM, defined in clause 17, and use an aSlotTime value of 9 μs. This number is not chosen at random, but rather is the result of an agreement among 802.11 experts on the time it takes for a receiving circuit to perform channel sensing (aCCATime), process what was detected (aMACProcessingDelay), and possibly switch from a receiving mode to a transmitting mode (aRxTxTurnaroundTime), as well as the time needed for any emitted signal to then reach another receiver in the BSS (aAirPropagationTime).[39] Thus, the aSlotTime is the minimum time that another STA in the BSS is expected to need to sense the channel, conclude that it is free of other transmissions, and start its own transmission.

When a STA (in this section, a STA is an AP or non-AP STA) wants to transmit, it first needs to listen to the medium to detect possible other transmissions. If the channel is concluded to be busy, the STA can wait for one aSlotTime, then listen again. This process repeats in a loop as long as the channel is busy; many implementations either listen to the channel until it is clear, or know (see the section "DCF-Based Access" a little later in this chapter) when the channel should be free and wait for that time Directly. The goal of this detection function, called carrier sensing,[40] is to make a Clear Channel Assessment (CCA). There are two ways to detect whether the channel is available ("idle") or busy. The

---

38 See, in particular, See 802.11-2020, clause 10.3.7.6, 17.3.9.6, 18.4.7.5, 19.3.18.6, 21.3.1.7.3, and 22.3.17.3 for the PHYs of primary interest for this book.
39 See 802.11-2020, clause 10.3.7
40 See 802.11-2020, clause 10.3.2.1

802.11 standard indicates that one of these methods is operated by the PHY layer, and the other by the MAC layer.[41] Practically, the PHY layer is a prerequisite for both methods.

The first method, called Signal Detect (SD) and sometimes preamble detect, is to effectively detect other 802.11 transmissions. While awake, the STA is required to listen continually for a transmission and detect it almost as soon as it starts. "Detect" in this context needs to be qualified. At its core, detection means that the STA receiver is set to the target channel. Each clause defines, for each channel width, the minimum receiving sensitivity expected for each modulation (BPSK, QPSK, QAM). For OFDM and 20 MHz, –82 dBm is the expected minimum receiving sensitivity. As the PPDU's preamble includes the L-STF (sent over 20 MHz), with symbols that are always modulated at using BPSK, a STA should, when listening to the channel, be able to detect an L-STF that marks the beginning of a transmission, if the preamble is received at –82 dBm or more. The STA should thus detect the signal, process it, and within a short delay (now implementation-dependent but traditionally set to 4 μs, or about half the aSlotTime) conclude that a transmission is starting.[42] Once the 11a/g Signal field is reached (and/or the HT-SIG or RLSIG for later PHY generations), if the Signal/SIG field is correctly decoded, then the PPDU duration is known and the CCA busy indication is elongated for the duration of the PPDU. Alternatively, if no transmission is detected as starting, the STA can repeat the process for the next aSlotTime. This requirement expects that the STA will be able to detect a preamble, not the transmission of the payload of the frame. In practice, the Signal field is modulated at the lowest possible data rate. In contrast, the MAC header and the frame payload can be modulated at one of many available Modulation and Coding Schemes (MCSs)—namely, the one deemed by the STA to be the most efficient (least amount of consumed airtime for maximum amount of reliably transmitted data, based on the current RF conditions). A receiving STA may be able to demodulate the PHY header, but not the rest of the PPDU (and thus it would only perceive energy for those fields). Therefore, the ability to detect the preamble is critical for proper 802.11 operations.

It should be noted that –82 dBm is the minimum threshold here. A device with better receiving sensitivity may apply a different threshold. For example, enterprise APs often have a receiving sensitivity much greater than the minimum threshold. As a result, they may be able to detect an OFDM preamble, and thus a transmission, as low as –95 dBm, and even –100 dBm or weaker for DSSS BPSK signals.

The 802.11 standard also recognizes that many of the bands where 802.11 is found are allocated for Industrial, Scientific, and Medical (ISM) operations, and that other (non-802.11) devices may be operating on the same channel. Other bands are allowed for 802.11 (unlicensed) use, but may have other incumbent technologies using the same frequencies. A STA would not be able to detect an 802.11 preamble from these transmitters, yet their energy may be sufficient to prevent 802.11 frame reception from being successful. When energy is detected on the channel with a power at least 20 dB above the STA minimum receiving sensitivity thresholds, the STA can use a technique called Energy Detect (ED).[43] Practically, this means that the STA will deem the channel busy if an energy of at least –62

---

41 See 802.11-2020, clause 10.3.2.1
42 See 802.11-2020, clause 17.3.10.6. Just as for minimum receiving sensitivity, this operation should be successful at least 90% of the time for a STA to be deemed 802.11-compliant, although IEEE 802.11 does not test any of these performance values.
43 See 802.11-2020, clause 17.3.10.6

dBm (20 dB above the −82 dBm minimum) is detected. Just as with signal detection, the STA needs to make that conclusion typically within 4 µs after sensing the channel.

The CCA technique works well and is sufficient for individual PPDUs. However, one critical transmission in 802.11 is the response to a previous transmission. This response is key to enable or conclude a dialog (without which transmissions are often not successful). Therefore, the standard also requires a virtual Carrier Sense (CS) function, by which the STA determines that there is an ongoing transmission. Several methods may be used to make this determination.[44] For example, the STA might read a MAC header that includes the Duration field. The STA might also detect a Request-to-Send/Clear-to-Send (RTS/CTS) exchange, or a CTS-to-self exchange, and deduce the intended duration of the exchange.

In all cases, the STA should wait for the channel to be idle, as identified by direct PHY measurements of energy on the medium, the indicated length of any ongoing PPDUs, and the Duration field in the MAC header of any ongoing TXOPs. Only then should it initiate its backoff procedure.

DCF-Based Access

When the STA (in this section, understood as AP or non-AP STA) listens to a channel, it should not take more than 4 µs to conclude that the channel is busy. Likewise, the STA should take the same amount of time to conclude that the channel is idle. The STA should not decide to then transmit immediately. Some exchanges require an immediate answer (an ACK frame, for example), and priority should be given to those responses over new transmissions. Thus, the standard identifies different silence durations for which a STA should wait before beginning its transmission. If it is answering a frame with an ACK, for example, that is highest priority and the STA should wait the shortest possible time. This duration is called the Short InterFrame Space (SIFS), referenced as aSIFSTime and defined (for OFDM in 5 GHz) as 16 µs.[45] If a STA needs to start a new transmission, it should first wait for a Distributed Interframe space (DIFS), whose duration is DIFS = $1 \times$ aSIFSTime + $2 \times$ aSlotTime[46] (thus 34 µs for OFDM in 5 GHz).

After that DIFS, the channel is available for a new transmission. If the medium was busy when an MSDU reached the transmitting STA, the STA still needs to wait a random time, by picking up a random number (called the random backoff count) in a range.[47] The range selection might seem exotic on first encounter. Practically, each PHY determines the bounds of the range, called the Contention Window (CW), with CWMin and CWMax. The CWMin and CWMax default values, called aCWMin and aCWMax, are 15 and 1023 aSlotTimes,[48] respectively, for OFDM in 5 GHz. Thus, CWMin ≤ CW ≤ CWMax. Then, the STA picks the random backoff count, randomly, in the range [0, CW]; in other words, the STA can randomly pick 0 or CW, or any number in between. The standard also expresses that, for the first attempt for a given frame transmission, CW = CWMin; thus, the STA picks the

---

44 See 802.11-2020, clause 10.3.1
45 See 802.11-2020, Table 17-21.
46 See 802.11-2020, clause 10.3.7
47 See 802.11-2020, clause 10.3.3
48 See 802.11-2020, Table 17-21.

number in the range [0, CWMin]. The STA then counts down from that number, sensing the channel at each aSlotTime count. If, at any point in the countdown, the channel is busy, the STA waits until the channel is idle again for at least one DIFS, then resumes its countdown from where it had stopped it. When the STA reaches 0, it sends its frame.[49]

If the frame is multicast, broadcast, or unicast with a no-ack[50] policy attached, then the STA has no mechanism to ensure that the intended receiver(s) properly demodulated the frame and its content, and concludes with a success if the transmission completes. If the frame is unicast, the STA expects an acknowledgment from the intended RA (again, with the exception of no-ack frames). As soon as the transmission completes, the STA turns around within the SIFS to wait for the start of a PPDU containing the expected ACK. If the ACK does not arrive by that time, the STA waits a bit longer—specifically, AckTimeout,[51] the time needed for the RA to transmit the ACK frame, including the preamble, PHY, and MAC headers, at the lowest data rate.[52] It then concludes that the ACK is not coming.

The STA cannot just try again immediately. After an unacknowledged unicast frame, if the STA wants to try again, it has to wait for an Extended Interframe Space, EIFS (= aSIFSTime + AckTxTime + DIFS). After picking up a new random backoff count number, it tries again. This time, however, CW is increased.[53] The process repeats until CW reaches CWMax, and stays at that value until the STA's internal algorithm decides to stop retrying.

In practice, these attempts are called hardware retries. After a certain number of failed hardware retries, it is common practice to dequeue the MPDU. The software part of the Wi-Fi driver may also attempt several retries, replacing the MPDU back in the queue and restarting fresh with a new transmission attempt with CWMin. Transmission attempts stop only after the number of hardware retries (times the number of software retries) is exhausted. The 802.11 standard does not mandate minimum or maximum retry counts.

HCF and EDCA

The DCF mechanism[54] provides equal opportunities for all STAs in the BSS to transmit (the AP is a STA). This opportunity appears in the form of equality of statistical access to the medium (two STAs competing for the medium would end up, on average and over time, accessing the medium 50% of the time each), but not in terms of access interval. Again, if two STAs compete for the medium and send PPDUs of the same size, as they end up accessing the medium 50% of the time, the PPDU duration (and the number of competing STAs) will end up dictating how often a STA can send its PPDUs. However, if one STA's PPDUs are much larger than the other STA's, the STA with the larger PPDUs uses the medium comparatively more often than the STA with smaller PPDUs.

---

49 See 802.11-2020, Figure 10-6.
50 See 802.11-2020, Table G-1.
51 See 802.11-2020, Equation 10-7.
52 See 802.11-2020, Table 10-8.
53 With the formula new CW = ascending integer powers of 2, minus 1. So, for example, if CW = 15 ($2^4 - 1$), next CW = 31 ($2^5 - 1$). See the end of 10.3.3.
54 Another mechanism created at the same time as the DCF is the point coordination function (PCF). It is rarely, if ever, used, so it is ignored for the purposes of this section.

> **Note**
>
> A toy example may help visualize this relationship. Oversimplifying the interframe spaces by supposing no wait between transmissions, suppose that STA1 transmissions are 1 ms long and STA2 transmissions are 500 µs long (half the duration of STA1's), and they each get a perfect 50% of the airtime and transmit in turn. STA1 transmits, then waits 500 µs for its next transmission. STA2 transmits, then waits 1 ms for its next transmission. Therefore, STA2 receives less airtime than STA1, and transmits less often.

This effect soon found its limitation when the 802.11 designers realized that different traffic types had different sensitivities to delays (in this context, absolute channel access delay for a given frame) and jitter (the variation of delays). For example, voice packets are usually small (e.g., 160 bytes for a G711 codec), but expected at very regular intervals (e.g., one packet every 20 ms). The receiving end has a buffer that permits a tolerance to delay and jitter, but to the scale of one or a few hundreds of milliseconds: As the number increases, the perceived quality of the call decreases. By comparison, the quality of reception of an email does not suffer if it is delayed, even by 500 ms or longer. Thus, providing equal access opportunities to STAs without accounting for the traffic they sent soon proved suboptimal, especially in enterprise contexts. 802.11e-2005 introduced (among others) provisions to solve this limitation, by introducing a Hybrid Coordination Function (HCF) in addition to DCF. HCF comes in two flavors, but only one is widely used and of interest for our discussion: Enhanced Distributed Channel Access (EDCA).

EDCA recognizes that applications making socket calls to the operating system, with the goal of transmitting some data, can qualify the type of traffic they intend to send. The inner workings of operating systems are irrelevant to 802.11, but an outcome of this qualification is that the packet, at Layer 3, can be marked with a Differentiated Service Code Point (DSCP) label that indicates the traffic type, as well as its sensitivity to jitter and delay.[55] When this information is available, the Layer 2 process can convert this DSCP value into a User priority (UP) value. The UP can then be associated to an 802.11 Access Category (AC). Each AC can be given different rules to access the medium, removing the limitations just described. For example, it can provide more opportunities to voice packets than to email packets. We will see in the next chapter (and in Chapter 7, on enhanced SCS) that this mechanism also suffers from some limitations. However, it provides a better traffic-type awareness to the access method.

With EDCA, four access categories are defined:

- AC_VO for voice traffic (UPs 7 and 6)
- AC_VI for video traffic (UPs 5 and 4)
- AC_BE for best effort traffic (UPs 3 and 0)
- AC_BK for background traffic (UPs 2 and 1)[56]

55 See RFC 2474, 2475, and 4594.
56 See 802.11-2020, Table 10-1.

With DCF, each incoming packet is converted to an MPDU and added to the STA's frame queue (and the DCF rules apply to the next frame to send). With EDCA, the STA maintains four parallel queues, one for each AC. The queues are parallel, in the sense that the STA counts down frames in each queue. It also accepts that two or more queues may have frames ready to be transmitted (i.e., they reach 0 at the same time).

Each AC operates under different channel access rules. The CS and CCA mechanisms are the same as before, but the backoff timers change. Each AC is allocated an Arbitrated Interframe Space Number, AIFSN[AC]. Then, instead of waiting one DIFS (remember, DIFS = 1 × aSIFSTime + 2 × aSlotTime) before declaring the channel idle and available for the next transmission, the STA needs to wait one AIFS[AC] = 1 × aSIFSTime + AIFSN[AC] × aSlotTime.[57]

AC_VO and AC_VI are assigned an AIFSN of 2, AC_BE receives an AIFSN of 3, and AC_BK receives an AIFSN of 7. These numbers seem to indicate that DIFS is assigned the same statistical access chance to the medium as AC_VO or AC_VI. However, EDCA also implements different CWMin and CWMax values for these ACs. By default, CWMin and CWMax are the same as DCF for AC_BE and AC_BK, but they are much shorter for AC_VO [CWMin = (aCWMin + 1)/4 − 1, CWMax = (CWMin + 1)/2 − 1] and somewhat shorter for AC_VI [CWMin = (aCWMin + 1)/2 − 1, CWMax = aCWMin]. This way, because the range within which CW is picked is smaller, AC_VO and AC_VI get a relatively better statistical chance to access the channel than they would under DCF.

EDCA brings an additional advantage: the transmission opportunity (TXOP). With DCF, when a STA gains access to the medium, it can send a frame, then has to contend again for the next frame. With EDCA, the STA can instead continue sending frames of the chosen AC,[58] for up to a maximum amount of time (the TXOP limit). This way, a STA can empty its buffer while it has access to the medium. This mechanism also limits the "dead air"—the amount of time wasted during which no STA transmits (they are all counting down).

Table 1-4 summarizes the differences between the ACs, and between EDCA and DCF.

TABLE 1-4   802.11 EDCA and DCF Parameters*

AC	CWMin	CWMax	AIFSN	TXOP Limit
AC_VO	3 (aCWMin + 1)/4 − 1	7 (aCWMin + 1)/2 − 1	2	2.080 ms
AC_VI	7 (aWCMIN + 1)/2 − 1	15 (aCWMin)	2	4.096 ms
AC_BE	15 (aCWMin)	1023 (aCWMax)	3	2.528 ms
AC_BK	15 (aCWMin)	1023 (aCWMax)	7	2.528 ms
DCF	15 (aCWMin)	1023 (aCWMax)	2	0

*This table assumes OFDM in 5 GHz. Times are in aSlotTime units. See 802.11-2020, Table 9-155, for other PHYs.

57 See 802.11-2020, clause 10.23.2.4
58 See 802.11-2020, clause 10.2.3.2 d

Summary

In this chapter, we reviewed the basics of the 802.11 standard. A STA (non-AP STA or AP STA) can contend to access the medium with a low probability of collisions. A non-AP STA can discover APs and associate to one of them, then apply one of many security schemes. The STA can then send frames while providing better statistical opportunities to traffic with a higher sensitivity to jitter and delay. However, the schemes necessary to achieve these goals are associated with a notable set of constraints: The STA is a function, and the non-AP STA can associate to only one AP at a time. The AP is also a STA, and all STAs in the BSS have the same tools for transmission, leaving very few opportunities for time-sensitive traffic (think AR/VR) or deterministic traffic to be scheduled exactly when it needs to be sent. When a STA reaches the edge of the BSA and needs to roam, the STA has no solution other than wasting time on sequential dialog with two APs. All of these elements present serious challenges for Wi-Fi 7 design ideas. In the next chapter, we will see that efficiency mechanisms were designed, before 802.11be, to make QoS and roaming more efficient, but they also came with an additional layer of constraints and challenges.

Chapter | **2**

Reaching the Limits of Traditional Wi-Fi

association: the service used to establish a mapping between an access point (AP) or personal basic service set (PBSS) control point (PCP), and a station (STA) and enable STA invocation of the distribution system services (DSSs).

802.11-2020, Clause 3

The long history of 802.11 has led the standard to incorporate new scenarios and new mechanisms as it grew in maturity and complexity. But each time a new possibility emerged, it had to be balanced with the need to "not break what was already there." A new device, incorporating new methods for better and faster services, could be positioned near an older device; however, that older device should not be stripped of its ability to operate for the sole benefit of the newcomer.

These complex and sometimes contradicting requirements have led the 802.11 designers to build careful pyramids of techniques that enable new features without destroying the experience of STAs (non-AP stations) supporting only older versions of the protocols. In turn, owing to this need for backward compatibility, some of the methods integrated into 802.11 to solve particular scenarios relevant to 802.11be provide only imperfect solutions. This forced the 802.11be designers to consider their limitations and explore new avenues.

The Burden of Roaming

From its first versions, the 802.11 standard included the notion of an Extended Service Set (ESS), because it was designed around the idea of mobility. A STA should be able to roam—that is, to move from one Basic Service Set (BSS) to another (supporting the same Service Set Identifier [SSID]). However, the standard did not describe how the STA would determine that it is reaching the Basic Service Area (BSA) edge, and would choose to go to this AP instead of the next AP. In some ways, the discovery of the next AP was considered the same problem as the discovery of the first AP: The STA would scan (see Chapter 1, "Wi-Fi Fundamentals"), then would select the next AP in the same way it selected the first one.

This approach soon proved insufficient, and observing a STA perform this operation is enough to understand why. As the STA moves away from the AP, its data rate shifts downward. The STA has no idea of distance to the AP. The STA examines specific messages from the AP (usually the beacons), and has a vendor-defined threshold below which it considers that the Received Signal Strength Indicator (RSSI) of these messages indicates the edge of the BSA. At that point, the STA is set to move to a discovery mode for the next AP. In parallel, the STA exchanges data frames through the AP. This operation is usually different from the BSA edge and discovery processes. In some implementations, a rapid degradation of the message exchange performances (caused by fast rate down-shifting, fast increase of retry counts, or failure to receive an acknowledgment for a certain number of transmissions attempts at the lowest possible rate) can also trigger the roaming process.

Next-AP Discovery

In all cases, the STA needs to start discovering other BSSs for the same SSID, which usually means leaving the current channel. However, the AP may be sending frames to the STA while it is away. There is no mechanism in 802.11 for the STA to tell the AP, "Please wait; I am scanning other channels"— even if this is a natural operation for all STAs, in all BSSs, of all Wi-Fi networks on the planet.[1] A common workaround for STA implementations is to send a null frame,[2] whose purpose is to provide updated parameters to the AP. In the null frame header, the power management bit[3] is set, indicating that the STA should go into power save mode and, therefore, be unavailable to receive frames from the AP. At that point, the STA can leave the channel and start exploring (scanning) other channels for other APs. Once it has assessed the presence of other APs on all channels, the STA can return to its active channel and send another null frame, this time with the power management bit unset. The unset bit indicates to the AP that the STA is out of the power save mode, and can receive frames if the AP has any to send.

In some cases, the STA might be in the middle of an exchange when this scanning process needs to start, or it might receive frames in its buffer (from the operating system) while it is scanning on other channels. These contradictory requirements—scan other channels, or stay on the channel to exchange with the AP—can lead the STA to make only short incursions on other channels, then come back briefly to the active channel to send or receive one or a few frames, before going away again to explore a few more channels. This back-and-forth can happen multiple times. The whole scanning process can then take several seconds, especially in bands with multiple channels (in the FCC domain, the traditional 5 GHz band includes 25 channels). This scenario becomes all the more difficult when the STA is searching for a better AP. Having reached the edge of the current BSA, the STA's data rate is likely low (so each segment of data takes a long time to send), and its retry rate high (increasing even

---

1 In this chapter, the footnote references point to the clause of the 802.11-2020 standard, or the 802.11be draft standard, where you can find more details, or the exact text, of the concept discussed here.

2 See 802.11-2020, Table 9-1

3 See 802.11-2020, clause 9.2.4.1.7

more the time needed to successfully transmit each data frame). These constraints mean that the STA should maximize its time on the channel to prevent its buffer (or that of the AP) from overflowing with untransmitted packets (which come in faster than the radio can successfully transmit). Yet this is the time when the STA is asked to leave the channel for a prolonged period of time. The whole process does not seem efficient.

802.11k

The 802.11 designers realized early on that the roaming process needed to be improved. In 2002, they created the 802.11k group, tasked with working on radio resource management exchanges. These exchanges could be used to help the STA and the AP "see through each other's eyes," by asking one side to perform some measurements and report to the other.

One such report is the Neighbor Report. The STA starts by asking the AP (to which it is associated) for a neighbor report.[4] This request indicates the SSID in which the STA is interested; it is typically the SSID the STA is using, but the process allows in theory any SSID value, or a null value to ask for "any" SSID. The AP then sends a neighbor report response,[5] which includes zero or more neighbor report elements.[6] Each element describes an AP of the ESS, with (among others) its BSSID, its channel, and whether the AP is reachable (i.e., likely in range of the STA, because it is in range of the reporting AP). The standard does not say how the AP builds this list. In many implementations, the APs detect each other over the air by a form of radio resource management process that prompts each AP to occasionally scan other channels and, over time, discover its AP neighbors within beacon signal detection range.

> **Note**
>
> The process described in the preceding paragraph is the neighbor report. The 802.11k amendment also allows the AP to query the STA, and have it scan one or more channels either dynamically (active scan report) or passively (passive scan report), or report on the APs that it has detected earlier and that are still in the STA memory (table report). The exchange is the Beacon Report— because in theory the STA detects the AP beacons, although in practice it can also collect probe responses. Do not confuse this process with the neighbor report: One is the converse of the other.

The 802.11k neighbor report was designed to be useful for the STA roaming scenario. Instead of scanning possibly tens of channels, the STA would receive only a subset of BSSIDs and channels of interest (e.g., half a dozen), which would speed up its discovery process. One difficulty lies in the idea of trajectory. Imagine 5 APs in a line (e.g., in a long corridor), named sequentially (AP1, AP2, …, AP5). If the STA associates to AP3 and asks for a neighbor report, AP3 might indicate AP1, AP2, AP4, and AP5. However, if the user of the STA is moving along the corridor toward AP4 (and then AP5),

---

4 See 802.11-2020, clause 9.6.6.6
5 See 802.11-2020, clause 9.6.6.7
6 See 802.11-2020, clause 11.10.10 for the general choreography, and clause 9.4.2.36 for the element.

providing information about AP1 is not very useful; neither is information about AP2, from which the STA might have roamed. It is also possible that AP3, positioned on the ceiling, would detect AP6 in another room, which the STA, near the floor level, might not detect. It is also possible that the STA might detect an AP7 that AP3 might not detect from its position.

In other words, the 802.11k neighbor report mechanism might shorten the discovery process in a roaming scenario, provided that the AP and the STA have the same view of the RF environment, and provided that the AP has some form of knowledge of the STA movements. Theoretically, the AP could receive a neighbor report request, respond with a beacon report request, collect the view of the STA, and then know what the STA sees from its position. This process would, of course, defeat the purpose of the neighbor report: The STA does not need the AP's help once it has scanned by itself to produce the beacon report. The AP could also use the beacon report from one STA to infer the view of another STA at the same position. However, this process would require the AP to sacrifice a first STA, by asking it to produce a beacon report when all the STA wants to do is to regain good connectivity. The process also presumes that the STAs would be willing to share their location. Notably, 802.11k allows the exchange of location information, but this process has privacy implications that make its support uncommon on STAs.

Beyond these procedural difficulties, another major limitation is that the STA still needs to leave its active channel to scan, from a position where the STA's traffic already suffers from losses and slow data rates. Thus, the 802.11k neighbor report was an improvement, but has many possible limitations (that are implementation-dependent).

802.11v

The 802.11v-2011 amendment was developed to improve the BSS management functions. The standard defines a series of elements that one side (the STA or the AP) may provide that could be useful to the other side. Among them, BSS Transition Management (BTM) exchanges allow an AP to instruct a STA to roam to a specific AP, or an AP in a set of preferred APs. The initial goal was load balancing, to improve the overall throughput (or QoS) of the ESS, by moving some STAs from one BSS to the other. Some of the explicit targets were sticky clients—that is, STAs that move away from one AP, get closer to a second AP that can provide a much better connection, yet stay associated to the first AP for a variety of reasons (see Chapter 4, "The Main Ideas in 802.11be and Wi-Fi 7"). The AP could instruct that sticky client to move to the other AP. As the STA has no reason to comply, the BTM message can include a disconnection timer (i.e., "If you have not roamed to that AP within x TUs, I'll disconnect you").

The 802.11v BTM process allows for an exchange,[7] in which a STA can first send a BTM query, asking the AP for the list of preferred next APs (in case the STA needs to transition to another BSS). The AP responds with a BTM Request, indicating the list of preferred APs to roam to, optionally with a maximum delay by which the STA should have roamed. In many cases, this list consists of a single target AP. The STA can respond with a BTM Response that indicates whether the STA will roam. If the

---

7 See 802.11-2020, Figure 6-18 and clause 11.21.7

STA intends to roam (within the delay mentioned in the request), then it can indicate to which AP (if the request list includes more than one AP). If the STA does not intend to roam, it can indicate why:[8] Perhaps the STA did not find the target AP(s) on the indicated channels, or perhaps these APs did not have the capacity to add the STA to their BSS. The STA can then indicate whether it prefers its session to be terminated (accepting that the AP would disassociate the STA), or whether it requests the AP for a bit more time.

This procedure is intended to facilitate BSS management and load balancing. However, a common scenario is a STA reaching the edge of the BSS and asking for the AP's help to find the next best AP. The STA sends the BTM Query, and the AP responds with the next best AP the STA could roam to. The process is delicate, because the STA is again at the edge of the BSA, in a configuration where its data communication to the current AP suffers. The AP's suggestion comes as a great help, because (in theory) the STA could just go to the channel indicated by the AP in the BTM request, find a great AP, and roam there. However, if the AP's advice proves wrong (the target AP is not reachable, or is not suitable for the STA's needs), then the process becomes wasteful. After having wasted a BTM exchange with the AP (instead of exchanging data), the STA leaves the channel to exchange with another AP (consuming even more time), just to realize that this other AP is not a candidate for roaming. Now the STA is just where it was when it first reached the edge of the BSA: in need of a next AP to roam to, with no indication of where to find it. The only difference from the beginning of the scenario is that the STA now has wasted time and energy on a discovery effort that was not useful, while traffic in its buffer continues to accumulate.

The BTM procedure is an improvement to the 802.11k neighbor report procedure in that the information provided is directly intended for the roaming process. The AP does not provide a map of the AP neighborhood, but rather points to an AP the STA can roam to. However, the BTM procedure suffers from the same limitations as the 802.11k neighbor report procedure. Without precise knowledge of the STA's position (and asking the STA for its location raises all sorts of concerns around privacy), the AP cannot easily provide the STA with information that the STA can use directly, without further scanning. The next-AP discovery process might be shortened, but it is not suppressed. If the STA cannot provide very detailed information to the AP (about its location), then the STA must be able to discover the next AP without stopping the data exchange with the current AP. At the time when the 802.11be design started, this requirement was impossible to meet.

A Limited QoS Scheme

When 802.11be was in its design phase, the issue of QoS also became a more prevalent concern. Well-known categories of applications like voice and video had been in use for many years, but new applications were making an appearance and demanded specific treatments. For example, how does one distinguish an enterprise Wi-Fi network video-conferencing for a business-relevant application from a video call on a personal application? Both send voice and video, but if congestion occurs, the business-relevant application should be prioritized over the personal application. The same issue occurs with

---

8 See 802.11-2020, clause 11.21.7.4

newer applications. How should the video flows sent by an augmented reality (AR)/virtual reality (VR) headset, which demands very low delay rates (users get headaches if the delay between motion and AR/VR screen refresh is too long; see Chapter 4), be treated, in comparison to a video-conference flow? The AR/VR demands 20 ms delay, while video conference tolerates 200 ms delay—yet both belong to the video category.

ADDTS

Recall from Chapter 1 that Enhanced Distributed Channel Access (EDCA) defines four access categories (AC).[9] All ACs use the same medium access principles, but different countdown numbers. As a result, AC_VO obtains a statistically better access to the medium than the others, followed by AC_VI, A_BE, and AC_BK (in descending order of advantaged access to the medium). The treatment of congestion should not be the same for all ACs, because different types of traffic have different needs.

In a wired network, where switches and routers have configurable buffer management capabilities, congestion can be addressed with policing or shaping. With shaping (Figure 2-1), traffic that enters an outgoing interface and that is in excess of the interface's configured transmit capability for that traffic type (the service rate) is delayed. The packets are kept in the buffer (usually with a target maximum amount, the line rate) and sent through the interface over time. This approach is useful for traffic that can be partially delayed, such as a web page sent to a browser, or a streaming video flow that will be buffered in the device before being played.

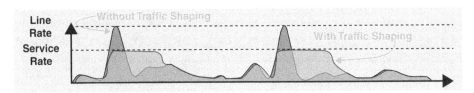

FIGURE 2-1   Traffic Shaping Principles

With policing, traffic that enters the interface buffer in excess of the service rate is simply dropped (deleted). This approach is useful for traffic that should not be delayed. A typical example is real-time voice. When a voice packet is delayed, there is no guarantee that it will reach the intended recipient in time to be played. When a voice packet is missing on the receiver side, the application usually plays an intermediate sound between the previous packet and the next packet. As each packet is typically a short segment of audio (e.g., 10 ms or 20 ms), the user does not perceive any issue, at least not until multiple consecutive packets are missing. If the missing packet arrives later, it cannot be played and is deleted at the receiver anyway. Therefore, there is no point in delaying the packet transmission with shaping. If packets arrive in a sequence in excess of the service rate, they are deleted; the receiver is then in charge of filling the gap.

9 See 802.11-2020, Table 10-1

Of course, the best design would ensure that no interface gets oversubscribed. A good design practice for voice is to measure the number of concurrent calls that each network interface can forward (at the service rate), and allow only that number of calls. This idea was present in the mind of the 802.11e designers, and they complemented the EDCA structure with an Admission Control Mandatory (ACM) mode. In the AP beacons and probe responses, an EDCA parameter set[10] indicates the parameters for each AC (AIFSN, CWMin, CWMax, TXOP limit), and can also indicate if ACM is enabled for each individual AC. The standard does not expect ACM to be enabled for AC_BE and AC_BK.[11]

When ACM is enabled (e.g., for AC_VO), any STA that intends to send traffic in that AC needs to request admission from the AP. The STA first sends an ADDTS request action frame[12] that describes the intended traffic with a Traffic Specification (TSPEC) element[13] and zero or more optional Traffic Class (TCLAS) elements.[14] The TSPEC indicates elements such as the MSDU minimum and maximum sizes; the service intervals (the pace at which these MSDUs are expected); and the minimum, peak, and maximum data rates. The TCLAS elements, when present, can add further information helping to classify the flow, such as the source and destination IP addresses and ports, the DSCP value, and more. Using this information, the AP (or a network management platform behind the AP) validates whether the flow is authorized and is within the maximum admissible volume; it then returns an ADDTS response action frame[15] that either allows or declines the flow. When the flow is declined, the STA can request admission to another AC (if that other AC also uses ACM), or send the traffic to an AC that does not implement admission.

This scheme was efficient in 2005. At that time, the Wi-Fi devices that supported voice were specialized Wi-Fi IP phones, and network designers would implement voice-specific SSIDs. Unfortunately (for the ACM design), a few years later, the smartphone explosion brought voice support to most Wi-Fi devices, as over-the-top (OTT) voice apps (allowing free or cheap calls over the data network) started to appear for mobile phones, tablets, and laptops. In that environment, two difficulties emerged:

■ Specialized SSIDs started to disappear, while general SSIDs (e.g., "corporate") would accommodate all traffic. In such a scenario, implementing ACM became problematic, because some client devices would support the feature (the specialized Wi-Fi IP phones), but many would not (smartphones, laptops, tablets).

■ Most OTT application developers had no idea about 802.11 specifications, and were not able to provide (from the application to the 802.11 driver) the information necessary for the ADDTS request parameters.

As voice (over Wi-Fi) support became mainstream on all STAs, smartphones, and general devices (tablets, laptops), the need for Wi-Fi IP phones receded. With these constraints, the usage of ACM

10 See 802.11-2020, clause 9.4.2.28
11 See 802.11-2020, Annex K2
12 See 802.11-2020, clause 9.6.3.2.1
13 See 802.11-2020, clause 9.4.2.29
14 See 802.11-2020, clause 9.4.2.30
15 See 802.11-2020, clause 9.6.3.3.1

faded. ACM was soon removed from the networking best practices, because enabling ACM would prevent general corporate devices from accessing AC_VO. In the early stages of designing 802.11be, using ACM to provide better treatment to a corporate voice application over a personal voice application, or to prioritize an AR/VR flow, was no longer a viable option.

UPs versus ACs

Another direction that emerged was the smart use of User Priorities (UPs). It is often surprising to read that there are eight UPs, but that the AC determines the access to the medium. Why, then, are there eight UPs and not four (in essence, one per AC)? The question is valid, and the answer comes from the history of 802.11. When the EDCA system was designed (between 2000 and 2005, with the development of 802.11e-2005), QoS was undergoing deep transformations in all standard bodies. Specifically, it was evolving from a mechanism where one traffic stream would be given precedence over all others, to a method where classes would be defined, each with its own characteristics and needs. The IEEE 802.1p group, which was active at the end of the 1990s, had improved several of the 802.1 standards with the idea that different types of traffic should receive different types of treatments in network nodes.

> **Note**
>
> Recall that the IEEE 802.1 working group is concerned with topics like general 802 LAN/MAN architecture, and internetworking between 802 LANS, MANs, and WANs. The group can produce documents that other groups, such as the 802.11 group, can leverage for the needs of the particular medium they work with.

Treatment can represent many actions, from queueing depth to overflow management techniques, but also carry an idea of prioritization, which is of interest to us in this context. In particular, the 802.1p group introduced in the 802.1D standard (its 1998 revision) the idea of seven traffic types: network control traffic, voice, video, controlled load, excellent effort, best effort, and background.[16] The capability of a network node to handle these categories would depend on the number of queues that it could support. In the best-case scenario, the network node would have seven queues and assign one queue to each of these categories.[17] As the 802.1D standard defines the operations of a MAC bridge (a link between two 802 types of mediums or networks), it was a useful source for the work of the 802.11e task group.

The 802.11e TG then adopted the 802.1D seven categories, plus an eighth category (spare) that was listed in 802.1D but not defined. The spare category was used as an additional queue for traffic that would need specific treatment, and that would not be described in one of the other seven categories. This eighth queue was useful for 802.11e, intending to use 3 bits[18] to encode the user priority

16 See 802.1D-1998, clause H2.2.
17 See 802.1D-1998, Table H-14.
18 The UP is coded over 3 of 4 bits of the TID field. See 802.1-2020, clause 5.1.1.3 for the meaning of TID values 8 to 15.

value (3 bits allow 8 binary values, with the decimal equivalent of 0 to 7). This 802.1D system translates into the series of values displayed in Table 2-1.

TABLE 2-1 Traffic Types and Names

User Priority	Traffic Type	Acronym
1	Background	BK
2	Spare	—
0 (default)	Best Effort	BE
3	Excellent Effort	EE
4	Controlled Load	CL
5	Video	VI
6	Voice	VO
7	Network Control	NC

The early experiments performed by 802.11e participants revealed that access to the 802.11 medium was very stochastic. It was also too dependent on complex sets of conditions (number of stations, the types of frames they send, the size of these frames, their intervals, the data rates used by the STAs, and the retry rate, just to name a few) for a protocol to easily create eight queues that would provide clearly distinct and predictable (at the individual frame level) performances. Four queues seemed to be the best a stochastic medium like that delineated by 802.11 could support. Consequently, the 802.11e TG grouped these eight UPs into four ACs, and created access rules per AC.

This clubbing of UPs, two by two, into their respective ACs created philosophical challenges even in the early days of 802.11e. Nevertheless, the general idea that ACs would be the maximum level of granularity allowed the group to continue their work without resolving these challenges. However, these challenges re-emerged soon after—first in 802.1, and then in 802.11. They stem from the very definitions of the 802.1D traffic categories:

■ Network Control is intended for traffic that is needed for the network to operate (e.g., router-to-router exchanges). This traffic type makes sense on an Ethernet network, through which network nodes may communicate about paths to various destinations. But 802.11 is primarily an access technology, connecting end-devices to DSs via APs. It is very uncommon to see network traffic sent over this type of medium. Other specifications[19] make it clear that UP7 should never be used for data (user) traffic. Therefore, the UP exists in 802.11, but the value is by definition likely to be never used, outside of very specific cases (e.g., mesh backhauls connecting routers).

■ Controlled Load was intended for business applications that would need priority but would receive some form of admission control. It became part of the general video queue in 802.11. However, no video queue (in the days of 802.11e creation) had any form of admission control.

19 See, for example, 802.1-Q-2003, clause 8.6.5.

- Excellent Effort made sense, when grouped with best effort. Best effort is the service that any traffic receives by default, and excellent effort would be treated a bit better (it was called "CEO best effort" at some point).

- Background was intended for traffic of low importance—that is, traffic that would be sent only if nothing else (including best effort) was asking for resources. In other words, background traffic is sent after everything else, including best effort traffic.

- Spare was not used. 802.11e clubbed it into the background AC along with UP1.

In 2005, the 802.1Q standard (Virtual Bridges) started integrating elements of 802.1D (and eventually replaced it—802.1D was deprecated in 2014). The group realized that the position of the spare queue could not be near the background queue. If best effort is the service provided when traffic has no specific prioritization need, and if background is the traffic that is sent when nothing more pressing (including best effort traffic) needs access to the medium, what exactly would be the treatment of a "spare" category between best effort and background? That position was nonsensical. So, the group moved UP2 above best effort to create an additional priority above best effort. However, 802.11e, with two UPs in the background category, was already (on the verge of being) published.

So, from its inception, 802.11e came with eight UPs, of which two (UP7 and UP2) could not really be used. This issue was raised when the 802.11aa group (formed in 2008) was looking at video transport streams. The group was working on the observation that some flows in a given AC (video in particular) could be of the same general nature (i.e., both might be video flows) but of different importance. The group then recognized that most implementations used only a single UP in each AC (UP6 in AC_VO, UP5 in AC_VI, UP0 in AC_BE, and UP1 in AC_BK). The 802.11aa TG then created the notion of an alternative queue, to allow a flow to use the second UP in each AC, as a way to signal that the matching flow was special in the category.

For example, all videos would be marked UP5, but the CEO video would be marked the special UP4. The special marking would appear in the QoS field of the frame header, but could also be communicated between the STA and the AP in an intra-access category priority element.[20] This exchange would allow a STA, requesting the admission of an upstream traffic (see the "Stream Classification Service" section in this chapter), to signal to the AP that the flow would be special. This mechanism is present in the 802.11 standard, but underlines additional difficulties:

- The alternative queue cannot really apply to background traffic, because there is no traffic that can be "special" and require a careful treatment in a category where all traffic is of low importance.

- The alternative queue cannot really apply to best effort traffic, because best effort traffic is already using UP0, and special traffic in this category is already using UP3.

---

20 See 802.11-2020, clause 9.4.2.120

- The alternative queue applies to video traffic, but with a strange mapping. That is, traffic using the lower UP4 requires a better treatment than traffic marked with the higher UP5, in a hierarchical system where higher UP suggests higher-priority needs.

- The alternative queue cannot really apply to voice traffic, because UP7 should be used only for network control traffic—not for data, and therefore not for voice.

Thus, the alternative queue could really be used only for videos, with severe semantic limitations, and the ability to flag just "normal" versus "special," without any further details. At the time of the 802.11be design work, it was clear that the UP system was not by itself a sufficient vehicle to properly distinguish AR/VR from video-conferencing, or personal video from business-critical video.

Stream Classification Service

Another promising direction came in the form of the Stream Classification Service (SCS), also defined at the time of 802.11aa-2012 development. SCS allows a STA to request the AP for specific QoS treatment.[21]

In this process, the STA starts by sending an SCS Request (action) frame.[22] The frame contains one or more SCS Descriptor Lists,[23] each of which includes zero or more TCLAS elements that allow the STA to describe the traffic of interest. The goal (and the main difficulty) of the TCLAS section is to describe the traffic in a way that is precise enough that the AP will recognize the matching application. For this reason, more than one TCLAS element per application might be needed, and they might need to be complemented with a TCLAS Processing element[24] that describes whether each PDU should match all the criteria in the TCLAS or only some of them. If the TCLAS element would not be sufficient to properly describe the traffic, the SCS Descriptor List can include optional vendor-specific elements; these elements could, for example, provide an application name or identifier that both the STA and the AP would understand. The TCLAS element also indicates how the traffic should be classified—that is, which UP the AP should use when forwarding this traffic from the DS to the STA. Within the context of 802.11aa, the SCS Descriptor List can also include an intra-access category priority element,[25] which can indicate the requested alternative AC UP value for the traffic, along with a drop eligibility.[26] Additionally, the SCS Descriptor List includes a request type that clarifies whether the STA is asking for a new stream classification service, is changing an existing one, or removing the request (along with the flow). Finally, the SCS Descriptor List includes a Stream Identifier, called the SCSID. The request can include more than one SCS Descriptor List. For example, the STA might want to describe the audio flow and the video flow of a video-conferencing application.

The AP receives and examines the request, applying decision processes that are outside of the scope of the standard. In most cases, the AP compares the request to some network QoS policy configuration.

---

21 See 802.11-2020, clause 11.25.2
22 See 802.11-2020, clause 9.6.18.2
23 See 802.11-2020, clause 9.4.2.121
24 See 802.11-2020, clause 9.4.2.32
25 See 802.11-2020, clause 9.4.2.120
26 Setting the Drop Eligibility flag indicates that, if resources are constrained, a frame for this flow can be discarded.

The AP then returns an SCS Response.[27] This response indicates, for each SCSID of the request, a status value, showing whether the AP accepts or rejects the request. If the SCS request is accepted, then the AP starts marking the matching downstream flows as requested by the STA. In most scenarios, an unspoken expectation is that the STA will also mark the upstream traffic in the same way.

The SCS process is a great direction because it allows the STA to express that one or more applications require special treatment. One difficulty, however, relates to the identification of that application. The TCLAS elements might describe the application with IP addresses, ports, or other elements, but it is not guaranteed that the AP will be able to recognize the application from these elements alone. This recognition has to happen for the AP to decide whether the request is accepted or rejected, as well as for the AP to identify incoming flows (from the DS) and mark them accordingly. This issue was not easily solved at the time when the 802.11be design work started.

> **Note**
>
> This problem proved so dire that, in the 802.11-2020 revision of the standard, the Mirrored Stream Classification Service[28] (MSCS) was added. With MSCS, the STA sends a request with an MSCS Descriptor element, similar in principle to that of SCS. But the STA simply tells the AP, "When you see this stream in reverse (coming back from the DS), just mark it with this UP." The AP does not need to recognize the application—just identify the application parameters (e.g., server IP address and port) in its incoming flow to mark it properly.

Another limitation of SCS is that its goal is merely to mark the downstream UP. However, at the time of the 802.11be design, 802.11ax was well on its way and had defined scheduling capabilities for upstream traffic (see Chapter 3, "Building on the Wi-Fi 6 Revolution"). With SCS, the AP has no way to know which type of latency an AR/VR flow can tolerate and has no possibility to ensure that it has the bandwidth to forward the incoming downstream flow with a delay rate that matches the STA app's requirements. The AP also has no way to know how to help the STA send its upstream traffic. The STA applies the regular EDCA rules, and competes with all other STAs the same way. In short, SCS was promising, but insufficient for the 802.11be goals.

Coexistence Challenges

The 802.11 standard was built upon 20 years of improvements and accelerating communication speed. However, as it operates in unlicensed bands, Wi-Fi always expects that non-802.11 and older 802.11 devices might be present, and the new standards have to account for these possibilities.

Coexistence is not just a courtesy aiming at letting older protocol devices access the medium. The coexistence process is critical to avoid collisions.

---

27 See 802.11-2020, clause 9.6.18.3
28 See 802.11-2020, clause 11.25.3

ERP Protection

The early mechanisms allowed for both individual and BSS-wide protections. When 802.11g-2003 enabled OFDM transmissions in 2.4 GHz networks that were previously 802.11 and 802.11b only, the problem of collisions became obvious. 802.11g accommodated older preambles for compatibility purposes, but also new preambles incompatible with the older variants of 802.11. Additionally, the OFDM transmission scheme could not be understood by the older devices (those supporting only DSSS). Therefore, the 802.11g amendment described several protection mechanisms:

- **BSS-wide mechanism:** When an AP would detect an 802.11/802.11b device in the vicinity of the 802.11g BSS,[29] it would enable bits in an element of the beacon and probe responses called the ERP element.[30] One bit (nonERP_Present) would inform the STAs in the BSS that a non-ERP[31] STA was detected, and another bit (Use_Protection) would indicate that 802.11g STAs should use the RTS/CTS protection (discussed next). One problem with this scheme was that the detection mechanism was flexible, so the AP could report nonERP_Present upon detecting all sorts of frames, including probe requests from non-802.11g devices. In turn, neighboring APs could forward this warning (because a neighboring AP reported a nonERP device), causing waves of protection to be launched across entire deployments for one 802.11b device present on the floor (and not associated to any AP).

- **Individual frame protection mechanism:** When an 802.11g device was informed that non-ERP devices were present and that protection should be used, the 802.11g STA would start each frame transmission with an initial frame, the request to send (RTS). This control frame includes (in addition to the Frame Control field detailed in Chapter 1) a duration field, followed by the RA and TA fields (then the FCS). The duration field indicates the intended duration of the entire exchange (including the data frame that follows this exchange). The frame is sent at legacy modulation (1 Mbps), which means older devices can receive and demodulate it, read the duration value, and stay silent for that duration. The RA receives the CTS, and responds with a Clear To Send (CTS) response, which also includes the Frame Control field, and the Duration value field. This doubles the chances that nonERP devices will be made aware of the upcoming frame exchange. The CTS includes the RA, but not the TA (which is obvious, when the CTS comes in response to an RTS). In some cases, a STA might skip the exchange and directly send a CTS, where the RA is its own address (CTS-to-self), to accelerate the process. The STA would then send an ERP/OFDM frame. The nonERP STAs would not be able to demodulate and decode that frame, or even detect that a frame was being transmitted. However, the duration values in the RTS/CTS frames were sufficient to keep them from transmitting during that time.

The latter process was "heavy," because the RTS/CTS exchange would consume airtime. This additional overhead would more than halve the overall throughput of the BSS.

---

29 See 802.11-2020, clause 10.27.2
30 See 802.11-2020, clause 9.4.2.11
31 802.11g introduces an extended rate PHY, so that 802.11 and 802.11b devices are nonERP.

802.11n Protection

802.11n-2009 (Higher Throughput [HT]) also allows for the protection afforded by RTS/CTS, but removes the need for nonERP_Present and Use_Protection warnings. Instead, 802.11n introduces the idea of a legacy physical header. The 802.11n STA can operate in greenfield mode and use an 802.11n preamble and PHY header. Alternatively, it can operate in HT-mixed mode, where the frame starts with a legacy header (L-STF, L-LTF, or L-SIG, as described in Chapter 1), followed by an HT header, as illustrated in Figure 2-2.

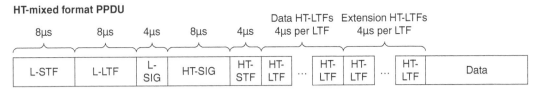

HT-mixed format PPDU

FIGURE 2-2   HT-Mixed Format PPDU

This format eliminates the requirement for an RTS/CTS exchange. As the STA starts its transmission, non-802.11n STAs read the legacy PHY header and understand that the medium will be busy. Even if they do not understand the HT header or the subsequent segments of the frame, the legacy STAs still know to stay silent for the duration of the exchange.

This protection scheme is much more efficient than the previous RTS/CTS-only method, but still not perfect. In particular, 802.11n allows operations over 40 MHz-wide channels, whereas previous versions of the standard allowed operations only over 20 MHz channels. Therefore, collisions could happen. For example, suppose an 802.11n AP is operating on channels {36, 40}, and a nearby 802.11a AP is operating on channel 40. The 802.11n AP could perform a CCA and conclude that both channels are free (because it is too far from the neighbor AP), even though the neighboring AP is actively sending and receiving on channel 40. The 802.11n AP could then send a frame over 40 MHz to a STA that, being closer to the 802.11a AP, would properly receive the transmission only on channel 36. The part on channel 40 would be undecodable due to a collision with the neighboring AP transmission. This scenario might not be extremely common, but it does happen. The legacy header did nothing to prevent this new issue.

802.11ac and 802.11ax Protection

802.11ac-2013 (Very High Throughput [VHT]) and 802.11ax (Extremely High Throughput [EHT]) also support the RTS/CTS mechanism, and also implement the idea of a legacy PHY header preceding the VHT PHY (or EHT PHY) header.

To protect the transmissions against the secondary channel collisions that affected 802.11n communications, 802.11ac introduced an additional idea: dynamic bandwidth operation.[32] With this type of operation, a VHT STA—for example, one operating a 80 MHz-wide channel—can send an RTS frame

32 See 802.11-2020, clause 10.3.2.9

that is actually a set of 4 PPDUs (one per each 20 MHz channel). The intended receiver can then respond with 4, or fewer, CTS frames (again, one per 20 MHz channel for which the responding STA has found available). This way, both sides can use the RTS/CTS exchange to express their view on the availability of the four 20 MHz channels, and negotiate the subset that is usable from both viewpoints.

This last protection mechanism nicely complements the others, and avoids the secondary channel collision issues that affected 802.11n transmissions. Such protection comes at a cost, however, and leaves some issues unresolved. A 40 MHz-wide channel consists of primary 20 MHz channels and secondary 20 MHz channels. An 80 MHz-wide channel adds a secondary 40 MHz channel to this scheme. Therefore, the position of the possible interferer matters. For example, suppose an 80 MHz-wide channel is set as 36 (primary 20), 40 (secondary 20), and {44, 48} (secondary 40). Suppose that some nearby system is found to communicate on channel 44 or 48, rendering them unavailable. A pair of devices hoping to exchange information over 80 MHz would have to revert to a 40 MHz exchange instead (over channels 36 and 40). But suppose that channels 44 and 48 are available, and the neighboring interfering system is on channel 40. The construction of the channel is {36}, or {36, 40}, or {36, 40, 44, 48}. Therefore, a pair of STAs hoping to exchange information over 80 MHz would be constrained to exchange only over channel 36 (20 MHz), even if channels 44 and 48 are free. The unavailability of the secondary 20 channel is equivalent to the unavailability of all channels beyond the primary 20.

Such a structure may have made sense in the construction of the channel, but it ensures that small interferences can have dramatic effects on the channel width available to all STAs. The 802.11be designers were well aware of these limitations, but redesigning the channel design would cause massive backward incompatibilities. Instead, they would have to explore other solutions.

802.11 Security

Security is another complex field covered in the 802.11 standard. In the early years of the standard, a weak security mechanism caused the whole technology to be called insecure. Since then, multiple generations of efforts have strived to make 802.11 security robust in multiple scenarios: at home, where simple schemes are required; in enterprises, where individual user and device authentication is needed; and in public places, where secure onboarding should be as seamless as possible. These diverse requirements have led the standard to support multiple security schemes,[33] all of which need to be incorporated into the 802.11be design.

Open System: With and Without RSN

Open System is the original method defined in the first 802.11 standard, and authentication is truly a link-level operation there: It validates that the STA is an 802.11 device, and does not introduce additional security concepts associated with identity validation. In the exchange, the STA starts by sending an 802.11 Authentication frame,[34] which is a management frame with 3 fields of particular interest: an

---

33 See 802.11-2020, clause 4.5.4.2
34 See 802.11-2020, clause 9.3.3.11

Authentication Algorithm field (set to 0, for Open System); an Authentication Sequence field, indicating the position of the frame in the exchange (set to 1, for the first frame in the exchange); and a Status Code (set to 0 for success, although the field has no real meaning when sent in that frame).

The AP validates that it can read the frame and, if successful, responds with an Authentication frame. The format is identical to the Authentication frame sent by the STA, but this time the Authentication Sequence field is set to 2, indicating the second frame in the exchange. The Authentication Algorithm is still set to 0 (Open System). The Status Code, still set to 0 (success), now indicates that the transaction completed successfully.

A STA can authenticate to as many APs as it wishes, thereby establishing an ability to communicate with them. A STA cannot associate to an AP without first authenticating.[35]

Once authentication completes, the STA attempts to associate to one of the APs it authenticated with, by sending an Association Request frame to that AP. This association frame, just like the authentication frame, is a management frame that can include several optional fields[36] and elements. It is essential, because it allows the STA to express to the AP its capabilities and the features it supports—from modulation and coding schemes (MCS) to security schemes, QoS, and more. The STA does not indicate these elements in a vacuum. In the discovery process, the STA should have scanned and obtained from the AP a beacon or probe response, both indicating the characteristics of the BSS. This information is used by the STA as a filter to decide if it can join that BSS (and which capabilities and features need to be enabled or can be supported).

The AP examines the request and the capabilities expressed by the STA, and responds with an Association Response frame.[37] The AP indicates in the response its own capabilities and support for various features; it is particularly useful if the STA suggests support for redundant schemes, and the AP selects one of them. The AP also adds a Status Code, confirming whether the STA association was successful (status 0) or not (non-zero status). The standard does not define the criteria by which an AP should accept or reject the association, as each AP administrator and/or their vendor might have different conditions for making that decision. However, it is clear that a mismatch in critical features can lead to such a rejection. If the association is successful, the AP provides to the STA an Association Identifier (AID), an integer uniquely identifying the STA in the BSS.

Part of the capability exchange indicates the two parties' support for security schemes. If none was indicated by either side, the RSN scheme is not in use, the AP and STA ports are uncontrolled, and both can start exchanging data frames as soon as the successful association frame response is received and acknowledged by the STA. If both the STA and the AP indicated support for RSN, the AP and STA ports are controlled and unauthorized. Part of this exchange then indicates support for an Authentication and Key Management (AKM) suite, representing the type of authentication algorithm to use, and a Cipher Suite, representing the type of encryption to use. In home networks, the AKM may be set to Pre-Shared

---

35 See 802.11-2020, clause 4.5.3.3
36 See 802.11-2020, clause 9.3.3.5
37 See 802.11-2020, clause 9.3.3.6

Key (PSK), referring to a passphrase configured on the AP and on the STA. In that case, the PSK is derived from the passphrase,[38] the PMK is the PSK,[39] and the STA and AP move to a 4-way handshake.

In enterprise networks, individual device identification is likely to be required. In that case, the 802.1X procedure is indicated in the AKM suite.

> **Note**
>
> 802.1X capabilities may also be useful in home networks. It is considered more robust than PSK, because the keying material does not need to be changed on all devices if a single device is lost. Unfortunately, support for 802.1X on home equipment is not very common.

Either side can proceed to the next step, but typically the STA triggers the dialog by sending an EAP-START frame. The AP continues with an EAP identity request frame, to which the STA responds with an EAP identity response frame. This exchange aims primarily to confirm which authentication method, within the EAP wrapper, the STA and the AS accept. The rest of the procedure depends on the method chosen. At the end of the procedure, both the STA and the AS will have generated the same MSK. The AS sends an access accept message to the AP as well as the MSK, and the AP forwards the success response to the STA to indicate an EAP success. At this point, the AP and the STA have the PMK, and continue to the 4-way handshake.

The 4-way handshake[40] is started by the AP, under the assumption that both sides have the same PMK. The AP sends message 1,[41] which is a management frame (subtype EAPOL-key) that contains, among other things, an identifier for the PMK (PMKID) and a random number, called Authenticator Number Once (ANonce).

> **Note**
>
> The PMKID can be used by either side in the future to reference the same PMK (e.g., during reassociation), thereby avoiding the need to reprocess the full PMK derivation.

While awaiting message 1, the STA also generates a random number, Supplicant Number Once (SNonce). With the ANonce, the SNonce, and the PMK, the STA can derive the PTK, using a Pseudo-Random Function (PRF[42]); its implementation details depend on the cipher suite agreed upon by the STA and the AP. At this stage, it is still possible that a malicious actor might have spoofed the AP MAC address and sent a fake message 1, but the malicious actor does not know the PMK. The real AP, if it is

---

38 See 802.11-2020, Annex J.4.1
39 See 802.11-2020, clause 5.9.2.2
40 See 802.11-2020, clause 12.7.6
41 See 802.11-2020, clause 12.7.6.1
42 See 802.11-2020, clause 12.7.1.2

the originator of message 1, knows ANonce (just like any observer seeing message 1). So, the next step is for the STA to reply to the AP with message 2[43] (type management, subtype EAPOL-key), which includes the SNonce. Here again, it is possible that a malicious actor might have seen message 1 and, impersonating the STA, may be attempting to send a fake message 2 to the AP. However, that malicious actor does not know the PMK. Therefore, as a way to authenticate its reply, the STA also uses the PTK to sign message 2 via a Message Integrity Code (MIC) process.[44]

Upon receiving message 2, the AP now has the SNonce (in addition to the ANonce and the PMK it already had). At this point, the AP can compute the PTK and use it to validate the MIC. The AP also verifies that the frame sequence number signaled by the STA matches its own—in case frames 1 or 2 had to be re-sent, or in case an attacker attempts to replay these frames. Moreover, the AP verifies that the cipher suite and AKM suite values, inserted in message 2, match the values that the STA signaled in its association request, as mandated in 8.5.3.3. This validation is not without consequences, because it means that the STA cannot use a security mechanism other than the one it requested in the association request frame, even if it discovered (e.g., in the AP association response) that another/more robust scheme was possible.

After these verification steps are completed, and if the MIC validation is successful, the AP installs the PTK and is ready to use it for unicast exchanges with the STA. The AP confirms this installation to the STA by sending message 3[45] (type management, subtype EAPOL-key). Message 3 also includes the GTK, encrypted with the PTK.[46] The frame is protected by a MIC as well.

Upon receiving message 3, the STA validates the MIC. If the verification is successful, the STA installs the PTK and GTK, and sends a confirmation response to the AP with message 4[47] (type management, subtype EAPOL-key). The message is also protected by a MIC. Figure 2-3 illustrates this choreography.

Shared Key

A deprecated Shared Key[48] authentication mechanism existed in the early versions of the IEEE 802.11 standard (pre-RSN). This scheme intended to validate that the STA would have knowledge of a key configured on the AP and all the STAs, the WEP encryption key, thereby verifying that the STA was valid and authorized for the BSS. Thus, the shared key sought to introduce a form of identity validation: The STA individual identity was not verified, but the scheme was supposed to verify that the

---

43 See 802.11-2020, clause 12.7.6.2
44 See 802.11-2020, clauses 12.5.3.3 and 12.5.5.3
45 See 802.11-2020, clause 12.7.6.4
46 This is the scheme described in the IEEE 802.11 standard and implemented in the Wi-Fi Alliance's Wi-Fi Protected Access 2 (WPA2) program. An older Wi-Fi Alliance program, WPA, defined a slightly different structure for GTK sharing. WPA is deprecated.
47 See 802.11-2020, clause 12.7.6.5
48 Do not confuse shared key schemes, which refer to the deprecated WEP exchanges, and pre-shared key schemes, which are newer home-centric schemes in which a passphrase is configured on the AP and the STAs.

STA belonged to the group of authorized devices, those on which the shared key was configured. Unfortunately, this scheme proved easy to attack and was deprecated.[49] However, it was the first scheme to insert into the standard the idea that 802.11 authentication was more than a validation of the STA's ability to transmit 802.11 messages.

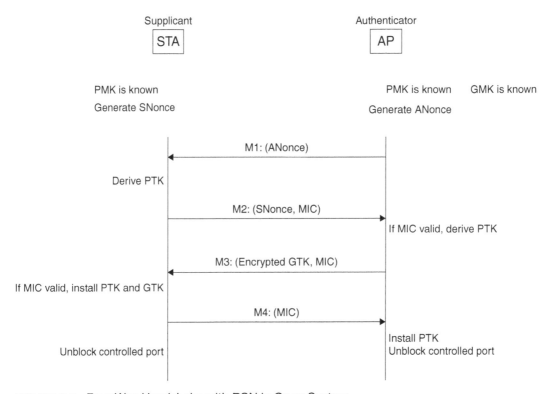

FIGURE 2-3   Four-Way Handshake with RSN in Open System

Shared Key authentication was a 4-frame exchange: The STA would start with an Authentication Request, receive a challenge from the AP in frame 2, respond to the challenge in frame 3, and conclude a successful authentication with a positive response from the AP in frame 4. This 4-frame exchange was followed by an association exchange similar to that of Open System. The STA and the AP could then start exchanging data frames immediately (no controlled port). Each data frame header would include elements that would allow the other side to deduce the key used to encrypt the payload, and therefore successfully decrypt the content. Unfortunately, an observer accumulating enough frames could also deduce the temporal and permanent keying material. The shared key scheme was declared deprecated in 802.11-2012.

49 See 802.11-2020, clause 5.1.2

FT Authentication and Association

The Fast Transition (FT) mechanism is defined in clause 13 of the 802.11 standard, integrating the 802.11r-2008 amendment. This scheme was designed to allow for early transfer of the keying material (and optionally QoS service obtained from the previous AP) for a STA between one AP and the next AP, thereby enabling a fast transition between BSSs. As soon as (re)association completes on the next AP, the STA is able to resume data exchanges. The scheme was not intended to perform an identity validation different from the 802.1X scheme, but merely to expedite the transmission of the keying material resulting from that validation from one AP to another as the STA roamed.

This requirement mandates that the key exchange must occur before association completes, whereas the standard scheme performs the 4-way handshake key exchange only after completion of a successful association. A direct consequence is that the STA needs to know if the next AP (here, called AP2 for simplicity) is able to retrieve key material from the current AP (here, AP1). This capability is expressed in both APs' probe responses and beacons, in the form of a Mobility Domain Information element[50] (MDE) that contains a unique 2-octet value identifying all APs in the ESS that can communicate keying material over the DS. The APs also express their support for FT, and also an FT policy, indicating whether FT negotiations should happen over the air or over the DS.

Naturally, sharing the same key over many APs presents a security risk. For this reason, FT implements a key hierarchy, with two levels. Each side implements both levels: the STA, called the Supplicant (S), and the infrastructure, called the authenticator or Resource (R). The first level is level 0, and the second is level 1.

As the STA associates to its first AP, the classical 802.1X process implemented in 802.11 sees a dialog between a STA/supplicant and an AP/authenticator, which results in a PMK on both sides. With FT, the dialog takes place between those same entities, but the outcome is a PMK of rank 0 (PMK-R0). Therefore, the STA/supplicant is also a PMK Key Holder, on the Supplicant side, of rank 0 (S0KH). Similarly, the AP/authenticator is a PMK Key Holder, but on the Resource side, of rank 0 (R0KH).

That rank 0 PMK is not intended to be used directly to generate PTKs. Instead, it is used to generate a second-level PMK (rank 1) used by each AP (and the STA) to generate PTKs. As the STA is the entity moving around, it is S0KH and S1KH, and therefore knows the PMK-R0 and the PMK-R1s. The first AP to which the STA associates is R0KH and R1KH (supposing that the STA effectively uses that AP to send and receive traffic), which also knows the PMK-R0 and its own PMK-R1. Each time the STA moves to another AP within the mobility domain, the PMK-R0 will be used to generate a PMK-R1 for that AP and the STA. Subsequently, each AP (and the STA) will use each individual PMK-R1 to derive the local PTK. Each key holder in the hierarchy is identified with a unique ID. Typically, the R1KH-ID and S1KHID are the MAC addresses of the AP and the STA, respectively. The S0KH-ID is also the MAC address of the STA, and the R0KH-ID can be any string, as long as it is not longer than a MAC

---

50 See 802.11-2020, clause 9.4.2.46

address.[51] In most implementations, it is either the MAC address of the first AP or the MAC address of a central entity (e.g., a WLAN controller) managing the APs in the ESS.

Thus, the FT authentication and association sequences are different from the Open System sequence with 802.1X or PSK. Likewise, the process is different on the first AP and on subsequent APs in the same ESS.

In the initial exchange (between a STA and its first AP), the first authentication exchange is Open System. The STA then indicates in the association request the MDE (it detected during the scanning phase) and a request for FT mode (in the AKM). In its response, the AP repeats the MDE for confirmation, and also indicates other key FT parameters (R0KH-ID and R1KH-ID).

These elements are information, in that they do not prove anything; instead, they merely inform the STA about the entities at play. As the goal is to pass keying material, RSN is expected so keys can be generated. Following association, the classical 802.1X authentication takes place (or is bypassed if PSK is used). The STA uses R0KH-ID as the identifier of the AP/authenticator, with which it exchanges during this phase. At the end of the procedure, both sides have the MSK and the PMK, in this case named PMK-R0.[52] The AP and the STA also derive the local PMK-R1 from the PMK-R0 value as well as the STA and the AP MAC addresses[53]—called S1KH-ID and R1KH-ID, respectively. The PTK is then derived from the PMK-R1 value, using the same logic as the 802.1X/PSK processes with 4-way handshake.

At this point, both sides have a local PMK. The logical next step is a 4-way handshake to generate the PTK. This exchange is similar to the one used with Open System and RSN, with the difference being that message 2 includes the same MDE and FTE values that the STA mentioned in the association. Message 3 from the AP includes these values as well. In a timeout interval element,[54] message 3 also provides a reassociation deadline or a key lifetime value, which informs the STA on how long this and other APs in the MDE will cache the keying material. Within that delay, the STA can start a FT process with another AP, then reassociate there. Beyond the delay time span, the FT exchange performed with another AP will have timed out, and the STA will need to perform the association steps from scratch. Figure 2-4 illustrates the FT initial association exchanges.

At some point in time, the STA reaches the edge of the current AP (AP1) BSA and decides that it is time to roam. The FT procedure allows the STA to make sure that a PTK is agreed upon with the next AP, so that the STA can resume data traffic immediately after the reassociation procedure. To achieve this goal, FT allows the STA to start the dialog with the next AP (AP2), in two possible ways:

- Directly Over The Air (OTA) with AP2. AP2 will then contact the R0KH to be able to generate a PMK-R1 (specific to AP2).

- Through AP1. AP1 will then establish a dialog with AP2 over the DS, pointing it to the R0KH and enabling it again to obtain a PMK-R1.

---

51 See 802.11-2020, clause 13.2.2
52 See 802.11-2020, clause 12.7.1.6.3
53 See 802.11-2020, clause 12.7.1.6.4
54 See 802.11-2020, clause 9.4.2.48

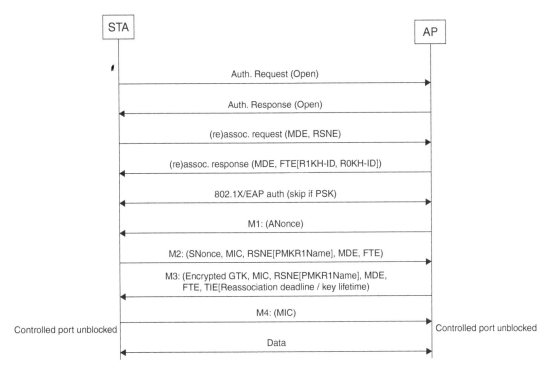

FIGURE 2-4 FT Initial Association

Both mechanisms competed with each other in the early days of the 802.11r group, and ended up being adopted together. Each of them has theoretical advantages and limitations:

- The OTA method allows the STA to pick an AP of its choice and directly start the dialog. It is therefore a logical consequence of the STA scanning function—that is, pick the best AP after scanning, likely the one with the strongest signal, and start the OTA dialog. However, the STA is still associated to another AP, probably on another channel. Any millisecond that the STA spends away from the channel where data communication happens may be detrimental for the STA's time-sensitive traffic. This concern might be particularly important when the STA asks AP2 to generate the PMK-R1, and then has to wait an unspecified time until AP2 responds (while AP2 contacts the R0KH).

- The over-the-DS method does not require the STA to leave its active channel. It can continue its data traffic while interleaving FT negotiations for the next AP roam through its current AP. However, the STA supposes that, because both APs have the same MDE value, they can reach each other. The STA also supposes that AP2, which was discovered through scanning, is still the best next AP as the FT exchange completes (even if the user is moving). However, being on AP1 channel, the STA has no good way of ensuring that AP2 is still the best next AP.

The mode (over-the-air or over-the-DS) is mandated by the AP, and expressed in its beacon, probe response, and association response. The FT procedure then depends on the mode chosen. In the OTA mode, the STA directly addresses AP2, and starts with an Authentication Request frame. The format of the request is similar to the authentication request frame used in Open System, except that the algorithm number is FT, and the frame includes additional elements. These elements include an RSN element that holds the name of the PMK-R0 key holder (derived from the PMK-R0 and a few other parameters[55]), the MDE, and an FT element (FTE) that includes the R0KH-ID and a SNonce.

AP2 uses these elements to locate R0KH to build PMK-R1. Meanwhile, AP2 also responds with an Authentication Response. Its algorithm number is also FT. The frame also includes an RSN element with the PMKR0Name, the MDE, and an FTE with the ANonce, the SNonce that the STA sent (for confirmation and to avoid session hijacking), R1KH-ID, and R0KH-ID.[56]

At this point, both sides have the core elements needed to build the keying material, with two restrictions:

- AP2 needs to contact R0KH and provide its R1KH-ID so that R0KH can build AP2's PMK-R1. Meanwhile, the STA, knowing both PMK-R0 and STA2's R1KH-ID, can derive AP2's PMK-R1 directly.

- The STA has, at most, the reassociation deadline time (discussed earlier) to attempt to reassociate to AP2. Past that delay point, AP2 will delete the values obtained in the authentication exchange. The STA would have to reinitiate an authentication exchange to re-create them before roaming to AP2.

Once the STA is ready to roam to AP2, it can terminate its connection to AP1, then send a Reassociation Request to AP2. In addition to the elements present with the Open System, the request includes, in the RSNE, what the STA thinks the PMKR1Name is. The PMKR1Name is derived from the PMKR0Name, the R1KH-ID, and the S1KH-ID—all elements that the STA obtained during the authentication exchange. If the STA is correct, as it can compute not only the PMK-R1 value, but also the PTK from the PMK-R1 and ANonce, it is ready to install the PTK. The request also includes the MDE and an FTE, which contains the ANonce and the SNonce that the STA received and sent (respectively) and used for the PTK generation, as well as the R1KH-ID and the R0KH-ID, all protected by a MIC.

AP2 validates these elements, generates and installs the PTK, and responds with a Reassociation Response. This response contains the PMKR1Name in the RSNE, the MDE, and the FTE with the same elements as in the association response (ANonce, SNonce, R1KH-ID, R0KH-ID), and the GTK, all encrypted and protected by a MIC.

At this point, both sides have a PTK, and the 802.1X controlled port is unblocked. The STA and AP2 can start exchanging data frames. You can see this sequence of exchanges in Figure 2-5.

---

55 See 802.11-2020, clause 12.1.7.6.3

56 All of these elements are either seen in the STA first authentication message or generated locally on AP2. Thus, AP2 can send this response even before it has reached R0KH.

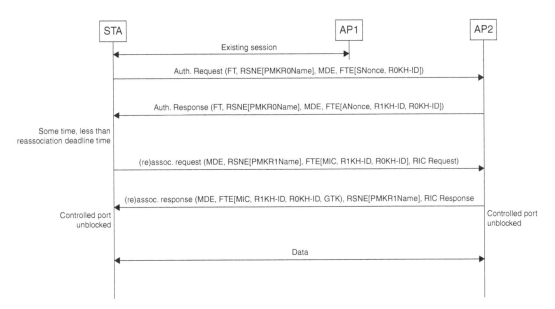

FIGURE 2-5   FT Over-the-Air Protocol

The over-the-DS process is similar in spirit, with the major difference being that the STA engages in dialog through AP1. Thus, it cannot send an authentication frame, because AP1 would not know if the goal is to relay the frame to another AP or to authenticate on AP1 again. Instead, the STA sends to AP1 an action frame called an FT Request. This frame includes the target AP MAC address, the STA MAC address (to be used as S1KH-ID), the RSNE with the PMKR0Name, the MDE, and the FTE that includes the R0KH-ID and a SNonce.

AP1 uses the target AP MAC to identify AP2, and relays the FT Request, through the DS, to AP2. The standard does not describe how this relaying happens or is secured, as it is not in the 802.11 realm (and is usually vendor-specific). AP2 receives and processes the request. It then responds with another action frame, the FT Response, that includes the STA MAC, its own MAC[57] (target AP MAC), the RSNE with the PMKR0Name, the MDE, and the FTE with the ANonce, the SNonce the STA sent, and R1KH-ID and R0KH-ID. AP1 relays this frame to the STA.

At this point, the STA can compute PMK-R1 from PMK-R0 and R1KH-ID, and can compute the PTK from PMK-R1 and the nonces. Just as with the OTA method, the STA needs to decide to roam to AP2 within the reassociation deadline time. When doing so, the STA terminates its connection to AP1, then sends a Reassociation Request directly to AP2. This request, just like the AP2 Reassociation Response, is similar to the OTA case. Figure 2-6 illustrates the over-the-DS FT exchange.

---

57 In case the radio MAC would be different from the Ethernet MAC, and AP1 would be dialoguing with several candidate roaming APs for the STA.

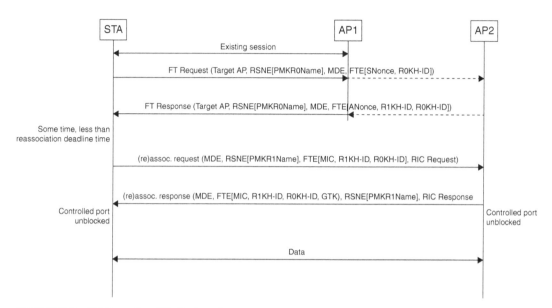

FIGURE 2-6   FT Over-the-DS Protocol

The FT protocol was also built to continue provisions devised in the 802.11e-2005 standard, which enable a STA to request resources from its AP for QoS purposes. In that 802.11e procedure, the STA sends an Add Traffic Stream (ADDTS) request[58] to its associated AP, which includes a description of the traffic the STA would like to send (including the number of frames per interval, their size, and the minimum and maximum acceptable data rates, among other information). The elements commonly used for that description are the Traffic Specification (TSpec) and the Traffic Classification (TCLAS) elements. The AP examines its available resources, and either accepts or denies the request.

When the STA roams with FT (OTA or over the DS), it can use the Reassociation Request frame to include, in the FTE, a Resource Information Container (RIC) Request. The RIC request includes a TSpec and one or more TCLAS elements describing the traffic that the STA would like to send through AP2. In the FTE of its Reassociation Response, AP2 provides a RIC Response with its conclusion for each intended traffic flow (accept/deny). In theory, AP2 can reject the association if it cannot accommodate the STA's traffic. In addition, in theory, STA2 can conclude that AP2 is not the best candidate if it rejects its service request. In practice, the RIC Request is done late (in the reassociation exchange), so it comes too late for the STA to find a better AP candidate on the spot; to do so, it would have to restart the scanning/FT authentication from scratch, with another AP. At the same time, the RIC exchange cannot be done earlier, because the AP cannot guarantee, at the time of the FT authentication exchange, that the resources that it has available at that time will still be available when the STA decides to roam, which may possibly be much later. In practice, however, the RIC process is seldom used.

58 See 802.11-2020, clauses 11.4 and 9.6.3.2

Simultaneous Authentication of Equals

The Simultaneous Authentication of Equals (SAE) scheme was initially intended for mesh APs,[59] as part of the 802.11s-2011 amendment. It was later ported for STA-to-AP exchanges, in particular during the Wi-Fi-Alliance development of WPA3. SAE brings the 802.11 authentication exchange back into the realm of security, as it is intended to prove knowledge of a shared password (known by both sides). As a side effect of this exchange of proof, both partners establish a cryptographically strong key.

The SAE authentication exchange consists of four frames.[60] As with the other methods, after scanning, the STA is the starting side, sending an 802.11 Authentication message. With SAE, that message is called the STA 802.11 SAE Authentication (Commit) message. In addition to the Authentication Algorithm field (set to 3, for SAE), an Authentication Sequence field indicating the position of the frame in the exchange (set to 1, for the first frame in the exchange, indicating the commit phase), and a Status Code[61] (set to 0, success), the frame includes a Finite Cyclic Group field (9.4.1.42) indicating a group, a Scalar field (9.4.1.39) containing the scalar, and a Finite Field Element. It may contain additional fields as well, such as an Anti-Clogging Token field (if the message is sent in response to an Anti-Clogging Token field request).

The AP responds with an SAE Authentication (Commit) message of similar structure, with its own Scalar and Cyclic Group field values. Both the AP and the STA need to agree on which cyclic group to use. At this point, both sides have agreed on the computation methods they will use to confirm that the other side has the right password, and that they exchanged the elements needed to verify this knowledge. The process might sound mysterious, but a simplified explanation can be sufficient to dispel the mystery. SAE[62] uses finite cycle groups, in which computations are made modulo a number q[63] that is known by both sides. One side (e.g., the STA) picks two random numbers (a and A) and computes their sum modulo q: $Sa = (a + A) \bmod q$. The STA also knows the shared password, and computes its hash to obtain a string that is a number, called the Password Element (PE). The STA then elevates that number to the $-A$ power: $elementA = PE^{\wedge}(-A) \bmod q$. What the STA sends in the commit message, beyond the agreements on the prime number and the way the numbers will be calculated[64] (this is the cyclic group), is Sa and $elementA$. The AP operates along a similar logic, picking two random numbers b and B, computing $Sb = (b + B) \bmod q$, $elementB = PE^{\wedge}(-B) \bmod q$, and sending in the commit frame Sb and $elementB$.

Each side can then use the knowledge of the password and the data sent by the other side to compute the same number. For example, the AP knows PE, knows b, and received from the STA Sa and $elementA$.

---

59 Hence the name—the protocol does not fix the "initiator" and "responder" roles, and any side can start the exchange.
60 See 802.11-2020, clause 12.4.5.1
61 See 802.11-2020, clauses 12.4.7.3 and 12.4.7.4
62 802.11-2020, clause 12.4.2 provides a complete, albeit dry, explanation.
63 Modulo is a ceiling. For example, modulo 5 uses 5 as the maximum number, thus allowing 0, 1, 2, 3, 4, 0, 1, …. In this environment, $(2 + 2) \bmod 5 = 4$, $(2 + 3) \bmod 5 = 0$, $(2 + 4) \bmod 5 = 1$, and so on. q is usually a large prime number.
64 The $(a + A)$ addition operates along a curve, instead of being in the flat line space in the previous example. 12.4.4 provides a more complete explanation. Operating along a curve (that is, not linear) makes it mathematically much more difficult to find a and A from the $(a + A) \bmod q$ value.

The AP can compute a new value, ss, defined with these numbers as ss = (PE^(Sa) × element$A$)^(b). A simple algebraic decomposition shows that this complicated expression can be simplified. As element$A$ is PE^(–A), ss can be expressed as follows:

$$ss = (PE^{\wedge}(Sa) \times PE^{\wedge}(-A))^{\wedge}b$$

Grouping all the powers together:

$$ss = (PE^{\wedge}(a + A - A)^{\wedge}b) = PE^{\wedge}(ab)$$

As the AP knows PE and b, it can then easily find a, and therefore A. Thus, it can verify that the STA, when it provided element$A$, did really know the password hash PE, so the STA really knows the password.

The STA performs the same operation in reverse. It knows a and PE, and received PE^(–B) and Sb, so it can compute the same ss value, with the following combination:

$$ss = (PE^{\wedge}(Sb) \times PE^{\wedge}(-B))^{\wedge}a = (PE^{\wedge}(b + B - B)^{\wedge}a = PE^{\wedge}(ab)$$

At the end of the commit phase, and after their computation, each side knows that the other side has the right password. But they can do more, and use the result of the computation they just ran to derive a session key.

They share that confirmation in the next two Authentication frames.[65] The STA first sends a second SAE Authentication message, this time of type Confirm (the Authentication Algorithm is still 3, SAE, but now the Authentication Sequence is 2, to indicate Confirm) and with the status set to success. This message also contains a Send-Confirm field and a Confirm field, whose values are the output of another function.[66] The other side can use the result to generate a new session key.[67] The AP responds with a message of similar structure and its own confirm value.

The authentication phase thus includes 4 frames instead of 2 with Open System. It is then followed with the standard 2-frame association exchange. Finally, there is a 4-way handshake in which the PTK is built from the PMK and the GTK shared by the AP with the STA, through a process identical to the one described earlier. Figure 2-7 illustrates this SAE authentication exchange.

65 See 802.11-2020, clause 12.4.7.5
66 See 802.11-2020, clause 12.4.5.5
67 Each side uses an expansion of the PSK, derived from the password, called the key confirmation key (KCK), and described in 802.11-2020, clause 12.7.1.3. For example, the STA computes and sends Confirm$A$ = Hash(KCK | scalar$A$ | a | element$A$ | element$B$), while the AP computes and sends Confirm$B$ = hash(KCK | scalar$B$ | b | element$B$ | element$A$). Each side uses the other side's commit, along with the PE, to generate a session key, which becomes the PMK. The operation details are explained in 802.11-2020, clause 12.4.5.4.

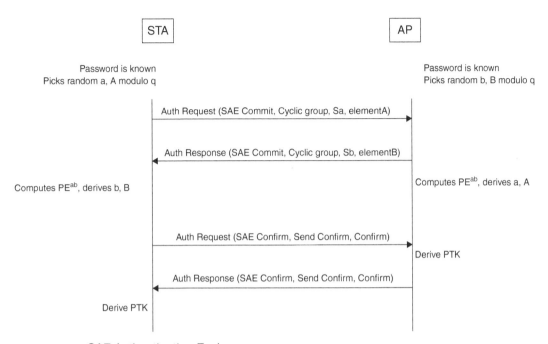

FIGURE 2-7 SAE Authentication Exchange

FILS Authentication and Association

Fast Initial Link Setup (FILS), introduced in the 802.11ai-2016 amendment, was initially designed to solve the "Tokyo Station problem." In this type of scenario, a large number of STAs attempt to join the Wi-Fi network within a small time window—for example, the mobile devices of hundreds of passengers on a subway train arriving at a station. The goal with FILS was to expedite the association process (including 802.1X and 4-way handshake) by reducing the number of frames needed for its completion, from 8 or more to 4.[68] In such a time-constrained scenario, the AP would not have time to facilitate the exchange between the STA and a RADIUS server using EAP (a process that can consume more than 10 frames). Therefore, to establish conditions similar to those used in the 802.1X exchange, the protocol had to reduce the identity validation scope and mechanism.

FILS can be used to validate that the STA has a valid password[69] or a valid certificate (in a PKI context). The AP might not be able to verify these elements alone, so it can turn to a Trusted Third Party (TTP) for this validation. The TTP is expected to be connected to the AP securely over the DS, and to be close enough that the exchange will be fast. As 802.11ai was developed in the 2010s, the idea that observing many authentication exchanges could be sufficient to run a statistical attack and find

68 See 802.11-2020, clause 4.10.3.6.1
69 Reintroducing the old vocabulary, the password is called a shared key, as in the WEP days (not a pre-shared key, as in 802.11i-2004 and later).

the static password had become widespread. The standard allows for more advanced mechanisms in which the STA and the AP (or TTP) first generate temporal certificates to protect their exchanges, and never exchange anything directly built from the shared password. Compromising the password would therefore not allow an attacker to attack current sessions. This process is called Perfect Forward Secrecy (PFS).

Note

The IEEE standard uses the acronym PFS for *perfect forward secrecy*, but sometimes incorrectly expands it (e.g., in clause 4.10.3.6.1) as *perfect forward security*. Perfect forward secrecy is a property of some key agreement protocols by which compromising a long-term secret does not also compromise the (shorter-term) session key. In the context of FILS, this is achieved by using a Diffie–Hellman exchange to derive a session-unique key from the long-term shared key. In other words, the AP and the STA derive ephemeral private/public key pairs (from nonces and other domain parameters) as part of their initial exchange, and pass the public part to the other side, which is then used to protect further exchanges. The operation is computationally complex, which means that an attacker obtaining the long-term shared key cannot easily recompute the particular session public/private keys that were derived by the STA or the AP.

PFS is optional for shared key authentication, but is always used with public key–based authentication. In the latter case, the exchange relies on finite cyclic groups algorithms. To limit password exposure (with or without PFS), the shared key process verifies that a returning STA either has a valid PMK that is still cached on the AP, or has a Reauthentication Root Key (rRK) that was derived from the MSKS-Nonce as defined in IETF RFC 6696. When the AP queries the TTP for validation, it uses EAP-RP as defined in IETF RFC 5295.

The FILS process then uses the 2-frame authentication exchange to establish the keying material. Next, in a 2-frame association exchange, the parties confirm to each other that they have the keys[70] and also potentially solve upper layer issues, such as IP address assignment.

In that scheme, the STA starts with an 802.11 Authentication frame (of FILS type), whose format depends on the FILS subflavor in use: shared key without Perfect Forward Secrecy, Shared Key with Perfect Forward Secrecy, or public key with Perfect Forward Secrecy.[71] In all cases, the frame includes the Authentication Algorithm field (set to 4 for shared key without Perfect Forward Secrecy, 5 for Shared Key with perfect forward secrecy, and 6 for Public Key with Perfect Forward Secrecy), the Authentication Sequence field (set to 1 for the first frame in the exchange), and the Status Code (set to 0 for success). When using Shared Key without Perfect Forward Secrecy, the frame can then also include a nonce (in a FILS Nonce element) and a random FILS Session value (in the FILS Session element). If the STA is coming back to an AP to which it associated before (and for which it has a cached key), it adds a FILS

---

70 See 802.11-2020, clause 12.11.2.1
71 See 802.11-2020, Table 9-41

Wrapped Data field that contains an encoded version of the EAP-initiate/reauth packet (by which the STA signals that it is returning and identifies its previous security association ID). When using the shared key with PFS and public key with PFS schemes, the STA adds a Finite Cyclic Group field that indicates which finite group the STA is targeting, and the ephemeral public key, encoded in the Finite Field Element field.[72]

The AP forwards the STA material to the TTP (if applicable), which in turn generates a response. The AP waits for that response before sending its Authentication Response to the STA (this is why the TTP is expected to be close to the AP). At this point, the AP has concluded the STA authentication (in a security sense) with a single exchange with the TPP (or the validation of the STA certification). Upon receiving the response from the TTP, the AP is ready to use the material transmitted by the STA to establish session keys.

The AP Authentication Response includes the same Authentication Algorithm Identifier as for the STA Authentication frame. The Authentication Sequence field is now set to 2, and the Status is set to 0 if the authentication is successful. The rest of the frame has a similar structure to that of the STA, including the TTP (or TTP public key) material and the AP-constructed ephemeral public key (when PFS is used). The STA validates the AP material, to confirm that the STA side (or the TTP behind the AP) has the same password (or a valid certificate). At this point, both sides have everything they need to generate session keys,[73] which will be generated during the association phase.

The STA then continues with an Association (or Reassociation) Request.[74] In addition to the capability informational elements similar to the Open System association frame, the STA includes not only the FILS session or FILS public key that the STA mentioned in the Authentication Request, but also the FILS Key Confirmation encrypted element, which includes a SNonce. Upon receiving this frame, the AP generates its ANonce, then uses it along with the SNonce and the keying material exchanged during the authentication phase to compute the PTK. The AP returns an Association Response that also includes FILS Key Confirmation element, with the AP ANonce (and the GTK). Similar to the AP, the STA uses the nonces and the keying material exchanged during the authentication phase to generate the PTK, then installs the PTK and the GTK. With the PTK installation complete, each side's controlled port becomes authorized and data communication can start immediately.

Naturally, data communication is not just a PHY and MAC layer problem. The STA also needs upper-layer elements—for example, an IP address. To address this concern, the association request can include a FILS IP Address Assignment element,[75] in which the STA can either indicate the IPv4 or IPv6 address it is requesting to use (for static IP address assignment and returning STAs that want to reuse their previously allocated address) or request a new address. The Association Response can also include the FILS IP Address Assignment, through which the AP accepts or declines the STA request (when the STA requests a specific address), or else returns an IP address to the STA (when the STA

---

72 See 802.11-2020, clause 12.11.2.3.2
73 See 802.11-2020, clause 12.11.2.5.3
74 See 802.11-2020, clause 12.11.2.6.2
75 See 802.11-2020, clause 9.4.2.184

requests a new address). In the second case, the AP can indicate other parameters, such as the network mask, gateway, or DNS server IP addresses.

The Association Request can also include one or more FILS HLP (Higher Layer Protocol) Container elements,[76] containing source and destination MAC addresses, and an HLD packet field. The AP is expected to forward this element to the DS, and from there to the target destination MAC (the STA is expected to be the source MAC). This process ensures that a server providing an upper-layer service (beyond DHCP) can respond and provide the elements that the STA needs. Figure 2-8 illustrates the FILS authentication exchange, in the case where no PFS is used.

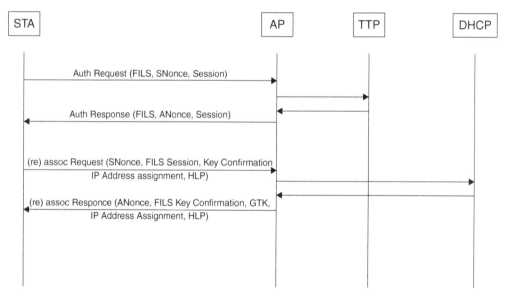

FIGURE 2-8   FILS Shared Key Authentication (No PFS)

One key difficulty with the HLP and DHCP containers is that upper-layer services, in standard networks, are usually applications running on servers. These applications need to receive queries, queue them, process them, and then answer—all operations that take time. In most implementations, the timeout value for a server nonresponse is expressed in seconds, a time scale incompatible with a STA expecting to receive an Association Response within a few hundred milliseconds.[77] Therefore, these upper-layer functions assume either the presence of fast servers, close to the AP, or the ability for the AP to cache some information (e.g., some IP addresses) for onboarding STAs. Collectively, these requirements make the adoption of FILS more difficult.

---

76 See 802.11-2020, clause 9.4.2.183
77 The standard does not mandate a maximum duration for the association response. The MLME allows each configuration to implement its own timer, in 802.11-2020, clause 6.3.4.2.2. However, the timeout value is in units of TUs, and most implementations conclude that a lack of response within a TU means that the request was not processed successfully.

Summary

802.11be considered the need to roam more efficiently. Unfortunately, the methods intended to facilitate roaming (e.g., 802.11k neighbor reports and 802.11v BSS transition management) all came with practical limitations. Without knowing the location of the STA (and sharing such information is problematic, because it raises all sorts of privacy risks), the AP cannot really provide in full confidence information about the best next AP. The AP can provide some indications that help the STA make faster AP discovery, but the STA still needs to get off channel to dialog with the next AP. This move negatively impacts the operations of the STA on its active channel—where it is already challenged, because it has reached the edge of the BSA.

Similar challenges were found when examining the need to provide differentiated service to applications in the same categories—for example, to provide better service to a business-relevant video-conferencing application over an application intended for personal video calls. The ADDTS techniques did not stand the test of time. The original 802.11e technique had 8 UPs, but many of them proved unusable to differentiate applications in the same AC. The 802.11aa Stream Classification Service seemed promising, but proved insufficient.

Meanwhile, the 802.11 group's long quest for optimal secure mechanisms for all environments has led to the development of four (and one obsolete) security schemes: some Open System schemes implementing the 802.1X technique for robust authentication and key derivation; a Fast Transition scheme that efficiently passes the keying material from one AP to the next at roaming time; a Fast Initial Link Setup that accelerates the initial secure onboarding of STAs entering the 802.11 domain; and a Secure Authentication of Equals scheme that offers a modern and robust scheme for password and PKI-based networks. All of these schemes needed to be incorporated into the 802.11be design.

Chapter **3**

Building on the Wi-Fi 6 Revolution

overlapping basic service Set (OBSS): A basic service set (BSS) operating on the same channel as the station's (STA's) BSS and within (either partly or wholly) its basic service area (BSA).

802.11-2020, Clause 3

IEEE 802.11ax, and its associated Wi-Fi Alliance certification, Wi-Fi6, paved the way for the improvements made possible with 802.11be. At its heart, 802.11ax aims at improving Wi-Fi performance in hyper-density scenarios, where many STAs compete for airtime. By allowing multiple concurrent uplink transmissions, and by designing a denser, yet more efficient carrier structure, 802.11ax ensured that throughput could increase, even in these high-density scenarios.

Channel Efficiency Improvements

One of the main features of 802.11ax is a new carrier design, made possible by the advances in 802.11 chipsets. Narrower and more densely packed tones allow for more resilience and flexibility in a multiuser scenario, leading to more efficient data throughput for a given channel width, even in dense STA environments. At the same time, slower symbols provide larger Inter-Symbol Interference (ISI) protection, making transmissions more robust not only outdoors, but also in reverberant indoor environments.

> **Note**
>
> This protection relates to the delay spread. Recall from the section on OFDM transmission in Chapter 1, "Wi-Fi Fundamentals," that signals can bounce against obstacles before hitting the receiver slightly after the main (line of sight) signal. This additional delay reaches a maximum value that depends on the distance to the farthest likely obstacle, and is called delay spread. An interval between symbols maximizes the chances that the reflected signal will reach the receiver at a time when no useful symbol is being received.

A New Carrier Structure

As described in Chapter 1, a standard OFDM scheme for 802.11 transmissions (until Wi-Fi 5) was to organize each 20 MHz subchannel into 64 subcarriers, each 312.5 kHz away from its neighbor. In a 20 MHz PPDU, the center tone (i.e., the direct current [DC] tone), the uppermost 3 tones, and the lowest 4 tones were kept null—the center tone so that DC offsets at the transmitter and/or receiver did not overlap active tones and could be ignored, and the edge tones to act as guards against overlaps to (and from) neighboring channels. In this 20 MHz-wide PPDU, 4 tones were used as pilots.

By contrast, 802.11ax organizes each 20 MHz channel into 256 tones, 78.125 kHz away from each other. Given that the transmitter and receiver might be up to ±40ppm apart at 6 GHz, the transmitter's residual DC might deviate as much as ±240 kHz from the receiver. For this reason, 802.11ax defines 3 to 7 DC tones at the channel center. At the edge, the lower 6 tones and the upper 5 tones are also set as null (DC). Figure 3-1 illustrates these principles.

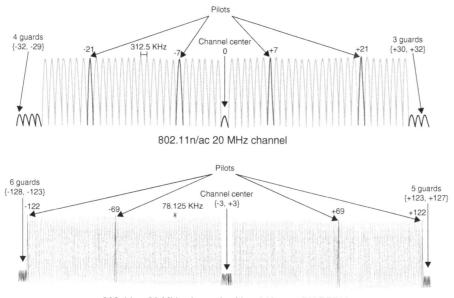

FIGURE 3-1   802.11n/ac versus 802.11ax 20 MHz Channel

Now that the tones are 4 times denser in the frequency domain, it is necessary to expand the core of the OFDM symbol duration by a commensurate factor of 4, so as to preserve the orthogonality of the tones. Even so, given the relatively fewer guard tones and pilot tone, the change is a net win because this subcarrier density is more efficient than the previous 64-tone per 20 MHz channel scheme. There is another benefit as well: Because the time spread of the wireless echoes between the transmitter and receiver is a function of the geometry and RF characteristics of the environment, the absolute length of the required cyclic extension (or guard interval) of each OFDM symbol does not *need* to increase. With the longer core OFDM symbol period, the relative overhead of the cyclic extension is smaller.

In practice, 802.11ax introduced a set of cyclic extensions with different trade-offs between efficiency and multipath/multi-uplink-user robustness. With 802.11n/ac, the core of each OFDM symbol lasts for 3.2 µs and is followed with an 800 ns (standard Guard Interval, GI) or 400 ns (short GI) that is a cyclic extension of the core (i.e., a repeat of the symbol's first 800/400 ns). With 802.11ax, the core lasts for 12.8 µs, and the GI is 0.8 µs (800 ns), 1.6 µs, or 3.2 µs. 802.11ax also allows up to 8 spatial streams (SS). This difference in tone, streams, and symbol densities results in the MCSs that Table 3-1 illustrates.[1]

TABLE 3-1   802.11ax MCSs (in Mbps), 800 ns GI

MCS Rate	Modulation and Code Rate	20 MHz 1 SS	20 MHz 8 SS	40 MHz 1 SS	40 MHz 8 SS	80 MHz 1 SS	80 MHz 8 SS	160 MHz 1 SS	160 MHz 8 SS
MCS0	BPSK ½	8.6	68.8	17.2	137.6	36.0	288.2	72.0	576.0
MCS1	QPSK ½	17.2	137.6	34.4	275.3	72.1	576.5	144.0	1152.0
MCS2	QPSK ¾	25.8	206.5	51.6	412.9	108.1	864.7	216.0	1728.0
MCS3	16 QAM ½	34.4	275.3	68.8	550.6	144.1	1152.9	282.0	2256.0
MCS4	16 QAM ¾	51.6	412.9	103.2	825.9	216.2	1729.4	432.0	3456.0
MCS5	64 QAM ⅔	68.8	550.6	137.6	1101.2	288.2	2305.9	576.0	4608.0
MCS6	64 QAM ¾	77.4	619.4	154.9	1238.8	324.3	2594.1	649.0	5192.0
MCS7	64 QAM ⅚	86.0	688.2	172.1	1376.5	360.3	2882.4	721.0	5768.0
MCS8	256 QAM ¾	103.2	825.9	206.5	1651.8	432.4	3458.8	865.0	6920.0
MCS9	256 QAM ⅚	114.7	917.6	229.4	1835.3	480.4	3843.1	961.0	7688.0
MCS10	1024 QAM ¾	129.0	1032.4	258.1	2064.7	540.4	4323.5	1081.0	8648.0
MCS11	1024 QAM ⅚	143.4	1147.1	286.8	2294.1	600.5	4803.9	1201.0	9608.0

As you can see from Table 3-1, the tone densification strategy pays off. Even at MCS 0 on a 20 MHz channel, 802.11ax can achieve 8.6 Mbps, whereas 802.11ac allowed for only 6.5 Mbps.

802.11ax also introduces the idea of Resource Units (RUs). RUs are useful in multiuser transmission schemes because they allow the AP to group a subset of the total available tones and allocate each group to a STA (see the next section, "Multiuser Transmissions"). Each RU has its own pilot tones— for example, 2 pilots for a 26-tone RU, 4 pilots for a 52- or 106-tone RU, 8 pilots for a (20 MHz) 242-tone RU, 16 pilots for a (40 MHz) 484-tone or (80 MHz) 996-tone RU, and 32 pilots for a (160 MHz) 1992-tone RU.

---

1 See 802.11ax-2021, clause 27.5 and in particular clause 27.5.6.

Multiuser Transmissions

Recall from Chapter 1 that 802.11n-2009 introduced the idea of Multiple Input and Multiple Output (MIMO), where a STA could send different signals out of different radio chains on the same channel and during the same PPDU. Two goals are possible in such a case:

- To send or receive the same data more reliably. For example, techniques like Transmit Beamforming (TxBF) can increase the signal at the receiver's antennas, and/or Maximal Ratio Combining (MRC) can use each copy at each receiver's antenna for increased diversity against any destructive interference occurring from multipath superposition.

- To send different segments of data via different radio chains (with Spatial Multiplexing [SM]), thereby increasing the data throughput to the receiver.

In all cases, there was a single target receiver and a single transmitter. Thus, the term was later renamed Single User MIMO (SU-MIMO).

802.11ac-2013 improved upon the MIMO scheme by introducing the notion of Multiuser MIMO (MU-MIMO). With this approach, not only could the AP send different data in the same PPDU via multiple radio chains, but the target of the different data could also be a different STA. For the first time, 802.11 allowed simultaneous transmissions to multiple STAs. This scheme made sense in the context of the 802.11 effort to improve efficiency. In many cases, a STA in an 802.11n BSS would benefit little from a stronger transmission with TxBF or MRC, because the STA would be close enough to the AP that it was already operating at a very high MCS. At the same time, the AP often would not have more than one frame to send to each STA. Appreciable overheads are incurred in sending individual frames and acknowledgments in their own, separate PPDUs with all PPDUs separated by SIFS. With MU-MIMO, the AP can better leverage its available radio chains and maximize the efficient reuse of preambles, SIFS, and the efficient retrieval of acknowledgments by sending different frames (or different aggregations of frames) to different STAs simultaneously.

MU-MIMO

The 802.11ac MU-MIMO scheme was limited to downlink traffic (AP to STA, DL-MU-MIMO). A primary reason behind this limitation was the difficulty of coordinating uplink transmissions. If several STAs were to send at the same time (toward the AP), their clocks should be accurate enough that they could synchronize exactly the time at which they would all start transmitting together. In the same way, their carrier frequencies needed to be tightly aligned. In an 802.11ac context, meeting these requirements was challenging.

802.11ax (also called High Efficiency [HE]) overcame this limitation by requiring the clients to precisely synchronize their behavior (in time and in frequency) to the immediately preceding PPDU (see the "Scheduling and Multiuser Operations" section). This approach allowed uplink MU-MIMO (UL-MU-MIMO). Given the tight sync, the process is conceptually similar to SU-MIMO, where each user's data is transmitted via a set of one or more spatial streams. However, with UL-MU-MIMO, each set of spatial streams originates from a different STA.[2] This scheme was a great improvement, but introduced a new source of inefficiency. With SU-MIMO, the Data field of the PPDU was filled with user data until either there was no more user data or the PPDU got too long. With MU-MIMO, the AP scheduler tries to pack the data of all users into a single PPDU, where the AP can adjust the duration of the PPDU's Data field and the MCS and number of spatial streams per user. In some cases, the packing is inefficient.

As an example, consider two users, one with a lot of data and one with a little, both otherwise similar in terms of optimal MCS and maximum supported number of spatial streams. The AP can allocate the maximum number of spatial streams and optimal MCS to the first user, and calculate the duration of the Data field for its data—but then the AP cannot allocate less resources to the second user than one spatial stream. In this case, the second STA, which does not need the full Data field, still has to pad the remainder of the Data field with non-useful octets. As multiple STAs need to send simultaneously, each STA that does not need the full data field of the PPDU has to pad the remainder of the data field with non-useful data.

Additionally, multiuser transmissions (uplink or downlink) require that the transmissions from or to multiple STAs do not interfere with each other and the receiver, or at least that their mutual interference can be easily learned and suppressed. On the downlink, this requirement means that the AP, coordinating these transmissions, needs to discover exactly how signals are scattered on their way from the AP's antennas to each client's antennas. This measurement is performed through a sounding (Null Data Packet [NDP]) transmission from the AP to the client, to which the STA responds with a set of compressed sounding matrices that report how the STA received regularly spaced tones of the sounding transmission (see the "Sounding Procedure" section later in this chapter). The AP may then determine which STAs are compatible with each other (achieve maximal spatial separation) and calculate beamforming (or steering) matrices such that the following criteria are met:

- The signal sent to one client's antennas is almost completely free from interference from other clients.
- This holds for each intended client.

As the precise scattering is highly dependent on each STA's position with respect to the surrounding walls, furniture, and other obstacles, and given that by nature 802.11 STAs or the people around them are expected to move often, frequent sounding is necessary to maintain the quality of the exchanges. The process is onerous in terms of airtime.

On the uplink, the training information embedded in the preamble of each PPDU enables the AP to identify how each client's signal is scattered on the way to the AP. Here, separate sounding is not

---

2 See 802.11ax-2021, clause 27.3.3.2.1

needed, but there is another complication: The AP needs to know when clients have uplink traffic ready to transmit to the AP.

OFDMA

A more useful scheme is enhancing multiplexing further in the frequency domain, and allocating to a STA only the set of subcarriers it needs for its transmission during the PPDU. This scheme, called orthogonal Frequency Division Multiple Access (OFDMA), was first proposed to the Study Group for 802.11ac[3] and later revised and adopted by 802.11ax. With OFDMA, the subcarriers are grouped into Resource Units (RUs). A RU can include 26, 52, 106, 242, 484, or 996 adjacent[4] subcarriers, depending on the channel width. Like MU-MIMO, OFDMA can be used for both the uplink and the downlink. For the uplink, the STAs' needs are assumed to be already known at the AP. Based on these needs, the AP can determine, for the next UL-MU-MIMO TXOP transmission, the RU of the best width that would maximize the overall BSS throughput, and then trigger a number of STAs to send simultaneously. The RU structure depends on the channel width, and defines which RUs may be allocated and how many DC tones are present. In turn, that allocation defines the number of null subcarriers as well as the count and position of the pilot subcarriers. Figure 3-2 provides a few examples. Notice how the center 26-RU is spread around the DC center subcarriers in the 20 MHz case. This split is the exception. In all other positions, the RUs are formed with contiguous subcarriers.

Also notice in the figure that larger RUs are constructed from smaller RUs, such as the 106-tone RU being built from four 26-tone RUs. This example shows the reason for the null subcarriers: 2 null subcarriers must be added to the 4×26 tones to reach 106 tones. It is also interesting to note that the 80 MHz HE PPDU includes contiguous null subcarriers. Both examples in Figure 3-2 include 7 DC subcarriers at the channel center. However, this is not the only possibility. A 20 MHz HE PPDU with a 242-tone RU would include not only 3 DC tones at the channel center, but also $6 + 5$ guard tones at the edges of the channel, but no other null subcarriers within the channel. These adjustments are necessary to harmonize the subcarrier count for a given RU width, but they also make the structure quite complex.

To help visualize the structures, Table 3-2 summarizes examples of the null and RU structures for the various channel widths defined in 802.11ax.[5] This table does not show all the mix-and-match possibilities, such as one 106-tone RU, one 52-tone RU, and three 26-tone RUs in 20 MHz. Keep in mind that these more complex combinations are technically allowed, as long as the total number of used RUs matches the number of available RUs. However, they add to the AP channel plan complexity, and many APs may not implement them.

3 See mentor.ieee.org/802.11/dcn/10/11-10-0317-01-00ac-dl-ofdma-for-mixed-clients.ppt
4 The center RU is an exception.
5 See 802.11ax-2021, clause 27.3.2.3 for more details.

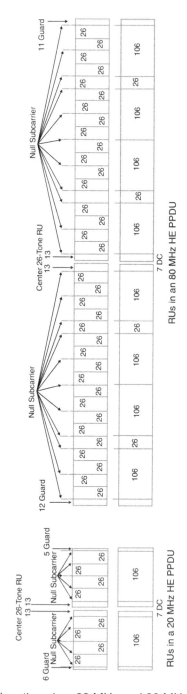

FIGURE 3-2   26 and 106 RU Allocations in a 20 MHz and 80 MHz HE PPDU

TABLE 3-2 HE PPDU RU Structure

Channel Width	RU Size (and Count)	Center DC Count	Left, Right Guard	Null Subcarriers
20	26 (9)	7	6, 5	±69, ±122
	52 (4), 26 (1)	7	6, 5	±69, ±122
	106 (2), 26 (1)	7	6, 5	None
	242 (1)	3	6, 5	None
40	26 (18)	5	12, 11	±3, ±56, ±57, ±110, ±137, ±190, ±191, ±244
	52 (8), 26 (2)	5	12, 11	±3, ±56, ±57, ±110, ±137, ±190, ±191, ±244
	106 (4), 26 (2)	5	12, 11	±3, ±110, ±137, ±244
	242 (2)	5	12, 11	None
	484 (1)	5	12, 11	None
80	26 (37)	7	12, 11	±17, ±70, ±71, ±124, ±151, ±204, ±205, ±258, ±259, ±312, ±313, ±366, ±393, ±446, ±447, ±500
	52 (16), 26 (5)	7	12, 11	±17, ±70, ±71, ±124, ±151, ±204, ±205, ±258, ±259, ±312, ±313, ±366, ±393, ±446, ±447, ±500
	106 (8), 26 (5)	7	12, 11	±17, ±124, ±151, ±258, ±259, ±366, ±393, ±500
	242 (4), 26 (1)	7	12, 11	None
	484 (2), 26 (1)	7	12, 11	None
	996 (1)	5	12, 11	None
160	26 (74)	7	12, 11	Null subcarrier indices in 80 MHz ± 512
	52 (32), 26 (10)	7	12, 11	
	106 (16), 25 (10)	7	12, 11	
	242 (8), 26 (1)	7	12, 11	None
	484 (4), 26 (1)	7	12, 11	None
	996 (2)	5	12, 11	None

A simple replacement logic surfaces the following points:

- A single 996-tone RU is composed from the tones of two 484-tone RUs, one 26-subcarrier (central) RU, and two DC tones.

- In turn, the 484-tone RU is composed from the tones of two 242-tone RUs.

- Each 242-tone RU is composed from the tones of two 106-tone RUs, one 26-tone (central) RU, and four null or DC subcarriers.

- A 106-tone RU is composed from the tones of two 52-tone RUs and two null subcarriers.

- The 52-subcarrier RU is composed from the tones of two 26-tone RUs.

Each RU includes data subcarriers and pilots, as shown in Table 3-3.[6]

TABLE 3-3   Data and Pilot Subcarrier Count for Each RU Type

RU Size (Subcarriers)	Data Subcarrier Count	Pilot Count
26	24	2
52	48	4
106	102	4
242	234	8
484	468	16
996	980	16
1992 (2 × 996)	1960	32

Preamble Puncturing

Large channels present the obvious advantages of providing more overall bandwidth. 802.11ac led the way by proposing channels larger than the 40 MHz in 802.11n—specifically, 80 MHz-wide and 160 MHz-wide channels. However, as 802.11ax was being developed, it started to become clear that the benefit of these wide channels might be lost in high-density environments, especially in mixed scenarios where 802.11ax wide-channel APs would coexist near non-802.11 systems, or near older 20 MHz-only or 40 MHz-only APs, or near 802.11ac/ax APs that, for some reason, would be statically configured with narrower channels.

The CCA mechanism implemented for 802.11ax allows a transmitter to check the availability of each individual 20 MHz subchannel of the wide channel before deciding to transmit (see the "Coexistence Challenges" section in Chapter 2, "Reaching the Limits of Traditional Wi-Fi"). The wide channel is constructed by adjunctions:

1. To a primary 20 MHz channel, a secondary 20 MHz channel is added, forming a primary 40 MHz channel.

2. To that primary 40 MHz channel, a secondary 40 MHz channel is added, forming a primary 80 MHz channel.

3. To that primary 80 MHz channel, a secondary 80 MHz channel is added (not necessarily adjacent), forming a 160 MHz, or more precisely 80 + 80 MHz, channel.

6 See 802.11ax-2021, clause 27.3.12.13 for pilot subcarrier indices.

With the dynamic bandwidth channel access mechanism introduced in 802.11ac, when the STA (after counting down using the contention window principles detailed in Chapter 1) detects that the primary 20 MHz channel is idle, it proceeds to sense the extended channels—first the secondary 20 MHz, then the secondary 40 MHz, then the secondary 80 MHz. The process works essentially instantaneously but we can conceive of it as an iterative process: If the next-widest secondary channel is idle as well, the STA expands its communication to the existing primary + new secondary, and thereby constructs a new double-width primary channel. Alternatively, if the next-widest secondary channel is busy—for example, because of a nearby system operating on a 20 MHz subchannel of that secondary channel— then the process stops and the transmitter's bandwidth stays at the existing primary channel width. In the worst-case scenario, when the competing system transmits on the secondary 20 MHz channel at a level high enough to cause that channel to be reported as busy, then the transmitter may use only the primary 20 MHz channel, even if the BSS is set to 160 MHz and 140 MHz out of the 160 MHz is available. This principle is illustrated on the left side of Figure 3-3 for an 80 MHz-wide channel.

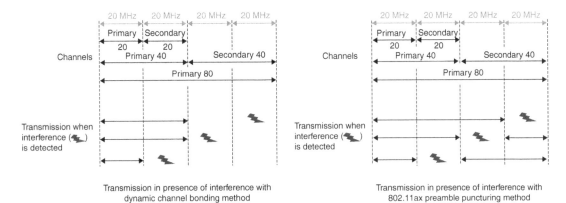

FIGURE 3-3 Transmissions in the Presence of an Interferer with Dynamic Channel Bonding and with 802.11ax Preamble Puncturing Methods

This system results in wasted bandwidth because a contiguous sequence of subchannels overlapping the primary 20 MHz must be clear for wide-channel use. To mitigate this issue, 802.11ax introduces the idea of preamble puncturing. With preamble puncturing, the AP detecting interference on one of the secondary subchannels can send a downlink OFDMA PPDU with that affected portion of the channel punctured. In other words, the AP does not send a preamble on that affected subchannel, and the RUs on that segment are also unused. However, the preamble is present on the other subchannels before and after the punctured channel, and the RUs on those subchannels carry data. You can see the difference on the right side of Figure 3-3.

When sending a trigger frame for UL OFDMA transmission (see the "Scheduling and Multiuser Operations" section later in this chapter), the AP merely needs to not schedule clients on the subchannels to be punctured. The STAs naturally send data only on their assigned RUs and send a preamble limited to the 20 MHz subchannels that actually span their RUs. In this way, the absence of triggering of any client on a punctured subchannel keeps the subchannel unenergized.

> **Note**
>
> 802.11ax adds certain constraints on the possible puncturing. This possibility arises because the user assignment information of 40 MHz and wider PPDUs is split across odd and even 20 MHz subchannels. Thus, at least one odd and one even subchannel within the primary 80 MHz must be unpunctured.

This mechanism allows for a better utilization of the channel. However, in 802.11ax (Wi-Fi 6), it suffers from several limitations:

- Support is required on the STAs, even for DL OFDMA transmissions from the AP, as the STA needs to understand which subchannels of the preamble are punctured so it can obtain its RU allocation from an unpunctured channel. Support is announced at association time in the HE Capabilities element in the (re)association frame. As support for this feature is optional (and implementation of the feature is complex at the chipset level), the AP must limit the use of this feature to the subset of STAs that support it.

- The AP is in charge of detecting the interference. In many cases, however, the interference affects some of the STAs and the AP does not detect the interferer. 802.11ax does not offer a remedy for this case.

- The process is tied to OFDMA transmissions. If a subchannel is busy, regular (non-OFDMA) transmissions still need to revert back to the primary 20/40/80 MHz channel using the standard dynamic channel bonding method. Thus, the gains offered by the method are limited to a single type of transmission.

HE PPDU Format

The new structure of the PPDU, with its narrower subcarrier separation, naturally means that 802.11ax transmissions cannot be understood by pre-802.11ax receivers. As usual, the protection mechanism consists of starting the transmission of most PPDUs[7] with a pre-HE modulated PHY header, with a Legacy Short Training field[8] (L-STF), a Legacy Long Training field (L-LTF), and a Legacy Signal field (L-SIG). This is followed by a repeated L-SIG (RL-SIG[9]) field, whose purpose is to differentiate the HE PPDU from all other non-HE PPDUs. (The latter also start with the L-STF, L-LTF, and L-SIG fields, but their next preamble field is quite unlike a repeated L-SIG.) The preamble continues with the

---

7 Exceptions at 2.4/5/6 GHz include the 1, 2, 5.5, and 11 Mbps PPDUs and the little-used HT-GF format, which is deprecated for modern devices. In the following discussions, these PHY formats are ignored for simplicity.

8 See Chapter 1, "Wi-Fi Fundamentals," for details on the structure of these pre-HE fields.

9 See 802.11ax-2021, clause 27.3.11.6

HE-SIG-A field and sometimes the HE-SIG-B field, too. These two fields are modulated with pre-HE data rates, but include the information needed to interpret the rest of the HE PPDU. In addition to the MCS used in the payload, the fields indicate whether the PPDU is uplink or downlink, is trigger-based or not, and uses dual carrier modulation or not (see the "Scheduling and Multiuser Operations" section). They also provide the BSS color (see the "BSS Coloring" section), the approximate duration of the TXOP, and the Spatial Reuse Parameters (SRP). After the HE-SIG-A/B fields, and beginning with the HE-STF field, the PPDU modulation switches to HE modulation, which is used for the remainder of the PPDU. Figure 3-4 illustrates this structure.

FIGURE 3-4 Modulations in the HE PPDU: HE MU PPDU Example

Some PPDUs do not follow this exact format. In fact, 802.11ax defines four PPDU format types: the HE Single User PPDU (HE SU PPDU), the HE multiuser PPDU (HE MU PPDU), a Trigger-Based PPDU for UL MU-MIMO or UL-OFDMA (HE TB PPDU), and an Extended Range Single User PPDU (ER SU PPDU). All PPDUs include the same legacy header and RL-SIG. The HE SU PPDU includes an 8 μs HE-SIG-A field, followed by a 4 μs HE-STF field. By contrast, the HE MU PPDU shown in Figure 3-4 also includes the 8 μs HE-SIG-A field, which is followed by a 4 μs (per symbol) HE-SIG-B field, and then the 4 μs HE-STF field. The HE TB PPDU does not include the HE-SIG-B field, and its HE-STF field lasts for 8 μs (instead of 4 μs, as in the HE SU PPDU). Lastly, the HE ER SU PPDU has a longer HE-SIG-A field (16 μs), no HE-SIG-B field, and a regular (4 μs) HE-STF. All four HE PPDU format types end with the HE-LTF, Data field and a (possibly zero-length) PE field.

In the HE-modulated part of the payload, recall that the core of each OFDM symbol lasts 12.8 μs. At the same time, 802.11ax allows for three core HE-LTF durations: 3.2 μs, 6.4 μs, and 12.8 μs. A longer HE-LTF allows for a better channel estimation, at the cost of a longer overhead. 802.11ax also allows three different guard intervals (GI): 0.8 μs, 1.6 μs, and 3.2 μs. The 800 ns GI is the most commonly used for indoor transmissions. The longer ones are available for the following purposes:

- Outdoor transmissions, primarily with IoT in mind, where reflection on farther objects (e.g., buildings) causes a longer delay spread

- UL multiuser transmissions, to account for small variations in ranges and transmit times at the different clients

Functional Improvements

The HE channel, with its high subcarrier density, allows for an increase in the PHY rate with minimal downsides. The RU structure also allows multiple concurrent transmissions via OFDMA (frequency-domain aggregation). MU-MIMO is another path to multiple concurrent transmissions via spatial-domain aggregation. Quite naturally, this organization works only if there is a form of coordination between potential senders. As you might expect, the PHY improvements are also accompanied by MAC improvements that allow for the HE orchestration.

Scheduling and Multiuser Operations

Downlink multiuser operations are somewhat easier to perform than their uplink counterparts, because a single device (the AP) is in charge of the transmission. When transmitting to more than one STA, the AP uses an HE MU PPDU, which contains the PSDU(s) for the different STAs, either in different spatial streams with MU-MIMO or in different RUs with OFDMA. The HE-SIG-B field contains the map of which part of the transmission is intended for which target STA (signaled by its Association Identifier [AID]).

A single-stream DL-OFDMA transmission can be sent as soon as the AP decides. A MU-MIMO transmission—DL MU-MIMO, UL MU-MIMO, or any multistream OFDMA transmission (i.e., DL MU-MIMO-OFDMA or UL MU-MIMO-OFDMA)—with its multiple spatial streams, requires sounding. The sounding process in 802.11ax is similar in logic to sounding with the previous 802.11 generations.[10] The AP sends a Sounding frame. The recipient HE STAs use this frame to determine the channel state information, and return a sounding matrix. The AP then uses the sounding matrices to identify STAs that can be grouped together for the next DL or UL MU-MIMO transmission. Sounding adds overhead, but both DL MU-MIMO and (single stream or multistream) DL OFDMA present the advantage of sending MSDUs simultaneously to several STAs. With these mechanisms, collisions are statistically reduced.

This phenomenon can be easily understood with a numerical example. Suppose 30 STAs are in the BSS, and a downlink transmission is being sent to 4 of these STAs. The AP first informs the 4 candidates (see the "Trigger Frames" section later in this chapter), so these STAs are waiting for the transmission and not contending for the medium. The AP contends for access to the medium for the initial trigger frame, against the 26 other STAs. If the AP wins the access, collisions are unlikely, except if one of the 26 STAs did not hear the trigger frame and sends at the same time as the AP. By contrast, without this MU mechanism, the AP sends one frame at a time to each STA; in other words, it needs to send 4 frames to reach the 4 STAs that the AP reached with one MU frame in the previous case. Each time, the AP contends with 30 STAs, including the target STA; the target STA, as it does not know that a DL frame is coming, may be attempting to send an untriggered UL PPDU. In essence, the AP contends against more STAs (30 instead of 26), and more times (4 times instead of once). The MU operation is the clear winner. Because 4 PPDUs are sent in one transmission, the MU operation also reduces latency and improves airtime efficiency.

10 See 802.11ax-2021, clause 26.7 for the details of the HE sounding protocol.

Sounding Procedure

The sounding procedure in 802.11 is as old as MIMO (and therefore 802.11n). 802.11ac added its own flavor, as did 802.11ax. Examining the 802.11ac procedure in detail is of limited interest for our purposes in this book. However, understanding the 802.11ax version will help you understand the enhancements brought by 802.11be. 802.11ax distinguishes trigger-based (TB), and non-TB sounding exchanges, depending on how many STAs are expected to respond. In all cases, the initiator of the sounding exchange (typically the AP, except in P2P scenarios) expects a feedback matrix from the responder(s), which are STA(s). The initiator is formally called the beamformer, and the responder(s) are formally called the beamformee(s). The feedback matrix can take three forms, depending on which report the beamformer is requesting:

- **Channel Quality Indication (CQI) feedback:** An HE Compressed Beamforming/CQI Report consisting of an HE CQI Report field; it shows the average SNR for each space-time stream at each RU.

- **SU feedback:** A single STA responds, providing an HE Compressed Beamforming/CQI Report with an HE Compressed Beamforming Report field.

- **MU feedback:** Multiple STAs respond, providing HE Compressed Beamforming/CQI Reports with an HE Compressed Beamforming Report field and HE MU Exclusive Beamforming Report fields.

The non-TB sounding sequence, with a single beamformee, is shown in Figure 3-5. In terms of the underlying principles, it resembles the previous 802.11 generations' single-STA-centric sounding exchanges. The HE beamformer sends a first frame, called HE NDP Announcement (NDPA), with the RA being the particular STA from which the beamformer requires feedback. The NDPA frame content helps the STA get ready for the subsequent sounding frame, and also includes information about the exchange—for example, a STA Info List field that includes the AID of the target STA, whether the exchange is SU or MU, whether the request is for CQI feedback or not, and so on. The beamformer waits one SIFS, then sends the HE Sounding NDP frame, which can be composed of one or more spatial streams. The beamformee responds (after a SIFS) with a report that includes the feedback matrix. With multiple streams and wide channels, the matrix can be fairly large, and 802.11ax allows the report to come in several segments (each being an HE compressed beamforming/CQI frame) if needed.

The TB sounding exchange seeks feedback from two or more STAs, as illustrated at the bottom of Figure 3-5. This exchange also starts with an NDPA frame, but this time sent to the broadcast address. The STA Info List contains two or more STA info fields, specifying which STAs are expected to respond. The beamformer, just as in the non-TB case, waits one SIFS, then sends the HE Sounding NDP frame. The feedback is different. Right after the HE sounding NDP (and one SIFS), the beamformer sends an additional frame instructing the target STAs that they should send their reports. The frame, called Beamforming Report Poll (BFRP) Trigger frame, indicates which STAs should send their reports and for which segments (i.e., which part of the channel, which spatial streams). The queried STAs wait one SIFS, then all send (the expectation is "at the same time") a common preamble, followed by their individual HE Compressed Beamforming/CQI Report, at the position of the channel (RU) indicated by the trigger. Here again, with wide channels and multiple streams (and multiple STAs), it may not be possible for all STAs to send their full reports in a single uplink transmission. Thus, 802.11ax allows the beamformer to query reports by segments and by STAs, with one or more BFRP triggers and subsequent reports, until all STAs have reported all streams and the entire channel width.

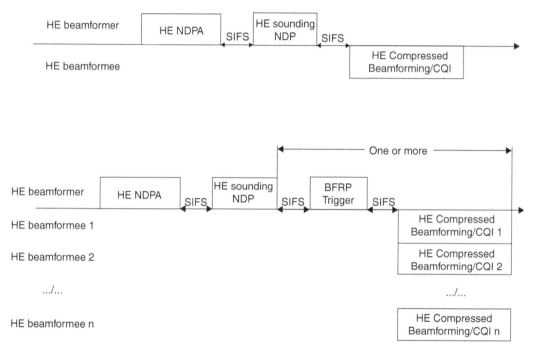

FIGURE 3-5   802.11ax Non-TB Sounding Procedure Example (top) and TB Sounding Procedure (bottom)

Trigger Frames

Sounding is useful for the AP to determine how to beamform DL frames to one or multiple recipients. As just described, when multiple responding STAs are expected, the AP uses a special frame (a trigger, called BFRP in this case) to organize the STAs' uplink transmissions. The Trigger frame is a new control frame defined in 802.11ax, and is used for many scenarios where the AP needs to coordinate STAs' simultaneous uplink transmissions. Figure 3-6 depicts the general format of the Trigger frame body.

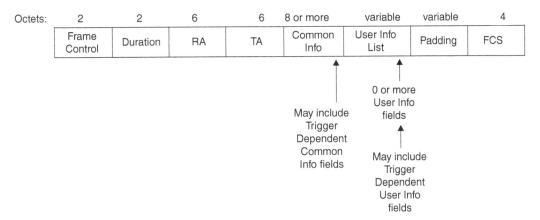

FIGURE 3-6 Trigger Frame Fields

The Trigger frame can be sent to a single STA or to the broadcast address, depending on whether it is intended for one or multiple STAs. The Common Info field expresses the general parameter and purpose of the trigger. As the expectation is that multiple STAs will send their transmissions together in an uplink frame, the Common Info field identifies the transmission parameters that the STAs should use, such as length, bandwidth, and the like. This field also indicates the trigger type; these types are listed in Table 3-4. The Common Info field can also include a Trigger Dependent Common Info field (in trigger type 5) that indicates the Block Ack Request parameters.

TABLE 3-4 Trigger Frame Types

Trigger Type Value	Trigger Type
0	Basic
1	Beamforming Report Poll (BFRP)
2	MU-BAR
3	MU-RTS
4	Buffer Status Report (BSRP)
5	GCR MU-BAR
6	Bandwidth Query Report Poll (BQRP)
7	NDP Feedback Report Poll (NFRP)
8–15	Reserved

Note

You will find many references naming a field within another field with the term "subfield." This semantic choice is absolutely correct. However, the 802.11 designers have noted that a subfield is a field, and in many cases do not bother with the underlying "sub" indicating that the field is within another field.

The User Info List[11] contains zero or more User Info fields. Each of these fields allocates RUs and/ or spatial streams to target STAs (identified by their AID) that are expected to transmit. The field also mentions the power at which all these joint transmissions are expected to be received at the AP level (see the "Spatial Reuse Techniques" section for details on its goal). A User Info field can also include a Trigger Dependent User Info field, whose content depends on the trigger type.

> **Note**
>
> The presence of the term "info" in the names of many fields nested into other fields also named "info" may be confusing if you do not look carefully at the field hierarchy. The Common Info field may include a Trigger Dependent Common Info field. The User Info List includes 0 or more User Info fields. Each User Info field may include a Trigger Dependent User Info field.

As all these elements have variable length, it might be necessary to add some padding bits to round the payload to the nearest octet.

The BFRP is a trigger frame with the special purpose of collecting beamforming reports from STAs, in preparation for a multiuser transmission. But the trigger frame can be used for many other purposes. The most basic purpose is to trigger multiple STAs to send an uplink flow together with OFDMA. Such frame is called the Basic Trigger frame. In addition to the Common Info field (uplink transmission characteristics) and User Info field (RU allocation for each STA), the Basic Trigger includes a Trigger Dependent User Info field that specifies which ACs are expected in the transmission and, based on a target MCS, the maximum length of each included MPDU. This field allows the BSS to ensure that the same access rules are used for the uplink transmission, as all transmitters use the same ACs. The STAs should respond one SIFS after the Trigger frame with their uplink transmission, at the RU(s) or spatial streams indicated. They also use the AP Tx power information (compared to the RSSI at which they received the trigger) to set their own transmit power. This ensures that all transmitted frames reach the AP not only at the same time (the timing tolerance is 0.4 μs), but also at the same power from the AP viewpoint. The AP receives the frame, and responds with a new Multi-STA Block ACK that indicates which frames were received properly.

The Basic Trigger frame supposes that the AP knows that the targeted STAs are ready to send, and that the whole channel is free of contention, especially from other BSSs around it. These conditions are not guaranteed, so the AP can also use the RTS/CTS technique described in Chapter 1 to further avoid collisions during a DL or UL OFDMA transmission. This type of RTS frame is called MU RTS.[12] It is sent to the broadcast address, using legacy rates, and on each 20 MHz channel of the expected MU transmission. Its duration value includes the duration of the expected exchange. The MU RTS frame carries the User Info List fields, with the AIDs of all STAs expected to participate in the MU transmission. All participating STAs are expected to respond simultaneously (after one SIFS) with a standard CTS frame. Recall that the CTS frame does not include the TA field, but rather the RA field—in this case,

---

11 See 802.11ax-2021, clause 9.3.1.22.1
12 See 802.11ax-2021, clause 9.3.1.2

the AP BSSID. As such, all STAs respond at the same time with the same frame. These different copies of the CTS frame may reach the AP (and other STAs in the cell) at slightly different times, just as multiple copies of a single normal CTS would reach a receiver in a multipath-prone environment. From the AP's (and other STAs') viewpoint, an AP sent an RTS frame and received in response (at least) one CTS. The exchange can then continue, with the trigger process for UL transmission. The AP can also send its frame to the STAs. Figure 3-7 illustrates these two forms.

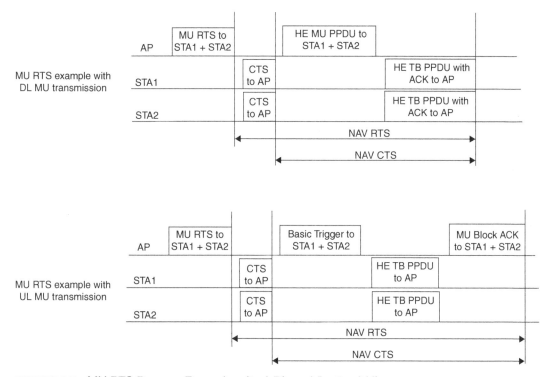

FIGURE 3-7   MU RTS Process Examples: (top) DL and (bottom) UL

With or without the MU RTS/CTS procedure, UL coordinated transmissions are challenging. In particular, a BSS can include hundreds of STAs—some in a dozing state, others fully awake. The STAs can change state (dozing/awake) at any instant. These STAs may also have bursts of traffic reaching their buffers at any time. One instant their buffer might be empty, and at the next instant they might have traffic to transmit. Then they transmit (possibly unscheduled), and their buffer is empty again. Organizing UL MU transmissions is therefore a complex task.

If the AP were to query each STA in turn to determine which ones have traffic to send, it would consume a considerable amount of airtime, would in many cases be unable to receive a response (if the target STA is dozing), and would list STAs in need of uplink scheduling that, by the time the AP round of querying is finished, would have long sent their traffic in an unscheduled manner. Most of the time, the AP's questions are simple, and can be answered with a single bit: Are you awake? (yes/no). Do you need uplink transmission scheduling? (yes/no).

To deal with these issues, the 802.11ax amendment resorted to inventing a more efficient polling mechanism. That is, the AP can send a single frame that many STAs will receive, but to which only some of the STAs will answer. This trigger frame is called NDP Feedback Report Poll[13] (NFRP). Its User Info field is quite specific. It contains a Starting AID field, which indicates the AID of the first STA that is allowed to respond, but also a Number of Spatially Multiplexed User field, which indicates how many STAs are multiplexed on the same tones in the same RU and allowed to respond. This structure is interesting, because more than one STA can be allocated to the same set of tones. However, the transmissions are set to be orthogonal, which minimizes the effect of one transmission to the other. When receiving the NFRP, a STA looks at the Start AID, computes (by using the Number of Spatially Multiplexed User value and the bandwidth of the transmission) the number of STAs that are queried (N_{STA}). It then verifies if it is in this list by checking if its AID is in the range (N_{STA} − Start AID). The STA responds only if it is awake, is in the list, and has traffic to send. This response is very short (it is an NDP Feedback Report) and is sent in the RU that matches the STA AID position. With a trigger and one MU UL response, the AP has the list of STAs that need scheduling. The AP can then acknowledge the response of all or some of the responding STAs, indicating to this subset that they are scheduled next. These STAs stay awake, waiting for the AP's trigger frame enabling their UL transmission. The resulting process is very light (2 frames for many STAs[14]) and very fast (the AP can query up to 36 STAs per 20 MHz channel, or potentially 296 STAs in a 160 MHz channel), making UL scheduling reactive enough to accommodate the changes in STAs' (and their buffers') states.

MU EDCA Channel Access

The OFDMA scheduling process for uplink transmissions provides a timing advantage to the participants. As it wins the channel access to send the trigger frame, the AP somehow shares this victory with all the scheduled STAs, by allowing them to send their MSDU right after the Basic Trigger frame (and one SIFS), without having to contend individually for the channel. These favored STAs would benefit from an unfair advantage (compared to the non-scheduled STAs) if they could, immediately after their scheduled UL OFDMA transmission, contend for the channel access just like any other (unscheduled) STA. However, the goal when scheduling uplink transmissions is to avoid excessive delay (see the "BSR" section that follows)—not to allow some STAs to benefit from more airtime overall than the others. When the STA has received an access advantage through a scheduled uplink transmission, it should have to wait a bit longer than the other STAs for its next unscheduled channel access. This would ensure any other (unscheduled) STAs have a chance to send first, and restore channel access fairness (where statistically the scheduled STA sent once, the unscheduled STA also sent once).

To enable this fairness, the AP advertises the regular EDCA parameters, including CWMin, CWMax, AIFSN, and TXOP for each of the four ACs (see the "Channel Access" section in Chapter 1), as well as

---

13 See 802.11ax, clause 9.3.1.22.9.

14 Contribution 11-16/1367 shows that this exchange consumes 260 μs, while a standard CSMA-CA process would consume, on average, 42 ms. Conceptually, the NFRP exchange could occur very often—for example, every 20 ms.

an alternative set of EDCA parameters (the MU EDCA Parameter Set[15]), which also indicates CWMin, CWMax, and AIFSN values for the four ACs. By default, the second CWMin/CWMax values are identical to the default values, but can be made to be longer. The MU EDCA Parameter Set also includes an MU EDCA timer for each AC. When it has successfully sent a trigger-based PPDU for one AC, a STA needs to apply the MU EDCA parameters for that AC first, before being allowed to continue and use the regular EDCA parameters. Practically, the STA examines whether it has received a trigger within the last MU EDCA Timer period. Such an event means that the STA was scheduled. In that case, the STA needs to apply the less-favorable MU EDCA parameters. Beyond that (MU EDCA timer) time, the STA goes back to the applying default EDCA parameters.

The AP can change the values of the MU EDCA Parameter Set at will. In particular, the element includes a version number that allows the STA to know if it has the latest values, and the AP can also send to the STA a MU EDCA reset frame to refresh the values for the STA. The MU EDCA Parameter Set also allows the AP to dynamically change the rules at any time, by causing the previously scheduled STAs to pause their transmissions. This second and more drastic mechanism is triggered by the AP setting, in the MU EDCA Parameter Set, the AIFSN value for one (or more) particular AC(s) to be 0 and communicating this new parameter set to the STA. If a STA receives the AIFSN[AC] = 0 message for an AC for which the STA was scheduled, it must pause its EDCA function (and so stop contending for channel access) for the duration of the MU EDCA timer. The STA can resume its normal operations at the end of the MU EDCA timer period. Figure 3-8 illustrates these MU EDCA contention principles.

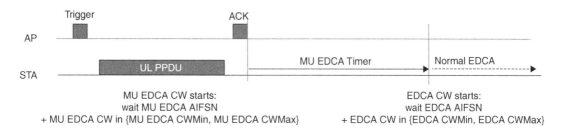

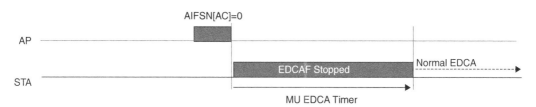

FIGURE 3-8 MU EDCA Contention Principles

---

15 See 802.11ax, clause 26.2.7

BSR

One key element that determines which STAs should be in the UL transmission pool is … their need to send frames! The AP should not just trigger STAs that are compatible with each other (from an RF standpoint). Rather, the AP should trigger STAs that have something to send. 802.11ax establishes three different methods by which the AP can gain that awareness.[16]

The simplest method is for a STA to send information about the amount of traffic waiting in its buffers, in an unsolicited manner. This method is called unsolicited Buffer Status Report (BSR). The STA sends either a QoS frame (carrying data) or a QoS Null frame, which contains a QoS Control field in its header. With pre-802.11ax transmissions, the QoS Control field primarily indicates the TID/UP of the frame. With the unsolicited BSR technique, the STA also sets the fourth bit of the QoS Control field to 1[17] to indicate that the Control field contains a Queue Size field[18] formed by the last 8 bits of the QoS Control field, and indicating the depth of the queue in the STA's buffer for that TID/UP. The queue depth can have a value of up to 64,768 octets. The field does not contain the status of other UPs, so the STA needs to send just one QoS frame per UP to inform the AP about all its UP queues.

A second possibility is for the STA to send a BSR for all ACs. In that scenario, the STA also sends a QoS frame, a QoS Null frame, or a management frame to the AP. The frame header includes an HT Control field. When the first two bits of the HT Control field are set to 1, the field becomes HE variant and includes an A-Control field. The A-Control field can include a BSR Control field.[19] In that case, the BSR Control field includes a Control Information field, which indicates the number of TIDs (1 to 7) contained in the report. An Access Category Index (ACI) Bitmap field represents the four ACs, and a Delta TID field further indicates which TID/UP in each AC has buffered traffic. The report is coarse, because it does not provide the buffer details for each individual TID. Instead, an ACI High field can point to the TID of highest importance (likely the one requiring urgent scheduling). A Queue Size High field indicates the depth of that critical queue. Another field, Queue Size All, represents the queue depth of the other queues (either their mean depth or the depth of the deepest queue, depending on the implementation). A Scaling Factor field indicates the octet units in use, thus allowing for an indication of up to 64,768 octets. Figure 3-9 illustrates this Control Information field.

	ACI Bitmap	Delta TID	ACI High	Scaling Factor	Queue Size High	Queue Size All
Bits:	4	2	2	2	8	8

FIGURE 3-9   Control Information Field in a BSR Control Field

Neither method is perfect. Notably, they rely on the STA sending the report in an unsolicited manner, which may be prohibited by the multiuser access parameters (MU-EDCA) under control of the AP itself.

---

16 See 802.11ax-2021, clause 26.5.5
17 See 802.11-2020, Table 9-10. The first 3 bits indicate the TID/UP, and in most cases the fourth bit is not significant.
18 See 802.11ax, clause 9.2.4.5.6
19 See 802.11ax-2021, clause 9.2.4.6a.4

As the AP is the entity needing that information for uplink scheduling, 802.11ax also allows the AP to send a BSR Poll (BSRP) Trigger frame to solicit BSR information from STA (solicited BSR). The Common Info field indicates the type (BSRP). The triggered STA then responds with BSR in a TB PPDU using one or more of the QoS Control field or HT Control field techniques. This way, the AP has a mechanism to learn about the traffic buffered on the STA; however, the rigidity of the BSR mechanism remains. If the STA uses the QoS Control field method, it needs to send as many frames as the STA has TIDs with buffered traffic. If it uses the HT Control field method, the STA sends an approximation of the buffered traffic in most queues (except the Queue High TID).

Additionally, BSR remains an *a posteriori* method. In an ideal world, a packet comes from the operating system into the 802.11 stack, gets encapsulated into a frame, and is sent to the medium immediately (because the medium is available). In contrast, in a congested environment, the frame is delayed and stays in the STA's buffers. At some point in time, the STA uses airtime to send a frame to the AP—not to send that buffered data, but rather to inform the AP that data is there that should be sent but continues to be delayed in the STA's buffer. Even if the AP immediately schedules airtime for the STA, the BSR method does not prevent congestion, though it limits its long-term effects. As explained in Chapter 4, "The Main Ideas in 802.11be and Wi-Fi 7," this type of reactive mechanism is not fast enough for near-real-time applications like AR/VR.

BSS Coloring

Another key issue that 802.11ax addressed was overlapping BSSs (OBSSs). As much as possible, neighboring BSSs are expected to operate on non-overlapping channels. However, with the increase of channel width (40 MHz with 802.11n-2009, then 80 MHz with 802.11ac-2013), and the reduction in the size of BSSs in most deployments,[20] it had become common to see neighboring BSSs on partially or completely overlapping channels. This overlap might not be problematic for the APs, but it could cause large losses for STAs positioned between the APs (irrespective of which APs the STAs are associated with). You can see this OBSS issue in Figure 3-10.

Suppose that STA1 is associated to AP1 (STA2, STA3, and STA4 are associated to AP2), and that both AP1 and AP2 are on the same channel. When AP2 transmits a broadcast frame or a unicast frame to STA2, STA1 detects that transmission and (because of CCA[21]) refrains from sending to AP1 at the same time.

Other STAs, like STA4, might be too far from STA1 to prevent STA1 from sending at the same time; however, the AP2/STA2 scenario is common. 802.11ax helps mitigate this issue with a multilayered spatial reuse mechanism. Spatial reuse allows two or more neighboring systems—BSSs in this case— to use the same medium by identifying signals from the overlapping system and applying interference management techniques.

---

20 In the days of 802.11n, a 6000 to 8000 ft^2 (550 to 750 m^2) BSS for basic data coverage was common. At the conclusion of the 802.11ac work, 2500 to 4000 ft^2 (230 to 370 m^2) BSSs were common.

21 See the section "Channel Access Rules" in Chapter 1 for details on CCA.

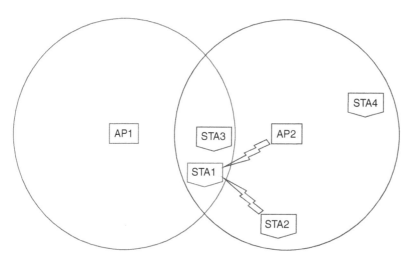

FIGURE 3-10   OBSS Issues

BSS Color Operation

The first element of the BSS coloring mechanism is to associate to each BSS a unique label, called *color*[22] (although there are no coloring concepts, just numbers). Upon startup, an AP implementing BSS coloring selects a number between 1 and 63 as its own "color." 802.11ax merely notes that the AP should consider, when selecting its own color, the colors of the existing BSSs nearby. In practice, in a controller-based environment, the WLAN controller will likely assign a different color to each AP in range of each other with primary channel overlap. Practically, the WLC would receive OBSS warnings and allocate neighboring APs to different channels. In an independent AP deployment, each AP operates independently. In that case, a booting AP may need to select its color by picking a random number between 1 and 63 to start operating, even before detecting other APs.

> **Note**
>
> Neighbor AP detection is a key feature of BSS color operations. With legacy techniques, the AP would need to receive a frame, decode the entire MAC header to extract the BSSID, and conclude that the sender is in an OBSS. With BSS coloring, the AP can identify the OBSS scenario directly from the physical header of any detected transmission on the channel, thus allowing for early detection of OBSS frames, and the subsequent action of ignoring them, as described further in this section.

The AP then mentions its color value in the HE Operation element in the beacon and probe responses. Each STA that associates to the AP adopts that color. For any frame sent in the BSS, the transmitter then indicates in the SIG-A field of the PHY header the BSS color associated to that BSS.

22 See 802.11ax-2021, clause 26.17.3

Color Collision Management

In a distributed system, it is possible that color collisions might happen; that is, two neighboring APs might pick the same color. With 63 values to choose from, the probability of such collision is a simple permutation calculation, in the form $1 - ((63! - (63 - n)!/ 63^n)$, where n is the number of overlapping BSSs. For example, with 3 APs in range of each other, the probability of two APs picking the same number would be $1 - (63.62.61 / 63^3) = 0.0471$, or 4.71%. The collision probability reaches 9.24% with 4 OBSSs, and 15% with 5 OBSSs. These numbers might look small, but they are statistically significant: 4.71% means that for 100 groups of 3 OBSSs (think of a medium to large shopping mall), 4 to 5 of them will display a color collision.

The collision detection event occurs when an AP detects a frame sent from a STA in an OBSS (i.e., another AP, or a STA that is not associated to the local AP) with the same color as its own. However, the AP is at the center of the BSA and may or may not detect the OBSS traffic. STAs that are closer to the other system are better reporters. Therefore, 802.11ax also allows a STA to report a color collision. Similar to the AP detection, the STA might detect a transmission with a SIG-A field matching its own BSS's color, but where none of the three addresses in the MAC header matches the BSSID with which the STA is associated. In that case, the STA can report the color collision with an Event Report frame.[23] This frame is an action frame inherited from 802.11v-2011 and enhanced for BSS coloring operations; it includes an Event Report element with an Event Report field that indicates the BSS color collision.

Upon concluding that a BSS color collision has occurred, the AP can start by disabling its own color for a while by setting the BSS Color Disabled bit in the HE Operation element to 1 (in beacons, association responses, and probe responses). All its associated STAs, upon receiving this beacon, would then understand that the current color is no longer valid. The AP can subsequently send a BSS Color Change Announcement, an information element that can be carried in beacons, probe responses, reassociation responses, or a specific HE BSS Color Change announcement action frame. The element contains the value of the new color, but also a color switch countdown value that indicates when (in units of TBTT) the new color will come into effect. As the message is repeated in each beacon (and the TBTT is a beacon interval), the color switch countdown acts as a countdown timer. When it reaches 0, the AP starts announcing the new color. At that time, the old color is no longer mentioned, and the BSS color disabled bit in the HE Operation element is set back to 0.

Spatial Reuse Techniques

After this operation, neighboring BSSs should operate with different color values. The next element in the spatial reuse mechanism is to apply interference mitigation techniques. Two techniques are implemented together: a technique to ignore the transmissions of the OBSS system (if they are below some target threshold), thereby allowing the STA to transmit while another OBSS is also transmitting, and a technique to reduce the transmit power, so as to minimize the disruptions to the neighboring OBSS. These two techniques come in two flavors: a parameter-free spatial reuse flavor and a parameterized spatial reuse flavor.

---

23 See 802.11-2020, clause 9.6.13.3

The parameter-free mode is called non-Spatial Reuse Group (non-SRG) OBSS Packet Detect (PD).[24] Recall from Chapter 1 that a STA should consider the medium to be busy if it detects an 802.11 frame at a power of –82 dBm or higher, or if it detects a signal at a power of –62 dBm or higher. With non-SRG OBSS PD spatial reuse, any transmitter in the BSS (non-AP STA or AP) can decide to lower its sensitivity threshold (the OBSS PD threshold) somewhere between the –82 dBm and –62 dBm values (called OBSS PD_{min} and OBSS PD_{max}, respectively). For example, a STA can decide to consider OBSS frames detected at –70 dBm or higher, thereby placing its OBSS PD threshold at –70 dBm. Any signal weaker than this value is ignored; thus, the STA starts transmitting, ignoring that weaker signal. When doing so, the STA needs to adjust its transmit power, so that its own transmission will not negatively affect the ignored OBSS transmitter too much. 802.11ax places a simple condition on this power reduction, by first defining a reference transmit power (TP_{ref}), which depends on the STA type and spatial stream count, as follows:

$$TP_{ref} = \begin{cases} 21 \text{dBm for STAs} \\ 21 \text{dBm for APs with 2 or less SS} \\ 25 \text{dBm for APs with 3 or more SS} \end{cases}$$

The transmit power of the STA (TP_{STA}) must then be chosen to satisfy the following constraints, where $\log_{10}(BW)$ is the bandwidth of the PPDU to transmit (20, 40, 80, or 160 MHz):

$$TP_{STA} \leq OBSS\ PD_{min} - OBSS\ PD + TP_{ref} + \log_{10}(BW / 20)^{25}$$

From this equation, you can see that a STA that sets its OBSS PD sensitivity at, for example, –78 dBm (thus ignoring only OBSS STAs that would be fairly far away) can send a 20 MHz signal at 17 dBm. However, a STA that set its OBSS PD threshold at –70 dBm (thus ignoring transmissions that are fairly close by) should set its power (for a 20 MHz signal) at 9 dBm. In other words, the STA decreases its transmit power as it ignores OBSS systems that are closer, which limits the STA's negative effect on the neighboring system. Just as in human communications in a quiet setting, the STA speaks more quietly if another conversation is occurring nearby, and it can speak more loudly (at normal volumes) if the other conversation is farther away. The neighboring system will likely operate in the same way. As a result, STAs that are between APs and strongly affected by the OBSS issue, such as STA1 and STA3 in Figure 3-6, will reduce their transmit power while ignoring the OBSS interference as much as possible. In contrast, STAs located farther away and potentially less affected by the OBSS issue, such as STA2, will reduce their transmit power only a little bit as they detect STA1 in the distance. Meanwhile, STAs on the other side of the OBSS issue, such as STA4, will not change their power, because they are not affected.

Within those limits, each STA can choose the OBSS PD value that best suits its needs, given that a STA usually aims at maximizing its access to the medium. The AP can also set some boundaries and prevent the STAs from setting their OBSS PD thresholds too high (too close to the –62 dBm OBSS

---

24 See 802.11ax-2021, clause 26.10.2.2
25 See 802.11ax-2021, Equation 26-5

PD_{max} value), by announcing a non-SRG OBSS PD_{max} value in the field of the same name, within the Spatial Reuse Parameter Set element in the beacons and association response frames.

The AP can advertise a single non-SRG OBSS PD_{max} value. However, a need may also arise—especially in enterprise managed networks (with an administrator in control of the deployment and managing the OBSS issue)—for different OBSS PD values for different neighboring systems. For example, the admin might configure AP1 to advertise a high OBSS PD threshold for traffic coming from the BSS of AP2 (which has a large overlap with AP1's BSA). The admin may also configure AP1 to advertise a lower OBSS PD threshold for traffic coming from AP3's BSS. Therefore, 802.11ax allows the AP to advertise SRG OBSS PD_{min} and SRG OBSS PD_{max} values. In the Spatial Reuse Parameter Set element of the beacon and probe response, the AP indicates a list of colors (in the form of a bitmap) or a partial list of BSSIDs, forming the SRG, and the SRG OBSS PD_{min} and SRG OBSS PD_{max} values for that group. The STAs in the BSS apply these thresholds to any traffic detected coming from an OBSS whose color matches the ones in the group. The STAs can then use the non-SRG OBSS PD_{max} value advertised by the AP (if it advertises the value) for any other color. If the AP does not advertise a default non-SRG OBSS PD_{max} value, the STA can use its own static threshold for traffic with any other color.

The Spatial Reuse Parameter Set element is central to this orchestration. You can see that element in Figure 3-11. In addition to the element ID (and its extension if the ID is larger than 255) and the element length, the SR Control field indicates which optional parameters are present. Note that the optional elements have a length of 0 or 1 octets, or 0 or 8 octets.

Element ID	Length	Element ID Extension	SR Control	Non-SRG OBSS PD Max	SRG OBSS PD Min	SRG OBSS PD Max	SRG BSS Color Bitmap	SRG Partial BSSID Bitmap
Octets: 1	1	1	1	0 or 1	0 or 1	0 or 1	0 or 8	0 or 8

FIGURE 3-11   Spatial Reuse Parameter Set Element

Instead of choosing a static OBSS PD threshold value, a STA can also dynamically set its transmit power by using the received power value of any frame coming from the OBSS as a reference threshold. In other words, as long as the received signal is lower than the OBSS PD_{max} value (i.e., weaker than −62 dBm or the threshold the AP advertises), that incoming signal is used as the OBSS PD value in the power equation. This way, the STA can set different transmit power values based on which OBSS STA is causing interference, and depending on the level of that interference.

The spatial reuse mechanism described so far is optimal for scenarios in which the STA sends frames without being scheduled. When the AP triggers several STAs to send their transmissions together an uplink frame, recall from the "Scheduling and Multiuser Operations" section that one requirement is that the frames arrive at roughly the same signal level at the AP. If STAs in neighboring BSSs also send at the same time, the AP might not be able to receive and decode some or all of the uplink components. To avoid this issue, the spatial reuse mechanism also comes in a parametrized flavor—called

Parametrized Spatial Reuse (PSR). Its goal is to inform STAs in neighboring BSSs how much they can transmit over triggered uplink frames without affecting the AP's ability to properly receive the frames.

The trigger frame sent by the AP includes a Common Info field. Recall that this field includes the AP Tx power and the bandwidth of the expected uplink frames that the STAs will send in response to the trigger. The field also includes, in its bits 37 to 52 (16 bits), the UL Spatial Reuse field, which indicates up to four PSR values,[26] called PSR_{input}. Depending on the bandwidth of the upcoming uplink frame (20, 40, 80, or 80 + 80 MHz), one to four of these PSR_{input} values will contain a number. The STAs in the BSS do not use these PSR_{input} values directly. Instead, the STAs receive these PSR_{input} values, and simply indicate them in the HE SIG-A field of the uplink frame they generate in response to the trigger frame. The real targets of these values are the STAs in the OBSSs.

The PSR_{input} values are computed by the AP based on its determination of an acceptable interference level, I_{AP}, which is defined as the maximum interference that the AP could tolerate while still receiving the uplink frame properly. You might remember that the AP, in the trigger frame, indicates a target RSSI (in addition to the AP Tx power), allowing the STAs to set their transmit power so that all the uplink transmissions reach the AP at about the same power level. The AP uses that target RSSI, with also taking into account the expected MCS (and the associated SNR at which that MCS can be safely demodulated) to calculate I_{AP}, with the addition of some safety margin (typically 5 dB or less). The PSR_{input} value indicated in the UL Spatial Reuse field is calculated as follows:

$$PSR_{input} = AP\_TX_{power} + I_{AP}$$

A STA receiving such trigger frame might observe that the sending AP is not the AP of the STA's BSS (i.e., the sender is a neighboring AP in an OBSS scenario). It can then use the signal level at which the trigger frame was received, along with the AP_TX_{power} and the PSR_{input} values, to determine the maximum power at which it would be allowed to transmit over the upcoming uplink frame. At that point, based on its OBSS PD parameter, the STA can decide whether transmission at that power is useful (or not).

A consequence of the spatial reuse mechanism is that STAs implementing 802.11ax fundamentally use two CCA mechanisms. One basic CCA mechanism, with its set of standard thresholds, applies to frames that are detected as intra-BSS (i.e., to and from non-AP STAs or the AP in the STA's BSS). Another CCA mechanism, which is adaptive, applies to inter-BSS frames. Each carrier sense conclusion affects whether the STA needs to update its Network Allocation Vector (NAV) and decides whether the STA needs to wait until the channel is no longer busy or whether the STA can continue its countdown until transmission. Therefore, as the STA maintains two sets of CCA thresholds, it also maintains two NAVs[27]: a basic NAV and an intra-BSS NAV. The intra-BSS NAV is updated when an intra-BSS frame is detected. The basic NAV is updated by an inter-BSS frame, or when a signal is detected at a high level but cannot be classified as intra-BSS or inter-BSS. Only when both NAVs reach 0 can the STA send its frame.

---

26 See 802.11ax-2021, Table 17-21, for the individual values of these PSR values.
27 See 802.11ax-2021, clause 26.2.4

Target Wake Time

Another issue that plagues high-density environments relates to waking after sleeping. As much as possible, STAs (often operating on batteries) attempt to switch to dozing mode when they are not sending or expecting incoming frames. In traditional 802.11 operations, the STA sends a null frame with the power management bit set to signal that it is entering dozing mode. The STA waking up from dozing mode needs to contend for the medium and contact the AP to retrieve any DL frame that had been buffered while the STA was unavailable. This method consumes airtime and requires some exchange overhead. When the STA's traffic follows unknown patterns (because, for example, the traffic peaks are dependent on user clicks and other actions, which the 802.11 stack cannot predict), the method is efficient. However, in many cases, the STA's traffic pattern is known. A typical scenario is an IoT object, such as one exchanging keepalive messages every 100 ms with a management application located somewhere in the infrastructure. Another example is a Voice over Internet Protocol (VoIP) phone during a call, where one packet is sent and one packet is received at known intervals (e.g., a G711 Codec sends one 160-byte packet every 20 ms in each direction). Signaling a dozing state and then contending to retrieve buffered traffic at each of these intervals is inefficient, and this inefficiency is multiplied by the number of STAs performing that same operation in the BSS.

To improve the efficiency of the dozing process, 802.11ax adopted a process derived from 802.11ah-2013 that is called Target Wake Time (TWT).[28] With TWT, the STA and the AP can agree on time periods, called Service Periods (SPs), during which the STA will be awake, with the understanding that the STA will likely be dozing between these service periods. As the AP knows that the STA is awake at precise points in time, the AP can directly send any frame to the STA without needing to verify that the STA is no longer dozing. Similarly, because the STA knows that the AP has allocated a time where the STA is expected to be active, the STA expects that it should be able to send a frame at that time, without having to contend with the AP. In this way, the TWT process solves the overhead of the dozing mode.

The TWT negotiation can be performed one STA at a time (resulting in individual TWT agreements), or the AP can create a TWT schedule that applies to a group of STAs (resulting in broadcast TWT schedules). In the individual case, TWT is initiated by the STA, sending a TWT Setup frame[29] to the AP. This action frame contains a TWT element with the parameters that the STA presents to the AP, such as the start time of the next SP, the interval between the SPs, and the duration of each SP. The TWT element also includes a TWT Setup Command field. This field is used by the STA to inform the AP about how the TWT parameters should be chosen:

- TWT Setup Command is set to *request* when the STA indicates that the AP should decide on the best TWT parameters.

- TWT Setup Command is set to *suggest when* the STA suggests some parameters, but may accept some variation suggested by the AP.

- TWT Setup Command is set to *demand* when the STA provides strict parameters and cannot accept different values.

28 See the "High Performance Channels" section in Chapter 5, as well as 802.11ax-2021, clause 26.8.
29 See 802.11ax-2021, clause 9.6.24.8

As you might expect, the AP, in response, can accept the parameters sent by the STA, provide alternative parameters, or flatly reject the request (in which case the dialog stops). In practice, more than one TWT setup exchange may be needed before the STA and the AP agree on a set of parameters. The result of this individual negotiation is that the AP can have multiple TWT agreements, each with a different STA. The service periods may then overlap, causing the STAs to have to contend with each other for channel access using the usual contention methods.

The AP can also have a shared TWT session with a group of STAs. This second mode is called Broadcast TWT. In this case, the AP creates one or more broadcast TWT schedules and specifies their parameters, and advertises each group identifier (the Broadcast TWT ID) and the matching TWT parameters in a TWT element of the beacon frames. A STA that wants to join one of the advertised broadcast TWT schedules makes that request by sending a TWT setup action frame that contains, in the TWT element, the Negotiation Type field set to 3 (broadcast) and the broadcast TWT ID. Here again, the AP can accept or reject the request, or suggest an alternative broadcast TWT schedule. During the setup phase, the STA can also optionally negotiate the next target beacon and the listen interval; that is, it can determine which beacons the STA should listen to, so that it can stay updated on the group TWT parameters. At any time, the STA can leave the broadcast TWT schedule by sending a TWT teardown frame.[30] The AP can send this frame to tear down either the STA membership to the group or an individual TWT agreement. The AP can create new broadcast TWT schedules (with different TWT parameters) or tear down existing groups as its scheduling, load, configuration, or other parameters require. However, the AP cannot delete a group that still has active STA members.

TWT is designed to be flexible. In particular, the service period is a key component, especially when the SP repeats at intervals. The entire system works only if the STA and the AP understand the SP boundaries in the same way. In practice, this requirement means that their clocks are aligned with enough precision that the STA will be awake when the AP decides that it is the beginning of the next SP. At that time, the AP can send a frame that the STA will receive. If the STA's clock is misaligned with the AP's clock, the transaction fails, defeating the whole purpose of TWT.

To accommodate all types of STAs and APs (those with high-quality clocks—and the others), TWT SPs can operate in different modes:

- **Implicit versus Explicit:**
  - In Implicit (also called periodic) TWT operation mode, the STA and the AP agree on a periodic schedule. The AP and the STA calculate the start time of the next SP based on this schedule agreement (helped by the fact that data communication can be observed during the SP). The clocks of the AP and the STA are expected to be fairly aligned.
  - In Explicit (also called aperiodic) TWT operation mode, during the SP, the AP sends information about the next TWT (in particular, the time at which the next SP will start) to the STA. This mode allows the STA to readjust its clock at each SP.

---

30 See 802.11ax-2021, clause 9.6.24.9

- **Announced versus Unannounced:**

  - In Announced TWT operation mode, the STA must send a PS poll or UAPSD trigger frame at the beginning of the SP, to signal to the AP that it is awake. This mode is virtually identical to the traditional power save operation. The only difference is that the STA announcement is expected to occur during a TWT SP, whereas in traditional mode the announcement can happen whenever the STA decides to exit dozing mode.

  - In Unannounced TWT operation mode, the AP expects that the STA is awake at the beginning of the SP, and can send frames without verifying that the STA is awake. This is the typical expected mode for TWT, as removing the announcement is one of the goals of the TWT mechanism.

- **Trigger-Enabled versus non-Trigger-Enabled:**

  - 802.11ax implements UL triggered transmissions, and you would rightly expect this aspect to be present in the TWT context. In Trigger-Enabled TWT operation mode, inside each SP, the AP must send trigger frames for the STA to be able to send UL traffic. This mode is well suited to broadcast TWT schedules, where the AP schedules multiple STAs for UL MU MIMO or UL OFDMA transmissions.

  - In non-Trigger-Enabled TWT operation mode, the STA can transmit using EDCA within the SP. In such TWT SPs, the AP is not expected to send trigger frames. This mode is well suited to individual TWT agreements.

Figure 3-12 illustrates individual and broadcast TWT agreements for the implicit and unannounced modes.

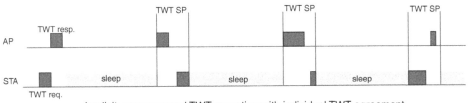

Implicit, unannounced TWT operation with individual TWT agreement

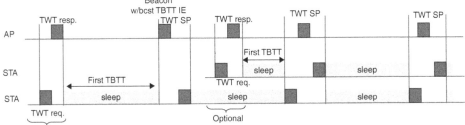

Implicit, unannounced TWT operation with broadcast TWT agreement

FIGURE 3-12 Individual and Broadcast TWT Agreements for the Implicit and Unannounced Modes

TWT can dramatically improve the efficiency of the operations by reducing the energy consumption and the overhead associated to the STA's dozing mode. However, this approach brings benefits only if the STA and the AP have clear parameters that they can negotiate. This, in turn, supposes that the STA's traffic pattern, a Layer 7 (Application) construct in the OSI model, is well understood by the 802.11 stack. This requirement is sometimes hard to meet, as application developers often do not understand the 802.11 stack, and the 802.11 layer cannot know all applications and their needs.

Wi-Fi in the 6 GHz Band

With BSS coloring, a more efficient subcarrier scheme, and performance mechanisms like TWT, 802.11ax and Wi-Fi 6 bring a lot of features that improve 802.11 efficiency, especially in high-density deployments. However, one issue that 802.11ax cannot solve alone is the requirement for backward compatibility. In an 802.11ax BSS in the 2.4 GHz or 5 GHz band, pre-802.11ax devices may appear that will limit the performance of the 802.11ax STAs for two reasons: because of the airtime that these pre-802.11ax consume, and because the 802.11ax devices need protection mechanisms (e.g., legacy headers) to coexist with these older devices.

Fortunately, by the time 802.11ax neared completion, and after years of discussion, several large regulatory bodies—the Federal Communications Commission (FCC), then the European Telecommunications Standards Institute (ETSI), then many others—announced their willingness to open the 6 GHz band for 802.11 operations. This new spectrum offered a fantastic opportunity for 802.11ax (with the Wi-Fi Alliance's Wi-Fi 6 E program) to operate in a band where backward compatibility was not an issue, and where optimal-efficiency rules could be implemented.

New Spectrum, New Power Rules

As you might expect, the 6 GHz band was not a totally unused space before 802.11 was authorized there. Several other systems were operating in various segments of this band. Authorizing 802.11 implied making sure that Wi-Fi would not disrupt the transmissions of the incumbents. Different regulatory domains had different incumbents. As a result, the 6 GHz band looks today like the 5 GHz band in the early days of 802.11 operations in that band. Some segments are allowed in some regulatory domains but not in others, and the power rules are not the same in all segments. These segments also have different names and designations depending on the regulatory domain. This book uses the FCC naming convention, because it seems to be defined clearly enough that any local nomenclature can be found from the FCC names.

Spectrum

In the FCC world, the 6 GHz band is divided in four segments of interest where each segment includes channels positioned 20 MHz from each other:

- **U-NII-5:** Includes 24 channels, from channel 1 centered on the 5.955 GHz frequency, to channel 93, centered on the 6.415 GHz frequency.

- **U-NII-6:** Includes five 20 MHz-wide channels, from channel 97, centered on the 6.435 GHz frequency, to channel 113, centered on the 6.515 GHz frequency.

- **U-NII-7:** Includes eighteen 20 MHz channels, from channel 117, centered on the 6.535 GHz frequency, to channel 185, centered on the 6.875 GHz frequency.

- **U-NII-8:** Includes twelve 20 MHz channels, from channel 189, centered on the 6.895 GHz frequency, to channel 233, centered on the 7.115 GHz frequency.

These channels can be grouped to form 40 MHz-wide channels, 80 MHz-wide channels, or 160 MHz-wide channels, as illustrated in Figure 3-13.

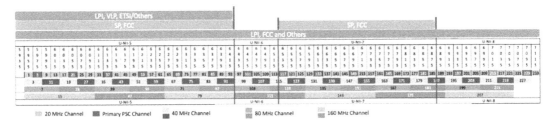

FIGURE 3-13 Channels in the 6 GHz Band

At the time that this chapter was written (mid-2024), not all channels were allowed in all regulatory domains. The FCC, followed by the regulatory bodies of several other countries, including Saudi Arabia, Brazil, Argentina, Peru, Colombia, and Canada, authorized the entire band in Figure 3-13, albeit with different power rules (standard, low, or very low power). Most of the ETSI countries and Australia have authorized the 5.925–6.425 GHz range, and are considering the adoption of the 6.425–7.125 GHz range. Other countries, including Russia, Sweden, Mexico, South Africa, Namibia, Kenya, Turkey, and many central European countries, have adopted the 5.925–6.425 GHz band but decided against allowing the upper part of the band (6.425–7.115 GHz). Still other countries continue to debate the adoption of one or more segments of the 6 GHz band. The difference between regulatory domains is important: In the United States, 59 new 20 MHz channels are accessible; in the ETSI domain, "only" 24 new 20 MHz channels are accessible. In all cases where 6 GHz is allowed, however, the spectrum available to Wi-Fi at least doubles, and sometimes triples.

Note

The Wi-Fi Alliance keeps track of the current state of authorization for every country, at this page: www.wi-fi.org/countries-enabling-wi-fi-in-6-ghz-wi-fi-6e.

General Power Rules

802.11 operations in the 6 GHz band come with interesting new power rules. In the 2.4 GHz and 5 GHz bands, the rules about maximum allowed power are built on the concept of effective isotropic radiated power (EIRP), a measure of the total quantity of energy radiated by the antenna of the transmitter (non-AP STA or AP STA). The maximum EIRP is fixed for 2.4 GHz and each sub-band of 5 GHz. This limitation means that an 802.11 STA transmitting a signal over an 80 MHz channel cannot transmit more energy overall than the same STA transmitting over a 20 MHz channel in the same band. As a result of this rule, the energy transmitted over each megahertz of a 80 MHz transmission is lower than the energy transmitted over each megahertz of a 20 MHz transmission. This idea might sound strange, but it is similar to the idea of the output of a water hose. Suppose your hose is allowed to deliver 1.5 liters of water per second. It will spray less water per unit of surface (square inches or square centimeters) if you spread the jet over a 5-meter-wide circular area than if you focus the jet, power-washer style, over a small (e.g., 1 square centimeter or 0.5 square inch) surface.

A consequence of the EIRP rule is that the functional area of an 80 MHz BSS is smaller than that of a 20 MHz BSS. In both cases, the beacons are transmitted over 20 MHz, so they can be transmitted at similar power and reach the same area. However, an 80 MHz data frame will contain less energy per megahertz than a 20 MHz data frame, with the consequence that the 20 MHz frame can still be properly received and decoded at a distance from the transmitter where the 80 MHz frame is no longer understandable. Another way to express this phenomenon is to note that the signal-to-noise (SNR) ratio over each megahertz of frequency decreases as the channel width increases.

This constraint was taken into consideration during the debate over the authorization of 802.11 in the 6 GHz band. The regulatory bodies that have authorized operations in the 6 GHz band have so far adopted another method for computing power, the maximum power spectral density (PSD). In this system, the maximum amount of energy that a transmitter can emit is regulated on a per unit of spectrum basis (i.e., per hertz). In the water hose analogy, the regulation would limit the amount of water spread per unit of surface, rather than the total amount of water sent by the hose. Practically, this rule means that a transmitter can send the same amount of energy over each megahertz irrespective of the channel width. As such, the 20 MHz transmission has the same useful range as the 80 MHz transmission. A natural consequence is that the total amount of energy sent over a 80 MHz transmission is larger than the total amount of energy sent over a 20 MHz transmission, at least for a frame of the same duration.

Standard, Low, or Very Low Power

One key factor driving the various 6 GHz adoption decisions of the different regulatory bodies is that the different segments of the band were already used by existing entities (the incumbents). In some cases, these entities had stopped using the band for historical reasons. In other cases, the entities were still actively using the band, and authorizing 802.11 meant ensuring a peaceful coexistence with these incumbents, by regulating the power that 802.11 entities could use.

LPI, VLP, and SP in Outdoor and Indoor Scenarios

The general power rules (based on PSD) apply to all segments of the 6 GHz band, but with different PSD maximums depending on the scenario and the segment of the band. Most regulatory domains distinguish indoor from outdoor scenarios. Most incumbents are outdoor systems, either fixed or mobile point-to-point and point-to-multipoint communication systems. An 802.11 radio operating at low power indoors presents limited interference risks for an outdoor incumbent. Therefore, many regulatory domains allow 802.11 to function with no restrictions indoors, as long as they use a power setting called Low Power Indoor (LPI). In the FCC domain, APs can operate in LPI mode in all four segments: U-NII-5, U-NII-6, U-NII-7, and U-NII-8. The AP must then not exceed 5 dBm/MHz. For a 20 MHz channel, this limit represents an AP set at 18 dBm. With the PSD rule, this limit also represents an AP set to 21 dBm for a 40 MHz channel, 24 dBm for an 80 MHz channel, and 27 dBm for a 160 MHz channel. In addition, the FCC rule clarifies that the AP maximum power in LPI mode is 30 dBm, alluding to the fact that the AP cannot be set to more than 320 MHz without reducing its power per megahertz. The ETSI also recognizes LPI, but allows this mode only in U-NII-5 and U-NII-6, and with a limit of 10 dBm/MHz, resulting in an AP power maximum of 23 dBm for a 20 MHz channel. However, the ETSI also sets 23 dBm as the AP maximum power. Therefore, the AP should also be set at 23 dBm maximum for a 40 MHz, 80 MHz, or 160 MHz channel. Practically, this means that the PSD rule finds its limit in the ETSI domain at 23 dBm EIRP.

Operation in LPI mode is allowed only indoors. Practically, this means that an AP leveraging LPI cannot be weatherized. The AP also cannot have an external antenna, a restriction meant to ensure that the gain of the antenna will not cause the AP to exceed the EIRP limits.

A 20 MHz channel set to 18 dBm (FCC) or 23 dBm (ETSI) in 6 GHz is called "low power," but keep in mind that the power envelope allowed in 5 GHz depends on the sub-band and can also be seen as "low." In U-NII-1, the maximum EIRP indoors is also 18 dBm in the FCC domain and 23 dBm in the ETSI domain. Other segments of the band allow for higher power—for example, 30 dBm EIRP in the FCC domain and the ETSI domain with U-NII-2c, with DFS and TPC (dynamic frequency selection and transmit power control, to avoid interference with airport radar systems). Therefore, 6 GHz LPI is not necessarily very "low." However, the propagation characteristics of the 6 GHz band (compared to the 5 GHz band) may result in a practically usable smaller cell in 6 GHz. (For Wi-Fi 6 and Wi-Fi 7 deployments, see Chapter 8, "Wi-Fi 7 Network Planning.")

Ultimately, proponents of 802.11 in 6 GHz requested the possibility for higher power, modestly called Standard Power (SP). The FCC granted this request, for operation both indoors and outdoors, but only in U-NII-5 and U-NII 7, and with the conditions detailed in the next section, "AFC Rules." SP enables the AP to operate at up to 23 dBm/MHz, but with a maximum EIRP of 36 dBm.

As you might expect, if the AP power is regulated, the non-AP STA power is regulated as well. One difficulty is that the non-AP STA cannot know if it is indoors or outdoors; by contrast, an AP is fixed, and an admin deploying the AP knows if the AP is deployed indoors or outdoors. Therefore, the non-AP STA is always designated as a client *under the control of an access point*. The non-AP STA must contact an AP first, before knowing if it can operate at the SP or LPI level. In both cases, the non-AP STA power limit is 6 dB below the AP maximum in the FCC domain, but the same level as the

AP in the ETSI domain. If the AP is not at maximum power in the FCC domain, it might be possible that the STA happens to operate at the same power as the AP. Even if it is at maximum power, the AP typically has highly sensitive antennas. As a result, the 6 dB difference might not cause dramatic asymmetry issues between DL and UL transmissions.

The ETSI has allowed, and the FCC is considering, a Very Low Power (VLP) mode, indoors and outdoors, with a PSD of –8 dBm/MHz and a maximum EIRP, for the AP and the non-AP STA, of 14 dBm. This mode is primarily intended for client-to-client communications (e.g., a smartphone to a smartwatch) or small mobile systems (e.g., a phone acting as an AP to connect a Wi-Fi laptop to the phone cellular network).

Table 3-5 summarizes these various power limits.

TABLE 3-5   6 GHz Power Limits

Mode	Domain	Max Tx Power EIRP		Max PSD EIRP	
		AP (dBm)	Client (dBm)	AP (dBm)	Client (dBm)
SP	FCC	36	30	11	5
LPI	FCC	30	24	5	–1
	ETSI	23	23	10	10
VLP	FCC, ETSI (under consideration)	14	14	–8	–8

AFC Rules

LPI is authorized indoors. Allowing an SP mode triggered a vigorous debate, because the notion of "indoors" itself is always ambiguous. An AP may be deployed indoors, yet close to a window with low RF absorption characteristics. Increasing the power beyond LPI meant increasing the risks that APs would interfere with incumbents, even if the AP is indoors. In the end, there may not be many differences in the interference risk if the AP is indoors (but close to the outdoors, from an RF propagation perspective) or directly outdoors.

Looking at the different segments of the 6 GHz band, the FCC concluded that there would be two key types of incumbents: some static (non-mobile) systems (e.g., person-to-person [P2P] radio links, from public safety dispatch to cell tower backhaul, but also satellite links) in U-NII-5 and U-NII-7 and some mobile systems (e.g., a TV news truck sending feeds back to the main station) in U-NII-6 and U-NII-8. Limiting interferences to mobile systems is very difficult, because these systems can appear anywhere, at any point in time. Therefore, the FCC decided to forbid 802.11 outdoors, as well as indoors operations at power beyond LPI, in U-NII-6 and U-NII-8. In U-NII-5 and U-NII-7, coexistence with non-mobile systems might be possible. Thus, the FCC allowed outdoors and indoors operations at the SP level, but with the conditions that an AP would not radiate energy upward (beyond 21 degrees above the horizon, to avoid disrupting satellite services) and that it would not use SP power on a particular channel before making sure that no incumbent (operating on that channel) would be in range.

The verification process operates through a query to a management system called the Automated Frequency Coordinator (AFC). Any incumbent can register its system, transmitter and receiver locations, frequencies, bandwidths, polarizations, transmitter EIRP, and antenna height, plus the make and model of the antenna and equipment used, with the FCC's Universal Licensing System (ULS). The incumbent is expected to register any new system, or any existing system whose transmission characteristics change. The incumbent must also update the ULS when any equipment is decommissioned.

Private entities can then set up an AFC system, which is a query/response service. An expected business model is that the AFC service will be offered to entities deploying APs for a nominal fee. Then, each time an AP is deployed and is configured to use SP in the 6 GHz band, the AP must first query an AFC, providing the AP location to some level of accuracy agreed upon by the FCC. (Typically, this level is within 100 m of the ground truth, with an uncertainty factor provided either as an ellipse with a specified center point and major and minor axis lengths, a polygon with specified vertices, or a polygon identified by its center and array of vectors.[31]) The AFC can then use this location information, along with the ULS information and the RF propagation models approved by the FCC (described in documents called the *6 GHz Report and Order*[32]), to compute the predicted interference of the AP with any incumbent nearby. The AFC then returns to the AP a list of channels in U-NII-5 and U-NII-7 that the AP can use without causing interference to an incumbent receiver of more than –6 dB, along with the possible EIRPs (from 21 dBm to 36 dBm). The AP must wait for this response before adopting SP in U-NII-5 or U-NII-7.

The AP can then operate normally at SP. The nonmobile systems are not expected to move. However, it is possible that new systems might be deployed, and these incumbents have spectrum access priority over 802.11. Therefore, at least every 24 hours (and each time it reboots), the AP needs to query the AFC again.

This system ensures peaceful coexistence between incumbents and 802.11 systems, while providing 850 MHz (500 MHz in U-NII-5 and 350 MHz in U-NII-7) of the spectrum to 802.11 where SP is allowed. This space represents forty-two 20 MHz channels, but only four 160 MHz channels (see Figure 3-13). In some areas of the FCC domain, it is likely that most channels in U-NII-5 or U-NII-7 will be available all the time, that an AP will operate in SP mode all the time, and that the only reason why an AP would change channels is because of some RRM decision. However, in other areas, such as dense urban environments, or in deployments where wide channels are in use, the number of SP-enabled available channels could potentially be very limited. In this scenario, some APs in a deployment might be allowed to operate at SP level, while neighboring APs on the same floor would be allowed to operate only in LPI mode. In such a case, a design decision needs to be made: allow neighboring APs to operate at 6 dB power differences, or design all BSSs to operate at LPI power level only.

---

31 See AFC System to AFC Device Interface Specification, www.wi-fi.org/discover-wi-fi/6-ghz-afc-resources.
32 See these FCC documents: FCC 20–51; 35 FCC Rcd 3852 (2020); 85 FR 31390 (May 26, 2020).

6 GHz Discovery and Special Features

Beyond the power complexity, a great advantage of the 6 GHz band for 802.11 is that it constitutes a greenfield. As no previous version of Wi-Fi was deployed in this band, there is no need for constraining backward-compatibility. Therefore, optimizations are possible that would not be allowed in other bands.

Short Beacons for AP Discovery

One element that had been bothering 802.11 designers for a long time is the issue of AP discovery. APs send beacons at regular intervals (typically every 102.4 ms or 100 TUs) to signal the BSS characteristics. The beacons contain a lot of information, so a beacon is considered as a frame that occupies a lot of airtime. Meanwhile, when a STA tunes onto a channel to discover APs, having to wait more than 100 ms to make sure that it hears all the beacons[33] is not desirable. A delay of 100 ms is a long time, especially if the STA has to scan more than 20 channels (in the 5 GHz band) or close to 60 channels (in the 6 GHz band in the FCC domain). A one-pass-per-channel scan operated at that scale will consume close to 10 seconds—an unacceptable delay. In the traditional spectrum (2.4 GHz and 5 GHz), there is no easy solution to this problem. Increasing the frequency of beacons would mean increasing the overhead on the channel.

One solution was to allow APs to send, between regular beacons, shorter beacons containing the minimum information that a STA needs to determine whether it wants to wait for the full beacon or move on to another channel. (See the "Scanning Procedures" section in Chapter 1, "Wi-Fi Fundamentals.") Another solution was to allow active scanning. In this approach, a STA sends a probe request, instead of waiting for the next beacon. The probe response includes the same information (in the context of AP discovery) as the beacon. However, many STAs performing probe exchanges end up adding to the channel overhead, ultimately consuming as much time as it would take if the AP were to send more beacons. In the 6 GHz spectrum, with 59 potential channels, the 802.11ax designers decided to implement new and more efficient discovery mechanisms.

Out-of-Band Discovery

One such mechanism relies on the idea that many AP devices operating in the 6 GHz band will be multi-radio, and also operate in another band (2.4 GHz or 5 GHz). A STA discovering APs will also scan the 2.4 GHz and 5 GHz bands—and will likely scan these bands first for the foreseeable future, as the STA has today more chances to find an AP in 2.4 GHz or 5 GHz than in 6 GHz. Therefore, Wi-Fi 6E mandates that multi-radio AP devices operating in 6 GHz and at least one other band must advertise their 6 GHz BSS in the other band(s). This advertisement uses an element of the beacon or the probe response called the Reduced Neighbor Report (RNR) element.[34] In its initial intent, this element informs the STA about neighbor APs in a compact form in a field called TBTT Information

---

33 In practice, the STA may even need to wait for two beacon intervals, to make sure to get the second-time beacons it may have missed the first time.
34 See 802.11-2020, clause 9.4.2.170; 802.11ax-2021 clause 11.49

Set. This field identifies each channel where the STA can find other APs, the TBTT offset that indicates the neighbor AP's beacon period offset with that of the local AP, and some optional information like the neighbors' SSIDs or BSSIDs.

> **Note**
>
> In the RNR, the SSID is the short SSID, a 4-octet hash of the SSID. Using the hash instead of the full SSID (which can be up to 32 octets long) makes the field shorter while still carrying relevant information the same way.

In Wi-Fi 6E, the RNR element is used by the 2.4 or 5 GHz AP to inform STAs about the 6 GHz AP. In that version of the RNR, the TBTT Information Set field still includes the TBTT offset (allowing the STA to know when to jump to the other channel with a minimum wait time before hearing the other AP beacon) and the optional BSSID/SSID fields. However, it adds a new BSS Parameters field (shown in Figure 3-11), with the following additional elements:

- **OCT Recommended:** Specifies whether the AP device recommends that the STA use on-channel tunneling (OCT) to communicate with the 6 GHz AP. OCT allows the STA to send a management frame (e.g., a probe) intended for a STA (and an AP in this case) through another STA (or AP) co-located on the same physical device. In other words, OCT allows the STA to use the 2.4 GHz or 5 GHz AP to communicate with the 6 GHz AP present in the same AP device.

- **Same SSID:** Indicates whether the 6 GHz AP supports the same SSID as the local AP.

- **Multiple BSSID and Transmitted BSSID:** Indicates whether the 6 GHz AP is part of a multiple BSSID set. The transmitted BSSID indicates whether the BSSID is that of the AP itself or that of the multiple BSSID set.

- **Member of ESS with 2.45 GHz Co-Located AP:** Indicates whether the other APs of the ESS are also multi-band. A value of 1 indicates that the ESS does not include 6 GHz-only APs, meaning that all 6 GHz APs in the ESS can be discovered while scanning 2.4/5 GHz.

- **Unsolicited Probe Response Active:** Indicates whether all the APs in the ESS also send unsolicited probe response frames every 20 TUs (or less). These probes allow the STA to discover the APs by just listening on a channel for 20 TUs. In such a case, there is no need to send probe requests and no need to wait 100 TUs.

- **Co-Located APs**: Indicates whether the 6 GHz AP described in this report is co-located with the reporting AP (i.e., the 6 GHz AP is in the same AP device as this 2.4/5 GHz AP).

In addition to the BSS Parameter field, the TBTT Information field includes a 20 MHz PSD field, which indicates the PSD of the 6 GHz AP. This information helps the STA determine whether the AP is operating in VLP, LPI, or SP mode (and therefore the maximum PSD of the STA when operating on that 6 GHz channel). Figure 3-14 illustrates the TBTT Information field in the RNR element.

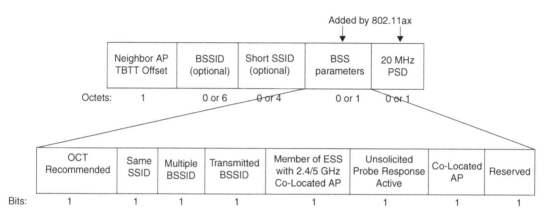

FIGURE 3-14   TBTT Information Field (in RNR Element) in 802.11ax

With these elements, the STA knows the main characteristics of the 6 GHz AP. It can go to the indicated channel, immediately send unicast directed probe requests to the 6 GHz AP, and obtain in the probe response all the operating parameters needed for the association.

Efficient Probing

The unicast probe request is not just a consequence of the RNR mechanism. In fact, Wi-Fi 6E does not allow a STA to browse each channel of the 6 GHz band and send its broadcast probe requests at will. The STA must obey scanning rules. First, a STA is allowed to send a unicast probe request on a 6 GHz channel. Such a message means that the STA knows the AP BSSID (the RA in the unicast probe request), and either is returning to the AP or has discovered the AP by other means (e.g., via an RNR in 2.4 or 5 GHz).

The unicast probe request method works well for multi-radio AP devices. Unfortunately, there are also 6 GHz-only APs, and those cannot be discovered through a 2.4/5GHz RNR. Therefore, broadcast probe requests may be necessary, but scanning 59 channels does not seem efficient.

To improve the scanning operation, the Wi-Fi 6E designers invented the idea of Preferred Scanning Channels (PSCs), depicted in Figure 3-13. In many deployments, the 6 GHz AP will use larger channels for the foreseeable future, because there is a lot of space in 6 GHz. The PSCs are the primary channels of these wider (likely 80 MHz) channels. Their list is known, so a STA just needs to scan the PSCs to discover the 6 GHz APs—in essence, scanning 4 times fewer channels than without PSCs. In the case of an ultra-dense deployment where the 6 GHz channels are narrower (40 MHz or 20 MHz), the

STA knows that it will find on the PSCs the APs that will advertise the RNR for the APs on the other 6 GHz channels. In all cases, the PSCs expedite AP discovery by reducing the need for broadcast probe requests.

At the same time, it is desirable to avoid broadcast probe requests, which consume precious airtime for the benefit of a single STA that does not want to wait for the beacon. In addition to unicast probing and PSC, Wi-Fi 6E allows for two other (mutually exclusive) mechanisms to limit the need for a STA to wait or send broadcast probe requests. One mechanism is the short beacon derived from FILS and detailed in Chapter 1. The short beacon is sent by the AP, typically at 20 TU intervals. The second mechanism is unsolicited probe responses, which the AP can also send at 20 TU intervals. As either of these mechanisms is expected to be enabled on the AP, the STA tuning to a new channel must first wait for a delay (called the probe delay time), measured in TUs, before sending a probe request. By that time, if an AP is operating on the channel, the STA should have received either a short beacon or an unsolicited probe response. If the STA has not received information about the AP after that delay, then the STA is allowed to send a broadcast probe request.

Summary

Designed during the years of the mobile phone explosion, 802.11ax (and its associated certifications, Wi-Fi 6 and Wi-Fi 6E) introduced many features aiming at solving high-density problems. A new carrier structure enabled narrower and more numerous subcarriers, with an associated throughput increase. OFDMA allowed simultaneous transmissions on subsegments (RUs) of the channel. Multiuser transmissions became possible for uplink flows. All of these features significantly reduced collisions, improving both throughout and latency for the BSS as a whole and for latency-sensitive applications in particular. At the same time, OBSS coloring and clever spatial reuse mitigated the issue of high-AP-density deployments with large channels, where neighboring BSS would overlap. Finally, operations in the 6 GHz band brought the opportunity for increased efficiency.

However, Wi-Fi 6E merely scratched the surface of what was possible in that new band. Meanwhile, by 2018–2019, as 802.11ax was nearing completion, new devices (e.g., AR/VR headsets) started to appear that had new requirements that Wi-Fi 6 did not solve. It was time for a new group to form, and build on the hyper-density solution to develop extremely high-throughput solutions.

Chapter 4

The Main Ideas in 802.11be and Wi-Fi 7

prioritized quality of service (QoS): [prioritized QoS] The provisioning of service in which the medium access control (MAC) protocol data units (MPDUs) with higher priority are given a preferential treatment over MPDUs with a lower priority.

802.11-2020, Clause 3

Each new generation of Wi-Fi both creates new market opportunities and solves challenges in existing deployments and previous Wi-Fi versions. Advances in the electronics design and manufacturing sector tend to drive market growth, with the need and the capability to achieve higher client density, higher throughput, smaller form factors, and lower power consumption. The previous chapters have touched on limitations of previous 802.11 generations. Some of these challenges were technical, as advances in some areas surfaced limitations in other areas. Others were operational, with needs for seamless roaming or higher QoS capabilities that were not answered fully by Wi-Fi 6 or the previous generations ("a Wi-Fi that just works, without the user needing to think about it because it just limited the user experience in some ways," as many protocol designers say). This chapter reviews the main drivers behind Wi-Fi 7, the directions that were explored, those that were retained, and those that were abandoned (at least temporarily).

802.11be versus Wi-Fi 7

The IEEE developed the 802.11be amendment to the 802.11 standard. During the development process, as it was clear that the market was eager to benefit from the improvements defined by the 802.11be group, the Wi-Fi Alliance adopted the 802.11be draft (v3.0) as a basis to develop a Wi-Fi 7 certification. As is customary with such a process, only a subset of the elements in the draft were penciled into the Wi-Fi 7 certification program—namely, those voted as most important and likely to receive widest adoption by the voting members of the Wi-Fi 7 group.

As the 802.11be amendment approached its final version, the Wi-Fi 7 group decided to start a Wi-Fi 7 R2 program. This program added more features to those that were part of the first Wi-Fi 7 program.

This chapter, like the others, uses the terms Wi-Fi 7 and 802.11be interchangeably when the context is not about the IEEE or the WFA work (e.g., "the Wi-Fi 7 era"). It uses 802.11be and Wi-Fi 7 specifically when referring to the amendment text (802.11be) or the certification program (Wi-Fi 7), noting that some elements of the amendment are not in the certification programs.

Pressure from the Market and Previous Generations' Limitations

Each time a group is formed to develop a new amendment to the 802.11 standard, a series of use cases are listed that summarize the main industry challenges that the new group will attempt to address. But as the group works toward new mechanisms to solve these challenges, new use cases and new issues are likely to arise. In most cases, these new issues cannot be integrated into the new standard under development, and have to be left unsolved for the subsequent generation(s) of the standard to address them. This scenario is particularly salient in the case of 802.11be, which was set to address issues that emerged during the development of 802.11ax.

802.11ax and Wi-Fi 6 Continued Trends

The main drivers that brought about 802.11ax and Wi-Fi 6 revolved around the idea of densification. As more and more networks adopted Wi-Fi as their primary access technology (including for business-critical operations, such as real-time production management or video-conferencing), the number of connecting devices and their overall throughput needs skyrocketed. The increasing deployment of Wi-Fi–based IoT solutions (for building management, safety, smart homes, and many other use cases) exacerbated the densification trend even more. Wi-Fi 6, which is based on 802.11ax (right-fully called, in its early IEEE inception, the High Efficiency Wireless [HEW] study group), met that demand with higher efficiency at scale. The 802.11ax amendment allowed for an increased number of supported clients per BSS, with technologies like orthogonal frequency-division multiple access (OFDMA), through which multiple clients could share the same uplink transmission time segments. Uplink scheduling also increased the density of data in each BSS, by reducing the risks of collisions or individual STA starvation. These mechanisms complemented, for the upstream, a densification effort that had started, for the downstream, in the Wi-Fi 5 generation. At the same time, 802.11ax allowed for denser AP deployments by limiting the detrimental effects of neighboring APs' operation on the same channel. With BSS coloring, adjacent APs (on the same channel) can detect each other, and suggest to their clients to partially ignore the noise coming from the neighboring BSS's traffic. Denser BSSs, more STAs per BSS—802.11ax "high efficiency" was primarily about "high density."

These trends are still valid in the Wi-Fi 7 era (Figure 4-1), and the need for densification continues. The accelerating growth of IoT devices and machine-to-machine (M2M) connections at home, in the office, and in the manufacturing/warehouse space, as well as the increasing need for virtual collaboration with traditional 2D or 3D virtual reality (VR) meetings, is bound to accentuate the requirement for denser and faster Wi-Fi.

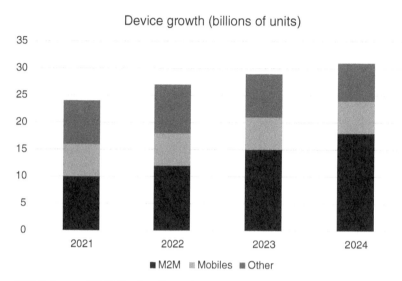

Device growth (billions of units)

■ M2M ■ Mobiles ■ Other

FIGURE 4-1   Wi-Fi Device Growth

Of particular interest in Figure 4-1 are the decreasing proportion of mobile devices (e.g., smartphones, tablets) and the increasing proportion of M2M/IoT devices (e.g., surveillance cameras, industrial robots). This trend signals the requirement to support the capacity and throughput needs of personal devices while simultaneously addressing the sheer number of dedicated IoT devices prevalent in all vertical segments.

The digitization of homes and businesses has, in part, led to a massive adoption of these IoT devices for home automation (smart homes), surveillance (video, smart doorbells, smart lights, and all sorts of sensors), and lifestyle management (e.g., fitness, with Wi-Fi–connected smartwatches, smart weight scales, connected treadmills, indoor bikes, and more). Similarly, in the operational technology (OT) space, businesses are adopting more and more IoT technologies, including industrial IoT (IIoT) solutions and related mission-critical applications where reliability or determinism is a key characteristic. Although these markets can be served by segregated wireless networks (e.g., 5G NR), the low cost point and ubiquity of Wi-Fi–based infrastructures make 802.11 suitable for a wide range of these applications. As the trend toward digitization continues, it is likely that these more traditional segregated or custom-engineered solutions will migrate toward converged standards-based platforms that can serve the IoT, OT, and IT needs on the same wireless LAN (WLAN).

Whether the application is for the home, office, home office, or your favorite warehouse, the same trends that drove Wi-Fi 6 continue apace and need to be addressed by the next generation: Wi-Fi 7.

More capacity (and use of more spectrum) is needed for the continued growth of IoT and mobile offload traffic. Determinism, or at least better reliability, is needed to tackle the emerging AR/VR and IIoT/OT applications with specific low bounds on latency. Higher capacity can be achieved with improved data rates, larger channels, and enhanced usage of the 6 GHz spectrum, especially via multi-link operation (MLO). Determinism, or reliability, can be solved with QoS enhancements that better allocate the transmission opportunities available in the BSS, to ensure that delay-sensitive applications are served within the bounds they tolerate.

Real-Time Interaction Trends

This requirement for bounded latency, better determinism, and higher throughput has taken its best expression in the fast deployment of applications that need real-time data in high volume. The idea of a high-resolution video flow is not new. However, until Wi-Fi 6, the generally accepted compromise was that 1080P or 4K videos would be reserved for one-way streaming, such as the CEO making an important video announcement on a "live" corporate TV channel. Although the flow would be near real time, it would practically be buffered on the receiving endpoint. Buffering means tolerable delays, and it would be acceptable to see the "live" stream with a 0.5-second delay behind its actual transmission. This tolerance was built on the idea that the flow would not be interactive, so the only constraint would be to ensure that all receivers would hear the same words (or see the same image) at roughly the same time (within a tolerance of a few hundreds of milliseconds), so that two receivers in the same physical space would have the same information when they wanted to react. However, no feedback was expected to the source.

For interactive video (video-conferencing, for example), a 0.5- or 1-second delay would cause awkward exchanges. For example, a listener might hear an apparent break and interject, only to collide with the talker, whose apparent break was only a brief pause. Therefore, the end-to-end delay should be shorter. However, the tolerance for losses was originally different for voice and video. Voice was the main message carrier (it is critical to understand what the CEO said). Video came in support of voice, but with a tolerance for loss (it did not matter if the side of the CEO's background was pixelized for a second or two). Video-conferencing adoption was also limited in the pre-COVID world.

As Wi-Fi 6 came to market, these trends changed. Video-conferencing became mainstream in the post-COVID world, with a requirement for increased performance. New collaboration and operations-oriented devices from the world of extended reality (XR), such as augmented reality (AR) glasses and virtual reality (VR) head-mounted devices (HMDs), also gained traction on the market. These XR systems not only redefined what network efficiency meant, but also demanded increased reliability or determinism: To maintain the illusion of reality or immersiveness, the video, audio, head, and hand movement must all be updated in synchronicity.

As you can see in Figure 4-2, the XR market is expected to grow exponentially going forward. A key trend is investment in AR systems that demand Wi-Fi connectivity and VR systems that need to be "untethered" via Wi-Fi to exploit their full market potential.

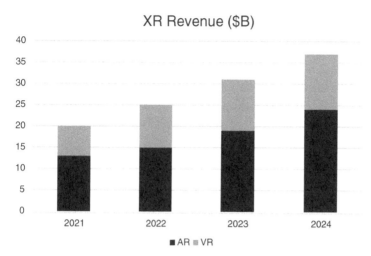

FIGURE 4-2   Extended Reality (XR) Systems Market Trend

Video-Conferencing

When the design process for 802.11be began in 2018, video-conferencing had become an essential part of business and personal communications. Getting on a Webex (video) meeting was becoming a normal daily activity for companies with workers spread all over the planet. With the ubiquity of smartphones, launching a video call to a friend was also becoming the new norm. Wi-Fi 6 was engineered for higher density and throughput, but the rapid development of video-conferencing changed the constraints put on Wi-Fi networks.

As shown in Figure 4-3, dedicated multi-monitor video-conferencing systems and multiple personal devices used simultaneously as video-conference endpoints are typical settings in business environments for team meetings. The challenge is that a single AP may be the only Wi-Fi connection point for all of these video clients, each possibly requiring up to 40 MBps of throughput. It is not atypical for a conference room to host 20 individual users, each with a laptop connected to the video-conference feed. With 3 additional video system endpoints, the BSS would need up to 23×40 or 920 Mbps of throughput in the worst case.

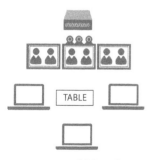

FIGURE 4-3   Video-Conferencing Room

Video-conferencing applications require a reliable connection to the network if they are to provide a consistent user experience. These applications are designed to account for network performance variability, and typically utilize adaptive video codecs to work around link impairments (reduced throughput or excess delay). Although the throughput needs are relatively modest (1–40 Mbps), any sustained deviation will cause "video down-speeding." In such a case, the application detects losses or network performance degradations, and immediately downgrades the video stream to a codec mode that consumes less bandwidth. This down-speeding occurs even in cases where the loss is very temporary (e.g., a few hundreds of milliseconds of congestion on the local BSS). The application takes a conservative approach to changes. The main driver is the notion that a loss of image is worse than an image of lower quality. Therefore, a perceived degradation of network conditions immediately causes a downgrade of the video quality in an effort to maintain a video stream, albeit of even lower quality. When the network conditions improve, the application waits several seconds (or tens of seconds) before deciding that the network conditions are good enough, and stable enough, to allow for higher bandwidth consumption. The application then slowly increases the quality (and bandwidth consumption) of the video flow. The asymmetry of these adaption mechanisms often leads to a degraded user experience, especially in Wi-Fi networks, where performance is stochastic (i.e., periods of good performance may be interleaved with times of congestion).

Thus, a primary goal when designing Wi-Fi for video-conferencing is to maintain a consistently high throughput so that the application maintains a stable color depth and frame rate. The effect of throughput inconsistency can be pixelization or resolution degradation for the video part, but can be clicks, drops, and silences for the audio part. These effects are, of course, detrimental to the call quality and should be avoided. The audio part typically does not require high throughput (often less than 1 Mbps), but delays are sufficient to make a packet arrive too late to be played. Here again, maintaining a consistent throughput is the key to a good user experience.

In low-density Wi-Fi deployments (e.g., single-family units), the probability of not being served the required throughout is low unless there are a significant number of personal and/or IoT devices (e.g., laptops, tablets, home security cameras) or a reduction of capacity due to a neighbor's unmanaged AP on the same channel. Conversely, in high-density dwellings, it is common to see several APs on the same channel, with devices in neighboring dwellings competing for airtime on the same channel. In Wi-Fi 6, BSS coloring was designed to address this type of problem. Unfortunately, the added bandwidth allowed by Wi-Fi 6 was expected by most actors. As soon as more bandwidth became available, gaming units, video-streaming services. and (of course) video-conferencing applications started offering high-resolution modes. This development created conditions for head-of-line blocking scenarios, in which a set of active applications consume most of the bandwidth that the AP can offer, when suddenly a new high-bandwidth-consuming application starts a new flow (e.g., a child starting an online interactive game while their parent is on a Webex video call). The new flow causes a sudden congestion, and therefore a degradation of the experience for the user of one or more of the other applications. In particular, the quality of the Webex video call in this example deteriorates.

In high-density Wi-Fi 7 deployments (e.g., offices), the probability of all video endpoints (e.g., 20 units in a video room, group of cubicles, or open-desk block) getting consistently served the best throughput (e.g., 40 Mbps) throughout the day is also low. As RF conditions fluctuate, as people move, and as new flows enter and leave the network, the throughput available to each device varies. With this variation, the quality of the user experience also changes.

Historically, the solution to this kind of QoS problem has been for an endpoint to mark its voice and video traffic appropriately with Differentiated Services Code Point (DSCP) and User Priority (UP) values. Each UP is then mapped to an AC, thereby providing access to prioritized Wi-Fi Enhanced Distributed Channel Access (EDCA) queues and allowing the STA (and AP) to prioritize these delay-sensitive flows.

This solution is satisfactory when the application and device density is varied. If one application (and one device) marks a flow with UP6, while the others use lower UPs and ACs, then it is easy to provide a statistical advantage to the voice flow (see Chapter 1, "Wi-Fi Fundamentals," and Chapter 2, "Reaching the Limits of Traditional Wi-Fi"). However, an online gaming application might also have a voice flow marked UP6. In enterprise environments, many applications have started offering real-time voice-exchange capabilities, and not all of them are business critical. In other words, the application performing the marking might or might not be critically important to that network (be it home or business). It is expected that a deployment might favor specific applications and wish to grant preferred access to those applications. For all these reasons, it was deemed necessary to move away from a single-sided marking logic and to implement a capability for the STA and the AP to dialog about the application and agree on a differential treatment, based on which flow was critical to the network.

AR/VR/MR/XR

Augmented reality (AR) was just the subject of market experiments at the time of Wi-Fi 6's development. But when the specification was officially released, many large companies had already started producing AR or VR headsets. Today, the world of augmented, mixed, and virtual reality (AR/MR/VR), which is collectively referred to as extended reality (XR), includes a wide range of applications with various levels of capabilities and complexity (Figure 4-4). The top left photo in Figure 4-4 pictures a simple heads-up display (HUD) application of AR in which an image is displayed on glass in the user's HMD, reflecting the state of one or more apps (e.g., machine monitoring app status, but also meeting notifications, today's weather, and so on). In general, these images have a low resolution and do not change much with respect to the HMD's orientation (or the user's pose). To reduce the compute complexity of the glass itself, the image rendering may be performed on a companion device (e.g., phone, watch) and then transferred to the glass via peer-to-peer (P2P) Wi-Fi.

HUD and HMD

A heads-up display (HUD) is a transparent display that can be positioned right in the user's field of vision. This allows the user to see overlayed information without having to "bring the head down" to look at the screen of a device positioned outside of the field of vision.

In contrast, a head-mounted display (HMD) is a display—typically a transparent screen—mounted on a headset or helmet. HMDs are usually, by design, HUDs.

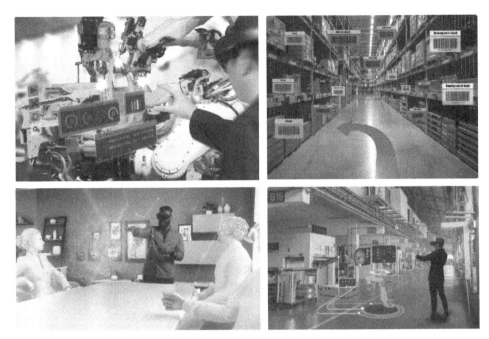

FIGURE 4-4 The Enterprise Metaverse

In the top right photo, a more immersive form of AR is shown, in which the image is full glass: It occupies the user's entire field-of-view (FOV) and is overlayed onto the real environment. In this scenario, the content shown is relative to the HMD orientation and location. Applications such as indoor/outdoor waypoint-based navigation and context-aware operations management are typical illustrations of this case. Just as in the previous case, rendering may be performed by a companion device, then transferred to the glass with P2P Wi-Fi.

The bottom right photo pictures a typical VR application, where a user controls a virtual robot, which might itself be a real robot in another location. Each movement of the user triggers a matching movement of the robot. The system needs to account for the HMD position and posture, but also other aspects such as the user's hands, feet, and so on. These elements are usually tracked by a built-in camera or other tracking sensors. The applications of this technology to training, industrial operations, and collaboration are practically limitless. Immersion requires that the virtual world keep pace with the real world (e.g., head/hand position, orientation) in real time. This constraint leads to a need for a motion-to-photon (MTP) or detect-to-project (DTP) control loop fast enough that the human brain does not sense any discontinuity. Such a degree of immersion and real-time operation requires significant rendering to compute, which is often offloaded to the nearby edge or companion compute resources. The video-rendering capabilities envisioned for VR (e.g., 4K/8K stereo with 120 frames per second) require significant throughput with low delays and, therefore, generally a direct Wi-Fi connection from the compute device to the HMD.

The complexity and capability of both AR and VR can be extended to mixed reality (MR), represented by the bottom left photo of Figure 4-4. In this virtual meeting, a user manipulates a virtual object. Some participants are remote and virtually displayed; other participants are physically in the room. In that case, cameras in the HMD itself capture live video and use it for both remote display/surveillance and contextual (object-recognition) purposes. This process implies that the input video stream can be part of the rendered image played back in the local HMD, thereby providing the same look and feel for the representation of the attendees physically in the room and the remote attendees. Another example is a game, where the user is in a virtual world, but can see in that world an object of the real environment, such as a table (at the object's real location in the environment). The object has to be captured by a camera and then displayed in the virtual landscape. The contextual-identification process can be simple (e.g., the user looks at a common and thus premapped object, such as a desk) or complex (e.g., multiple users, as in the MR video-conference example). Due to the combination of VR (virtualized objects), AR (augmented), and live image processing (context), MR represents the most complex application in the XR spectrum and is often realized with significant compute offload (e.g., edge computing) as opposed to simple AR companion pairing. Figure 4-5 shows the key components of an XR deployment in the enterprise setting.

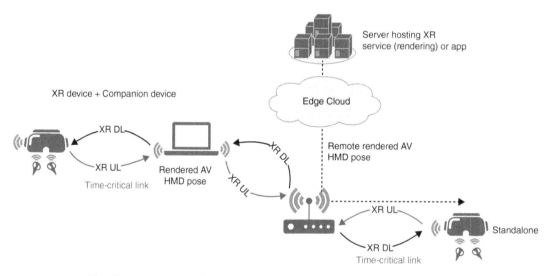

FIGURE 4-5   Key Components of Some Common Enterprise XR Use Cases

From a Wi-Fi standpoint, XR applications introduce hard constraints on the maximum delay considered tolerable for a set of frames to be transmitted over the WLAN. One key difficulty is that the delay must be consistently low. However, Wi-Fi is stochastic. The delay might be low "most of the time," but there might be periods of congestion when multiple devices compete for the medium, fail (because of collisions or a temporal interference), and retry within the same time window. When this issue occurs during a video-conference, a few glitches appear, such as pixelization of the image and small drops in the audio flow. When this issue occurs in an XR application, the user may lose their balance and fall or lose the object of focus. In other words, the issue is not tolerable to the same level as it was for older applications.

Wi-Fi 6 was not designed to eliminate these periods of congestion. The problem is naturally made even worse when the user moves, and roams from one AP to the other. The process of next-AP discovery, the roaming times, and other roaming overhead can worsen the delays, and in turn the user experience.

Promising Directions

As described in Chapter 3, "Building on the Wi-Fi 6 Revolution," Wi-Fi 6 dramatically improved the performance in dense client environments where tens or hundreds of devices or users are associated to a single access point (AP). Wi-Fi 6 also improved scenarios with high AP-density, and introduced an extreme low-power mode for those battery-powered devices in users' hands or mounted in their homes and offices. With the release of Wi-Fi 6E, the ability to use the new 6 GHz spectrum proved to be game-changing for almost all applications across all WLAN segments. Imagine doubling your WLAN capacity and speed by just replacing your Wi-Fi APs! Arguably, this has been the single most impactful Wi-Fi upgrade since the transition to 5 GHz channel-bonding and MIMO in Wi-Fi 4.

However, the challenges brought by the fast growth of the IoT, the explosion of real-time video calls (especially in the COVID and post-COVID worlds), and the apparition of XR applications with near-zero-delay requirements could not be solved only with the addition of more spectrum. One core reason is that a massive shift of all Wi-Fi clients to 6 GHz is unlikely to occur. The original 802.11 standard (in 1997) was released solely for the 2.4 GHz band. The 802.11a amendment of 1999 introduced the 5 GHz spectrum. Yet it took more than 10 years for 5 GHz to become mainstream and commonly supported on STAs and APs. The adoption of 6 GHz is likely to follow a different curve, because most vendors have been eagerly hoping for this spectrum addition. Nevertheless, for the foreseeable future, devices will still operate in the 2.4 GHz and 5 GHz bands. In other words, 6 GHz will be an augmentation, not a replacement. Given this reality, the limitations of Wi-Fi 6 in the 2.4 GHz and 5 GHz bands still need to be solved.

A Better Quality of Service for Mission-Critical Applications

A particularly acute problem was related to the standard QoS scheme inherited from the 802.11e-2005 days. Under this scheme, applications mark their class with a label (DSCP) that is translated into an 802.11 UP and AC, entirely within the system (STA or AP, depending on the flow direction). This mechanism brings up the risk of the duplication of labels mentioned in the previous section. Two applications might have the same label because they have similar sensitivity to the same parameters; however, in a given network environment, they might have different values for the IT manager or owner. In a home environment, the video of a Webex call might be more important than gaming or surveillance cameras. In complex environments, such as a warehouse, the application of choice might be the remote piloting of robots assembling a customer's order for delivery. In all cases, the network operator intends these applications to be treated differently than others, even if their traffic falls in the same DSCP or PHB class. In times of high network utilization, these applications should be given prioritized access in accordance with their specific needs.

Until Wi-Fi 6, this prioritization was achieved either via infrastructure duplication (i.e., an overlay of APs deployed to support key devices and their applications) or via some form of virtual segmentation (e.g., specially configured SSIDs, reserved for key devices, with targeted QoS settings). Duplication is economically and logistically challenging, and virtual segmentation often suffers from a lack of fidelity. This fidelity challenge arises because the current 802.11/Wi-Fi QoS mechanism relies on the AP correctly identifying and classifying the traffic flows belonging to these preferential applications (e.g., the robot control loop, the video-conferencing flows) among a multitude of flows from the same user device. The AP also needs to know which performance element is critical to the target traffic (e.g., low latency, high reliability) if it is to select the appropriate scheduling or channel access technique that will meet its needs.

Application-based traffic classification is a commonly deployed network capability; however, it is an expensive function that relies on deep-packet inspection (DPI) technology. In some cases, the mere detection of communication ports (Layer 4) or the snooping of DNS requests is sufficient to detect the application of interest. In other cases, these techniques are insufficient, and the DPI engine needs to observe the flow patterns to predict the most likely application. In a world where most traffic is encrypted, DPI performed on the AP is a resource-intensive operation that is, by its nature, subject to false classification (e.g., video content misclassified as web browser content due to encryption).

A more elegant approach was needed. The most natural direction ended up being a dialog between the STA and the AP. Wi-Fi 7 adopted a QoS management methodology in which each application on the STA can explicitly indicate its name, or parameters sufficient for the AP to identify it, using low-complexity (non-DPI) methods. This traffic classification (TCLAS) method allows the AP to know which traffic flows (e.g., IP tuples) belong to key applications and to schedule them properly (for upstream flows). An extension of this method, implemented in one of the Wi-Fi Alliance's QoS management programs, allows the AP to also instruct the STA on how to properly mark (DSCP and UP) that key traffic (or the other traffic). Thus, the critical applications can be marked as needed to benefit from the services they need, while the other applications will access the medium on a best-effort basis.

This new signaling method was a great step forward. However, the class, or category, is just one element of QoS. Another key consideration is the type of scheduling structure (or performance) that the application needs to perform optimally. For example, two applications might use interactive video streams as part of their multimedia service; under Wi-Fi 6, both of them would be classified and queued as video, with the same mapping to the AC_VI access category. However, one of those applications is web conferencing and the other is a VR application with head tracking. The VR application has a very strict latency requirement, because delays between the head motion, reported back to the server, and the matching new landscape view, sent and played back to the headset, may cause nausea or headaches in the user. Both applications belong to the video general category, but have very different requirements from a delay point of view. A 100- to 200-ms delay might be acceptable for the web conferencing application, but the AR/VR application requires a delay closer to 10 to 20 ms. Unfortunately, Wi-Fi 6 does not include any reliable mechanism to signal the differences between these applications.

A similar situation occurs for robotic control applications. These applications are not voice (AC_VO) and not video (AC_VI); by default, they are often classified as best effort (AC_BE). Even when they are positioned in the highest access category (AC_VO), that category does not fully express their needs. Each packet might be the difference between controlled and uncontrolled robot movement. Therefore, the application does not just need a statistical advantage for its access to the medium; it also needs very low latency and high reliability (actual delivery assurance with close to no loss). Unfortunately, this granularity is not provided by Wi-Fi 6 QoS management, as Figure 4-6 illustrates. In that example, a VR application does not meet its low-latency targets because of co-channel-interference (CCI) or overlapping BSS (OBSS) load.

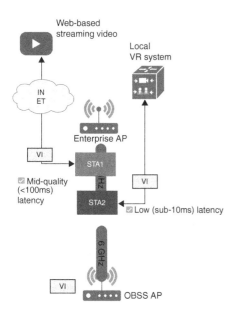

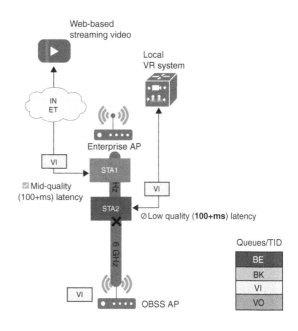

◆ STA1 with Internet video traffic and QoS Management R2 can access high-priority video (VI) queue/TIDs
◆ STA2 with LOCAL VR video traffic can access high-priority video (VI) queue/TIDs
◆ Low overlapping BSS (OBSS) video (VI) load allows the VR HMD of STA2 to experience low (<10 ms) latency

◆ High OBSS video (VI) load causes congestion with video (VI) flows of AP
◆ STA1 video quality is not impaired as it can already tolerate high latency
◆ STA2's VR HMD video quality is **poor** as it is very latency sensitive

FIGURE 4-6   Limitations of Wi-Fi 6 QoS Management: VR Use Case

To deal with these issues, the 802.11 standard designers looking into QoS problems for Wi-Fi 7 extended the QoS management mechanism further. Specifically, they augmented an exchange that was defined in a previous version of the standard, the stream classification service (SCS), with additional elements called QoS characteristics (QC). This mechanism provides an elegant way to not only clearly identify critical applications (using the TCLAS), but also let the application express its general scheduling needs. The STA can then request resources (e.g., uplink triggers) via QC for each direction of the flow (uplink to AP and downlink from AP). The key performance indicators (KPIs) expressible with SCS QC include the maximum delay, periodicity, minimum throughput, and reliability. This granularity solves the under-classification problem present in previous Wi-Fi generations and allows for better KPI delivery of real-time, mission-critical services.

In a typical implementation, the STA is configured with a list of critical applications. This configuration can be done through group profile management or other methods. When the STA needs to start a flow for a critical application, it sends an SCS request to the AP, with the QC element that describes the traffic and its characteristics. The description is flexible and provides different ways of identifying the target application. The STA can indicate its needs in different forms, such as the expected packet size and pace. The AP compares the request to its QoS policy configuration. Based on this configuration and the STA request, the AP responds to accept or decline the request. The mechanism is flexible and allows for negotiation. As the AP starts scheduling upstream transmission slots for the STA and its target application, it also permits the STA to send updates (e.g., "Please schedule a bit more" or "I am not using all the slots you schedule, so you can slow down"). In all cases, the exchange makes sure that the AP knows in advance the type of service that the application needs and allocates this service (or negotiates with the STA, if the BSS is congested, so that the application uses, for example, a more conservative codec). The frequent updates between the STA and the AP allow the AP to adapt its service to the peaks and troughs of the application traffic volume. Because the AP is also considering its own configuration for the QoS policy, applications that are not relevant to the business do not get privileged treatment if other, more critical applications are in motion.

A typical illustration is a hospital, where doctors and nurses may use real-time applications for patient care operations. Meanwhile, a patient (connected to the same AP, possibly a guest SSID, but still on the same channel) might want to play an online VR game. The patient headset might provide an SCS request with a description of the VR traffic. If there is enough airtime in the BSS, the AP can then allocate the requested slots. Every few seconds, the STA and the AP exchange updates on the VR flow. Then, a nurse starts a flow from a work application that needs a real-time exchange (e.g., a video-conference exchange with a specialist). The nurse device also makes the SCS request to the AP. As the nurse's traffic is business-critical, it gets the scheduling it needs, and the AP updates the first STA that the VR traffic will soon receive less bandwidth. The VR headset has the time to downsize to a more conservative codec and negotiate a new schedule with the AP.

Figure 4-7 illustrates this scenario, where the VR headset experiences poor quality under congestion with QoS Management R2, but requests a specific service level agreement (SLA) via QoS Management R3 with SCS. Due to the AP scheduling, it achieves its target latency.

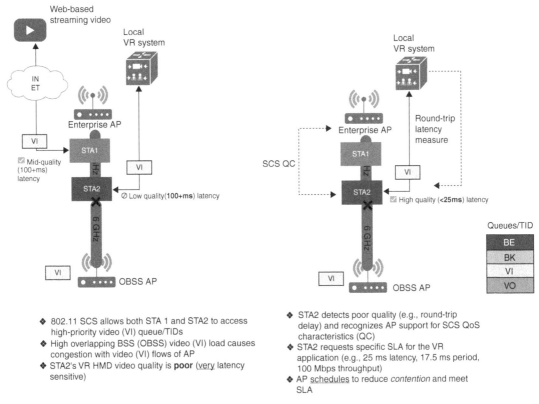

- ◆ 802.11 SCS allows both STA 1 and STA2 to access high-priority video (VI) queue/TIDs
- ◆ High overlapping BSS (OBSS) video (VI) load causes congestion with video (VI) flows of AP
- ◆ STA2's VR HMD video quality is **poor** (very latency sensitive)

- ◆ STA2 detects poor quality (e.g., round-trip delay) and recognizes AP support for SCS QoS characteristics (QC)
- ◆ STA2 requests specific SLA for the VR application (e.g., 25 ms latency, 17.5 ms period, 100 Mbps throughput)
- ◆ AP schedules to reduce contention and meet SLA

FIGURE 4-7   Wi-Fi 7 QoS Characteristics: VR Use Case

Client Roaming and Stickiness: Coverage Discontinuity

The QoS requirements are not limited to a single AP. The very nature of Wi-Fi is based on mobility. However, a STA often has very limited knowledge about its environment. The mobile device is torn between the need to conserve energy as much as it can (so your phone battery lasts one day instead of one hour) and the need to continuously scan the RF environment to find the best AP. As a result, even the simplest Wi-Fi deployment (e.g., multi-AP home or branch office) suffers from client stickiness challenges, where the STA does not roam to the next AP when that AP provides a better signal than the previous AP. The issue of underperforming roaming is experienced daily by almost all Wi-Fi users. This infamous "sticky client" issue is a form of "coverage anxiety." The STA attached to an AP does not know if it should let go and seek a potentially better AP, for fear of dropping the connection and inconveniencing the user.

In a trivial single-AP scenario, this is exemplified by the dual/tri-band challenge. The first AP radio that the STA detects (via scanning while approaching the venue) is often in the 2.4 GHz band. This band provides relatively low speeds. The rules of physics—in particular, the general principle that the antenna of a receiver should be half the size of the RF wave to receive—mean that the STA collects more energy from a signal in the 2.4 GHz band than from a signal in the 5 GHz or 6 GHz band (at the same distance, and sent at the same power). This is because the STA collects energy over a 6-cm-long antenna, compared to over 3 cm for 5 GHz and over 2.5 cm for 6 GHz.[1] A way to express this phenomenon is to observe that the 2.4 GHz BSA is often about twice as large as the 5 GHz BSA for the same AP radio power.

So, the STA will most likely detect the AP 2.4 GHz radio first, and then try to associate with this AP radio while entering the Wi-Fi coverage area, such as when transitioning from outside to inside a building. If so configured, the AP will likely simply accept the association. The STA will be served as best as possible, given the inherent performance limitations of 2.4 GHz channels, such as their limited capacity and low data rates. Once the STA is associated, and keeps approaching closer, the 2.4 GHz signal becomes even stronger, while the 5 GHz radios, then the 6 GHz radios start to appear to the STA scanning (if it chooses to scan). The STA can then choose to leave this strong 2.4 GHz channel and change to the new AP (and channel). However, this choice must be weighed against the risks—namely, the necessary process of de-association (from the old 2.4 GHz radio) and the re-association (with the new 5/6 GHz radio). Such a change will cause a definite connectivity outage, even with the fastest 802.11 roaming scheme, and a possible failure, possibly leading to a hand-back to the cellular system. The STA doesn't actually know if it can be accommodated on the other radio. The STA also does not know whether the other radio service is better. The STA just knows that the 2.4 GHz signal is strong and the 5 GHz signal is weaker. This is the classic "break before make" problem of wireless.

It is worth noting that the APs will advertise the other AP radios (e.g., the 5 and 6 GHz radios) over the current radio (2.4 GHz) so the STA at least knows which channels to scan. Nevertheless, for a single-radio STA, the "scan" process itself is disruptive to traffic flows and might be prohibited under certain scenarios, such as when a voice call is in progress. Figure 4-8 illustrates this discontinuity, where, even with the fastest roaming protocol (Fast Transition [FT]), the client needs to periodically start a connectivity-impacting scan before the roam itself.

Although some vendor implementations attempt to improve upon this situation further by providing "hints" as to when or if to move between radio bands of the AP, many client implementations tend to favor conservative "sticky" behavior. That is, if the AP signal strength is above a given threshold, and unless data throughput is catastrophic, the client stays on the current AP radio. This behavior results in overall low network utilization and poorer user experiences (e.g., low throughput). You can see this issue in Figure 4-9, where the speed of the connection varies from low to high, and then low again as the client moves.

---

1 The physical constraints of smartphones mean that this rule is often bent, but the principles remain the same: The collection surface for the 2.4 GHz band is larger than the one for the 5 GHz or 6 GHz band.

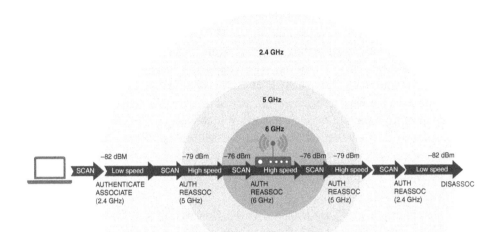

FIGURE 4-8   Standard 802.11r Roaming

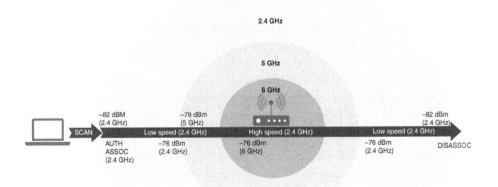

FIGURE 4-9   Roaming with Classic "Sticky" Client

In the more complex multi-AP scenario, the problem deepens, as now the STA faces a multitude of radios from a multitude of distinct APs. Just as in the single-AP case, each of the APs can advertise the radios of the other, non–co-located AP radios, over any of the other radios. This way, the STA detects a first AP radio, and can exchange with that AP to learn the channels (and BSSIDs) of all other surrounding AP radios. But the STA typically does not know at that stage which AP radios it will detect from its position, and at what levels their signals will be. The STA also does not know the type of service it can expect from each of these neighbors, or if connection will even be possible. A single-radio STA can still perform only sequential scanning—albeit at a risk to the current active data connection, as the STA radio needs to leave the current channel. In all cases, these refinements help in theory, but the simplest STAs will prefer to stay on the AP that they know is functional, even if it is not optimal.

This issue had been present in previous generations of Wi-Fi. However, in a world where 802.11be designers faced the requirement of maintaining high connection quality for applications such XR and voice, the problem had to be studied in more depth and solved.

First, it would be nice if a STA associated to a single AP with dual/tri-bands did not need to de-associate (break) before re-associating (make) to a different AP radio, but on the same AP device, simply to move from the low-speed 2.4 GHz radio to the higher-speed 5 GHz radio. But recall from Chapter 1 that a STA cannot associate to more than one AP at a time. Before Wi-Fi 7, a STA had no other choice than de-associating first. This requirement is difficult to break for backward-compatibility reasons. The 802.11 designers could not simply change the rules and decide that a STA would be allowed to associate to more than one AP. Opening that Pandora's box would wreak havoc on existing Wi-Fi networks, where STAs would potentially start associating to all APs in range, and no infrastructure would be able to handle a STA MAC address suddenly appearing everywhere.

Before Wi-Fi 7, an AP device also had no easy way to advertise, "These two radios are mine; this is just me on two frequencies." Even if the AP could make such advertisements, the principles detailed in Chapter 1 remind us that the AP is the device that informs the DS about how to join the STA (MAC address). But each AP radio has its own identity and its own MAC address. If the STA appears on two radios, even of the same physical AP, there will be plenty of implementations where the DS would see two possible paths to a single STA MAC address. Such a scenario would cause confusion, duplication of messages, and other problems.

For these reasons, the 802.11 designers could not simply loosen the "no association to more than one AP" rule. They had to invent new mechanisms—so they designed the 802.11be/Wi-Fi 7's multi-link device (MLD) architecture.

You will find the details of MLD in Chapter 6, "EHT MAC Enhancements for Multi-Link Operations." In essence, in the MLD architecture, a client station (called the non-AP STA MLD) can be simultaneously associated with all of the APs in a single co-located AP MLD (i.e., its 2.4, 5, and 6 GHz links). Note that the non-AP STA MLD and the AP MLD are different entities from the STA and the AP. The non-AP STA MLD (or AP MLD) exists in the upper part of the MAC layer, above the STA and AP functions. Thus, the non-AP STA MLD associates with the AP MLD (and could not associate to just "an AP"). The AP and the STA still exist at the link level—for example, one STA for the 2.4 GHz link, one STA for the 5 GHz link, and both under the same STA MLD entity. The AP MLD, by accepting the multi-link (ML) association, ensures that the STA will be accommodated. The AP MLD that cannot accommodate the STA and its multi-connection can reject the MLD association, causing the STA to directly consider alternative APs (instead of failing on the re-association to the other AP radio, before Wi-Fi 7). This dual/tri-association then makes it possible to seamlessly move (or steer) traffic from one link to another without ever disconnecting from the AP. There is no more "coverage anxiety" or "sticky client" issue (at least within the same physical AP).

You can see this behavior in Figure 4-10, where an inactive radio of the MLD can be used to scan for a new link to add while the active link still serves traffic. Then, seamlessly, the new link is added to the MLD and traffic can flow over both.

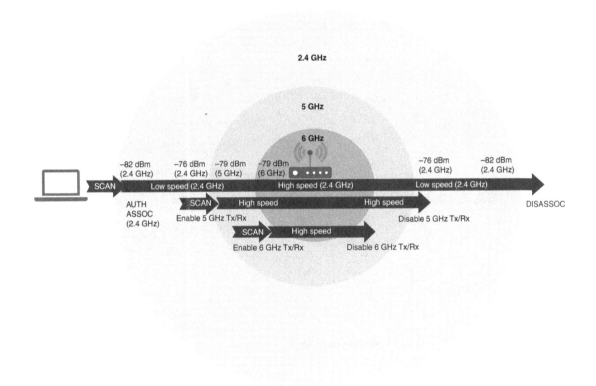

FIGURE 4-10   Wi-Fi 7 Intra-AP Roaming with MLO

Even in the simplest non-AP MLD implementations, where the client has only one physical radio, the movement between the AP's radios is seamless—simply a matter of selecting the active link using real-time signaling. For the majority of Wi-Fi deployments, the MLD concept will be invaluable.

The AP MLD resides in the upper part of the MAC layer, which means that both radios need to be within a single AP device. In Wi-Fi 7, you cannot have an AP MLD residing across two physical APs. In a multi-AP scenario, this limitation seems to offer no benefit for the classical roaming problem—that is, a STA moving from one AP radio to another physical AP radio. However, the expected design of Wi-Fi 7 clients might bring additional benefits for this scenario. As many clients start to support Wi-Fi 7, they will come with two or more radios that can be active at the same time, which may dramatically reduce the proportion of single radio clients on the market. Recall that the barrier to discovering or hearing a candidate AP for roaming is scanning (off channel), which requires sending the current, active radio off the current active channel. With a single Wi-Fi radio, a client needs to decide whether it can afford to drop potentially time-sensitive (e.g., voice) or mission-critical (e.g., robotics control) traffic in an effort to find a better AP. With a Wi-Fi 7 MLD-capable device, the client can associate to an AP MLD, temporarily use a single link for its data communication, and simply use its second radio to scan for candidate APs without interrupting the flow of important traffic—hence, roaming or handoffs can finally be made more predictable! Figure 4-11 illustrates this concept.

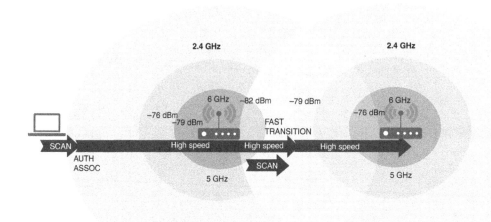

FIGURE 4-11   Wi-Fi 7 Inter-AP Roaming with MLO

Faster and Wider

Each generation of Wi-Fi has recognized that new applications demand more bandwidth and faster throughput. Wi-Fi 7 is no different. During the development of Wi-Fi 6, a common prediction was that, in the 2010s decade, video applications would become common on the Internet and, therefore, on 802.11 networks. This prediction proved valid. Real-time video applications invaded personal and professional exchanges, and people turned to streaming video as a way to consume content (sometimes, replacing completely cable-to-TV solutions). As 802.11be was being designed in 2018, the next prediction was that video applications would compete for bandwidth. At that time, new standards like

4K were emerging. As 802.11be development continued, the 25 Mbps bar for a single (UHD) video streaming was reached. By 2018, the main video-streaming content companies reported together that close to 10% of their streamed flows were at rates greater than 25 Mbps, the majority around 12 Mbps, and seldom less than 6.5 Mbps (in 2014, 1.5 Mbps was the standard). When the first XR applications appeared, their standard resolution mode was also commonly inscribed around 25 Mbps.

This evolution meant that more bandwidth was needed for the next Wi-Fi generation. Quite clearly, MLD was designed with this requirement in mind. If your STA connects to both the 5 GHz and 2.4 GHz (or 5 GHz and 6 GHz) radios, the bandwidth available to the STA increases. It does not directly "double," as the 2.4 GHz channel is typically narrower than the 5 GHz channel. Even when the connection is in the 5 GHz and 6 GHz bands, the presence of legacy devices in 5 GHz is likely to slow down the overall throughput. But in most cases, a double connection increases the STA throughput, if it needs it. Indeed, even before Wi-Fi 7 emerged, some vendors had started implementing proprietary solutions. That is, connecting a STA of brand X to an AP of the same brand would allow the user to enable a high-speed mode (designated with names like Dual-Band, Dual-Wi-Fi, and Turbo-Wi-Fi), by which the user's STA would establish a connection to the AP's 2.4 GHz radio and another connection to the 5 GHz radio. This mode was different from MLD, in that it would literally require two connections. The STA has one connection, one IP address, and one stream to a server through the 2.4 GHz radio, and another connection, another IP address, and another stream (possibly to the same server) through the 5 GHz radio. When the connection was to a single server, the server application needed to support this special mode for packet reassembly. By contrast, MLD is inscribed in the upper MAC layer, and the STA has only a single IP address (above the MLD). But both the proprietary mode and MLD were designed based on the idea that STA could consume more bandwidth than a single radio link could provide.

As the world of video continued to evolve, it became clear that the dual connection might be just a short- to mid-term solution. The designers knew that, in the 802.11 world, there are three ways to increase the available bandwidth:

- **Increase the number of spatial streams:** This approach is tempting, because if the AP sends (or receives) two spatial streams, then the AP can (at least conceptually) send and receive twice as much data as if it had only one spatial stream to work from. The issue, of course, is that the receiving end detects two signals on the same frequency, only one of which is of interest. The receiver needs to null one of the streams. This operation is complex, and makes the interesting signal weaker. A third spatial stream adds even more to the complexity, and makes the signal to noise ratio of the interesting stream even lower. Although 802.11-2020 described up to 8 streams in clause 21 (VHT, for 802.11ac[2]), no vendor in 2018 was implementing more than 3 streams, so the 802.11be designers decided that increasing the number of spatial streams was not a realistic route.

- **Increase the channel width:** Recall from Chapter 1 that when the channel width increases from 20 MHz to 40 MHz, the bandwidth more than doubles, because the number of pilot tones

---

2 See 802.11-2020, clause 21.5

decreases and the tones at the edge of the two merged channels are reused. As the channel width increases beyond 80 MHz, this additional gain disappears. In 802.11ax, a 160 MHz-wide channel is a set of two 80 MHz-wide channels. They are used together, not merged into a larger channel. However, the bandwidth of a 160 MHz channel is twice that of a 80 MHz channel, which is a great improvement. Naturally, the next step was to increase the width to 320 MHz. This was impossible in 2.4 GHz (because the whole band is 83.5 MHz wide), but also proved challenging in the 5 GHz band. In most regulatory domains, it proved impossible to find 320 MHz continuous segments in that band (see Chapter 5 for details). These considerations led the 802.11be designers to abandon the idea of a 320 MHz channel in 5 GHz. But in 6 GHz, such segments can be found. Therefore, in 6 GHz, 802.11be allows 320 MHz-wide channels. Three 320 MHz channels are possible in the regions that support the 6 GHz spectrum over 1200 MHz. The channel is made of two adjacent 160 MHz-wide channels in all cases, and can be centered on channels 31, 95, and 159. This set of 3 channels is called *320 MHz-1 channelization*. Alternatively, the channels can be centered on channels 63, 127, and 191. This set is called *320 MHz-2 channelization*.[3] Figure 4-12 illustrates the 6 GHz band 320 MHz-1 and 320 MHz-2 positions.

- **Increase the density of the modulation:** Wi-Fi 6 had increased the modulation density by allowing 1024 QAM. 802.11be allows for 4096 QAM (with both ¾ and ⅚ coding rates), where 12 bits are encoded for each target of the constellation. This new mode increases the PHY data rate by 20%. With such a constellation, where each target is very close to the other, the signal needs to be strong and the noise low for the receiver to properly identify the intended target (see Chapter 5, "EHT Physical Layer Enhancements," for details). The distance at which these modulations can be achieved is expected to be small at first. However, the designers also kept in mind that many higher-end devices need to perform far above the standard minimums, and it is very likely that APs and STAs will come to the market that can achieve these rates at a "normal" distance from the AP.

Table 4-1 showcases the maximum single-link achievable data rates for typical IoT devices (with one antenna and one spatial stream [SS]), consumer mobile devices with two SS, standard APs with four SS, and mesh APs with eight SS. For simplicity, the table considers only the 6 GHz band and use of up to 320 MHz channels, as well as the smallest guard interval (GI) of 800 ns suitable for indoor use.

TABLE 4-1   Maximum Data Rates for Typical 6 GHz Use Cases

Channel BW (MHz)	1 SS (IoT)	2 SS (Mobile)	4 SS (Indoor AP)	8 SS (Mesh AP)
20	172 Mbps	344 Mbps	688 Mbps	1372 Mbps
40	344 Mbps	688 Mbps	1376 Mbps	2752 Mbps
80	721 Mbps	1442 Mbps	2884 Mbps	5768 Mbps
160	1441 Mbps	2882 Mbps	5764 Mbps	11,528 Mbps
320	2882 Mbps	5764 Mbps	11,528 Mbps	23,056 Mbps

3 See P802.11be D5.0, clause 36.3.24.2

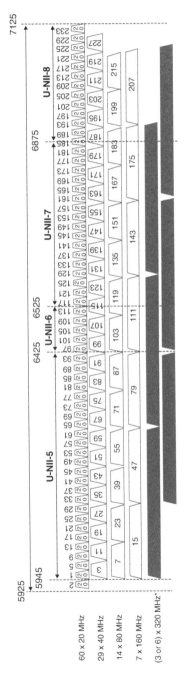

FIGURE 4-12   320 MHz Channels in the 6 GHz Band

Of course, with the introduction of MLO, an implementation might choose to aggregate two or three of these single links together to form an even higher-speed bundle. Theoretically, a STA might have all three bands active simultaneously (2.4, 5, and 6 GHz) and thus be capable of 344 (2.4 GHz/20 MHz) + 2882 (5 GHz/160 MHz) + 5764 (6 GHz/320 MHz) for a surreal total of 8990 Mbps or almost 9 Gbps! Indeed, there will be implementations that require such speeds and critically have no cost or power-consumption constraints (e.g., an autonomous vehicle with multiple 8K cameras and a large battery, a VR headset with low-complexity uncompressed 4K stereo video and a battery pack accessory). However, for the mass market of mobile devices that must interoperate in regulatory regimes without ubiquitous 320 MHz channels and single-AP coverage constraints, a more modest use of MLO is envisioned, with 344 (2.4 GHz/20 MHz) plus 2882 (6 GHz/160 MHz) for a total of 3226 Mbps being more likely. As cautioned earlier, these highest data rates are achievable only when the STA is very close to the AP.

Spectrum Sharing (and Interference Avoidance)

MLD is a major advance and adding bandwidth is even better; however, interference remains a problem, particularly in multi-dwelling buildings and high-density enterprise networks. By its very definition, the unlicensed spectrum where Wi-Fi operates is shared by different technologies and users. In some cases, the other users are your neighbors, also using Wi-Fi (the same technology as yours). In other cases, those users might be incumbent (e.g., aviation radars that take precedence over 802.11 operations in their band, such as UNII-2e). In still other cases, the other users may operate unlicensed technologies that interfere with Wi-Fi (e.g., ZigBee in 2.4 GHz), but have the same access right to the medium as Wi-Fi. In the case of a Wi-Fi interferer, the AP usually detects the overlapping Wi-Fi signal and decides to not use that channel or (if the signal is low enough) decides to share the channel with the other users (e.g., using techniques like carrier-sense [CS] or listen-before-talk [LBT]). In the case of an incumbency, the Wi-Fi user is typically regulated to not use that particular channel either as long as the incumbent signal is detected (with an additional time margin, such as 5 GHz DFS) or as instructed by an authority that the Wi-Fi system must query (i.e., 6 GHz AFC).

With the introduction of the 6 GHz spectrum, the channel sizes can be very large (160 MHz in Wi-Fi 6E and 320 MHz in Wi-Fi 7). Large channels are attractive because they carry a lot of traffic. In particular, with MU-MIMO or OFDMA, traffic can be sent to and from multiple STAs simultaneously, which improves the efficiency of the BSS. However, the presence of an interferer or an incumbent in any segment of the large channel forces the AP to abandon it. In some cases, the AP can look for another part of the spectrum where a large channel is available. However, in most high-density cases, such channels are few and far between, and the AP is forced to reduce the channel width, sometimes dramatically—to 40 MHz or 20 MHz. The impact on the Wi-Fi network efficiency is significant, especially for APs next to airports or military facilities where the need to vacate happens often and on all channels. The issue is all the more frustrating because the interference, or the incumbent signal, might affect only a small segment of the large channel. For example, the AP could be on 160 MHz channel 47 (in UNII-5, composed of the 20 MHz channels 33, 37, 41, 45, 49, 53, 57, and 61). An incumbent, or an interference, in channel 43 can force the AP to abandon all channels and fold back to channel 33, switching suddenly from 160 MHz to 20 MHz.

The solution introduced by 802.11be is to exclude Wi-Fi transmission and reception only on the portion of the wide channel affected by the narrow-band interference. Recall from Chapter 3, "Building on the Wi-Fi 6 Revolution," that 802.11ax introduced the idea of preamble puncturing; however, this process was limited to the AP's actions. With 802.11be, the AP detects the interferer, then advertises which segment of the wide channel cannot be used, as shown in Figure 4-13. In the preceding example, the channel stays 160 MHz-wide on channel 47. The AP informs the STA that the 20 MHz subportion, on channel 43, cannot be used, leaving 140 MHz available for the BSS. This subchannel exclusion includes the AP advertising the subchannels to avoid and, of course, not allocating affected resource units (RUs) in those subchannels to its associated clients (STA). This procedure is called *static preamble puncturing* in the 802.11be standard, because the preamble occupies the full channel in a normal transmission, even if only a portion of the channel is effectively allocated to the payload of a specific STA. But with preamble puncturing, the preamble (and the payload after the preamble) are removed, such that the entire transmission is punctured around the affected subportion.

Resource Unit Optimization

In Wi-Fi 6, an OFDMA mode was introduced to aggregate small frames from multiple STAs in the same multiuser transmission. Each STA would be allocated a single narrow portion of the channel bandwidth, called a *resource unit*. In Wi-Fi 7, the resource units can be any of the following:

- 2 MHz wide (26-tone RU)
- 4 MHz (52-tone RU)
- 8 MHz (106-tone RU)
- 20 MHz (242-tone RU)
- 40 MHz (484-tone RU)
- 80 MHz (996-tone RU)
- 160 MHz (2×996-tone RU)
- 320 MHz (4×996-tone RU)

With Wi-Fi 6, the AP would select a subset of STAs and a set of RU sizes. It would then allocate one RU per STA for the next downlink or uplink transmission.

This mechanism has a clear limitation: Each STA has an RU allocation, whose size is determined by the AP. If it needs only a portion of that RU, the STA would fill the rest with padding, so that the AP receives a full frame. This structure can be wasteful for small channel bandwidths and/or a few STAs because it is unlikely the padding can be minimized to zero bytes.

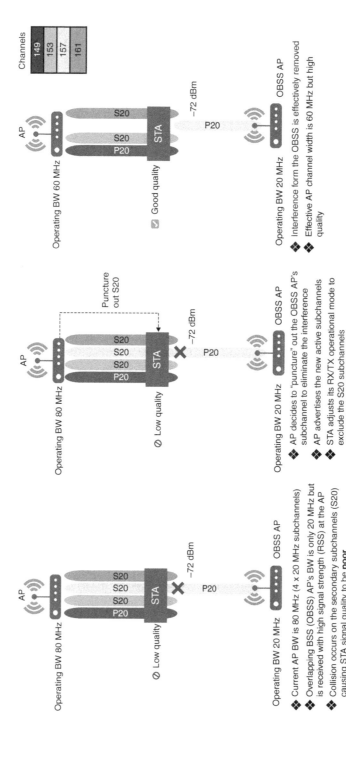

FIGURE 4-13   Preamble Puncturing in 802.11be

With 802.11be and the multiple RU (MRU) capability, an individual STA can be assigned two contiguous RUs of small size (e.g., 52 + 26-tone) or two or more possibly noncontinuous RUs of large size (e.g., 484 + 242-tone) in the same transmission to maximize utilization and, therefore, spectral efficiency. In the former case, better spectral efficiency is possible, especially for smaller channel bandwidths (e.g., 20 MHz) where under Wi-Fi 6 it might have been difficult to find the optimal size of RU for each STA. But in the latter case, the ability to allocate discontinuous RUs provides spectral diversity and hence reliability to wide channels (e.g., 320 MHz).

High-Performance Channels

Sometimes, interferences do not come from strangers, but rather from your own ranks. In enterprise networks in particular, it is common to see the corporate SSID available on all bands (e.g., 2.4 GHz, 5 GHz, and 6 GHz). Some device applications are highly delay sensitive, but others are not. Implementing scheduling and mechanisms to provide a statically better access to the medium for the devices running delay-sensitive applications is great, but another possibility is simply to reserve channels only for those delay-sensitive devices. This way, their traffic is never interrupted by the traffic of slower devices. This type of design is particularly attractive in scenarios where the slower devices send large frames, which consume a lot of airtime and significantly delay the next time-sensitive frame. For example, a real-time robot guidance system might be sent in a special set of 6 GHz channels, where only small frames for these specific applications are found. Even if there are multiple robots in a given BSS, they all send small frames that are forwarded with a very low delay-bound. Other, less-sensitive traffic operates in the 5 GHz band (e.g., a Webex video call). Time-insensitive traffic operates in the 2.4 GHz band.

This type of requirement can be fulfilled with specialized SSIDs: one SSID for each type of traffic, and each SSID only in one particular band. Unfortunately, sooner or later, in one location or another, this type of design ends up with all SSIDs needing to be supported on a given channel. For example, perhaps the 6 GHz band is not available in this location, and laptops and robots have to coexist in the 5 GHz band. If multiple SSIDs share the same channel, there is no longer any benefit in separating them for channel efficiency purposes. Therefore, a more common model is to attempt to use a single SSID (which also simplifies the WLAN management tasks), and steer the devices to the band that matches their needs.

In the 802.11 original design, any STA that has the right credentials can access the SSID on the radio of their choice. Recall from Chapter 1 that an AP must respond with a probe response to a probe request. Once a STA knows that the SSID is available on a given band, there is no good mechanism in 802.11 to convince the STA to join instead on another band. There are proprietary mechanisms—for example, the AP can reject the association when it occurs on the "wrong" band. However, those mechanisms are not standardized, and therefore ambiguous. In consequence, the STA might misunderstand the rejection and simply attempt to rejoin on the same channel in an endless loop until the user opens a support case with the label "can't connect."

Developing an entirely new mechanism would have been challenging. The 802.11 standard is based on the idea of backward compatibility, and a method that would require all STAs to understand a new approach to channel steering would have predictably failed, as legacy devices, ignoring the new philosophy of the WLAN, would have ignored the new messages. Therefore, a solution would need to use messages that had a chance of being already understood by some, if not most, STAs. This method would also need to steer the time-sensitive devices to a channel where legacy devices had low chances of being present. If you stop to think about this last condition, you might come to the realization that 6 GHz channels are good candidates, as they were opened only recently to 802.11. All that was left to do was to find a suitable mechanism.

A good vehicle turned out to be Target Wake Time (TWT). This system was designed at the time of 802.11ah (published in 2016, but in development since 2010) and also inserted into 802.11ax. With TWT, the STA and the AP can exchange on the time at which the STA is expected to wake up. This initial design solves a density problem. 802.11ah was conceived with the idea of a large field of IoT devices (e.g., sensors) waking up at intervals to report some value through the AP. A single BSS would be large (as these devices use simple and far-reaching modulations) and could include up to 4000 clients. If they were left to wake up at random intervals, the transmissions of these devices would collide with those of other devices waking up at about the same time. In large BSSs, they might not even detect these collisions—but they would occur at the AP, placed at the center of the BSS. To reduce this risk, TWT in 802.11ah allowed the STAs to signal to the AP the time (or interval) at which they need to wake up. In turn, the AP could divide the BSS in quadrants, and instruct the STAs in each quadrant to use a certain wake-up time matching the interval they need. This way, only a subset of STAs would be awake at any given time, and collisions would be avoided.

This scheme is also useful in a standard indoor BSS and for IoT devices, but also for non-IoT STAs. In many cases, STAs that do not have much data to send prefer to go into dozing (sleep) mode to conserve their battery life. However, 802.11 includes the notion of session timeout. A STA might have many reasons to stop exchanging (become "idle") with the AP. The STA might be sleeping, or it might have moved to other channels to scan or perform other operations. The STA might also have left the BSS without saying goodbye (disassociation frame). To avoid the burden of keeping in memory states for STAs that have left, 802.11 allows the AP to simply remove the entry (and AID) for a STA that has not been heard for too long. The definition of "too long" is implementation-dependent, and 802.11v-2011[4] introduced a mechanism for the AP to tell the STAs what this interval would be (a BSS maximum idle period IE inserted into the AP association response). To avoid being removed from the AP memory, a STA would need to send keepalive messages at intervals smaller than this maximum idle period value. This requirement can be burdensome for STAs that do not need to send data very often. Having to wake up just to send an "I am still here" message is a waste of power.

To deal with this issue, the 802.11ax designers adopted the 802.11ah TWT mechanism. It allows a STA to negotiate with the AP an interval at which the STA will wake up irrespective of the BSS maximum idle period value. The general concepts of quadrants and collision avoidance from 802.11ah are irrelevant in this approach: Client battery consumption optimization becomes the primary target.

4 See 802.11-2020 clause 11.21.13

This mechanism has interesting properties. In particular, the AP can group STAs with similar needs, as alluded to in 802.11ah. Therefore, in 802.11be, the mechanism is enhanced. Specifically, the STA can negotiate with the AP a membership setup, by which the STA expresses the TIDs in which the STA would have traffic to send and receive. For example, the AP can describe traffic characteristics that match those of the flows sent by auto-guided robot applications. This mechanism is called restricted TWT (rTWT[5]). A STA that does not have this traffic description can be pushed to another band with 802.11v BTM messages. This way, one band can be kept for the high-efficiency, highly time- and delay-sensitive traffic, while less sensitive traffic is kept on the other bands.

Challenged Directions

If you read papers on Wi-Fi 7 written in the early years of the development process, you might see great ideas that are not covered in this book. 802.11be, just like most other 802.11 amendments, started with a Topic Interest Group (TIG). The goal of a TIG is to study the use cases that the next generation of 802.11 protocols could solve. Its conclusions are used to form a Study Group (SG), whose goal is to explore how these use cases might be solved. Once the group has a general idea of whether 802.11 could offer solutions, a Task Group (TG; 802.11be, in this case) is formed to define the exact mechanisms necessary to solve these use cases.

At each step of the path, difficulties might arise that lead the group to conclude that this or that direction cannot be part of the envisioned update, at least for the time being. Thus, the initial phases of 802.11be explored many more directions than the ones that ended up being in the final standard. Most of these discussions stayed within the realm of the 802.11 TIG, SG, or TG; others seemed strong enough that early papers brought them up as likely features, although they ended up not being adopted in the 802.11 standard.

Multi-AP MLD

One such promising direction was the idea of implementing MLD on multiple APs. The 802.11be amendment calls for an architecture that allows MLD on multiple radios of a single AP device (see Chapter 7, "MAC Improvements of Wi-Fi 7 and 802.11be," for details). This idea sounds natural to anyone reading about the general concepts of MLO. If the MLD function sits above the AP radio, with the explicit goal to improve the user experience, increase the bandwidth, and facilitate "make before break" while roaming, it is only natural to imagine an MLD entity residing above two AP devices, and allowing a STA to connect to AP1 and AP2 at the same time. This architecture would increase its throughput by allowing for connection to more than one wide channel, allow "make before break" by associating to AP2 before dropping its link to AP1, possibly increase its transmission reliability by sending key frames to both AP1 and AP2, and more.

This idea was tempting and explored in detail in the early phase of the 802.11be group. Undoubtedly, the idea of a STA connecting to both radios of a dual radio-AP sounded like a first step, because this procedure would have the merits of increasing the throughput of the STA as well as improving the

5 See P802.11be D5.0, clause 35.8

robustness of the transmission. That is, if one link turned out to be congested, the STA could move some or all of its traffic to the other link. However, as this first case was studied in depth, building up to the multi-AP device case, a lot of questions arose, which proved challenging to solve.

Up to this point, an AP had been a function (see Chapter 1) that informed the DS about the location of the non-AP STA. Over 20 years of standard development, the AP had taken on multiple roles and functions that were still connected to its official function, but that also relied on the underlying idea that the AP was associated with a single physical device. For example, after association, the AP gives the STA an association identifier (AID). This AID is used in many scenarios where the AP needs to distribute buffered traffic or call out a list of STAs for group operations. If the STA (as a device) connects to one AP (as a device) with two radios, should it have two AIDs? Two AIDs would mean two identities, and likely a duplication of packets that need to be sent to each STA (though still as a device). In contrast, if the STA is seen as a single device with a single AID, which AP radio allocates that AID?[6] If the connection is spread over two APs, the problem becomes even more complex each time a non-AP STA joins an AP (seen in its 802.11 sense, as a function). In this scenario, the AP would need to query other APs to check whether the STA would be known elsewhere, so as not to allocate two AIDs. But what happens if the AID is already assigned by another AP, and the AID is another STA on the local AP? When the association happens (and then the MLD is an entity sitting above two physical APs), how do these APs share security keying material? How do they even generate the material that is supposed to be generated using (among other parameters) the AP MAC address value? How do they reconcile the frames that both receive from the STA? If one AP becomes congested, how would the other AP know? How would that second AP allow the STA to switch some of its flow to the second AP?[7]

The issue is already complicated on a single physical MLD AP hosting two or more 802.11 APs,[8] where the central AP host would have real-time visibility on the STAs connected to each of its radios. When this scenario was extended to two physical APs, the coordination, synchronization, and link removal or addition (while roaming) issues proved extremely complex to solve within the 802.11be time frame. As the 802.11be work continued, the topic was paused, then removed. This decision does not mean that the topic was abandoned—just that it was postponed, possibly to 802.11be's successor (a future Wi-Fi 8 generation).

Full Duplex

The idea of full duplex operations keeps coming back in 802.11. Recall from Chapter 1 that 802.11 is half duplex: A STA can transmit or receive, but it cannot do both at the same time. This idea makes sense conceptually. It seems difficult to imagine that a STA, while sending a signal at 20 dBm or more, would be able to hear at the same time a signal at –60 or –70 dBm, which is a billion times weaker than the signal the STA, transmits. However, the progress in electronics allows for better self-interference cancellations, which operate with a logic similar to noise-canceling headphones. Right before reaching the antenna, the transmitted signal could be forwarded to a circuit that applies this exact signal in

6 See, for example, 802.11-20/0770 on this topic.
7 See, for example, 802.11-20/292 or 802.11-20/0014 on this topic.
8 See, for example, 802.11-20/0396, 802.11-20/1639, and 802.11-21/0396 on this topic.

reverse to the receiving circuit entrance, "perfectly" nulling the emitted signal and allowing the STA to receive a signal at the same time.[9]

Realizing such a circuit at scale within the budget envelopes usually assigned for Wi-Fi receivers is, of course, quite challenging, especially with Wi-Fi systems that implement multiple spatial streams and multiple radios. If the Wi-Fi chip cost is multiplied—for example, by 10—the attractiveness of the added throughput might be limited by its cost. Another major difficulty is backward compatibility and coexistence. If two signals are sent at the same time in the BSS (one from the AP to STA1, and one from STA1 to the AP), then other, non-duplex devices in the BSS might not be able to understand that there is an ongoing 802.11 transmission. If the nulling effect is not perfect, the signal level (RSSI and SNR) at which the STA (STA1, in this example) can read the AP signal might also need to be higher than for STAs that implement regular half duplex rules.

In short, the full duplex scenario introduces a lot of complexity. The idea might be attractive, but the group ended up deciding that the design should be focused on mechanisms that could provide similar gains, but at a lower complexity cost.

Outdoor Networks

Today, most network traffic transits through Wi-Fi networks at the access layer. Wi-Fi is almost ubiquitous indoors. For example, when you walk out of your home and then drive from your house to your office, your phone likely switches from Wi-Fi to cellular during the transit, before switching back to Wi-Fi at the office. The data volume exchanged over the cellular network matches the time you spend away from a place that is familiar to you, and where you likely can connect your phone to a Wi-Fi network.

Just as the cellular world tries to extend its coverage to the indoor case, the Wi-Fi world often debates outdoor coverage. Several solutions exist, from Wi-Fi mesh networks (in 802.11s-2011) to long-range Wi-Fi in the TV whitespace[10] (802.11af-2013). Therefore, with each generation of Wi-Fi, the case of outdoor networks is brought to the discussion table.

Wi-Fi 6 and 802.11ax did consider the outdoor case, and implemented several features that could be useful for longer-range communications—in particular, for low-power IoT devices. For example, OFDMA allowed for a slower symbol pace transmission, increasing the usable range of the message. Dual Carrier Modulation (DCM) allowed a message to be repeated on different segments of the channel, increasing the chance that noise or interference would not prevent proper reception of the message. It was logical to think that 802.11be should also implement mechanisms for outdoor transmissions, and many use cases and directions for this extension were presented.[11] However, as noted in the previous sections, the main ideas dealt with Extremely High Throughput (EHT)—the very name of the TIG that led to the 802.11be group formation. In the end, the outdoor case proved to be less important than the

---

9 See the proposal in 802.11-13/1122, for example. In practice, the system would need RF analog canceling, more canceling before the ADC, and finally digital canceling (i.e., 3 canceling phases).

10 A set of frequencies in the sub-GHz range that were used by analog TV transmissions, and progressively abandoned as TV broadcast became digital.

11 See, for example, contributions 802.11-13/0872 and 802.11-13/514.

indoor one. This prioritization pushed the designers, in the time they were given to develop 802.11be, to focus primarily on indoor scenarios and only marginally on outdoor issues.

Sharing Knowledge

One key goal of the AP is to provide the best possible service to the largest possible percentage of the associated STAs. However, each STA operates independently and somehow "suffers in silence" when it does not receive the service its needs. In most cases, this issue is not the AP's fault. More often, the problem is that the AP does not know that the STA's needs have suddenly changed or are no longer being fulfilled. Over the years, successive generations of 802.11 amendments have introduced mechanisms for the STA to express these needs in a series of request/response exchanges. One recent proposal appeared in 802.11ax. It allowed the STA to express the amount of traffic sitting in its buffer and waiting to be transmitted toward the AP. With this buffer status report (BSR), the AP could allocate more airtime to the STA in the form of uplink scheduling.

However, these mechanisms are essentially reactive. The STA must first have a need, try to fulfill it, conclude there is a failure without the AP's help, and finally send a request to the AP for additional support. The BSR case is particularly illustrative. In an ideal scenario, a STA application sends a packet to the transmission layer ("socket call"); the 802.11 layer then receives the packet and transmits it to the medium without delay (beyond the usual random countdown). If the medium is busy and the STA has more frames to transmit than its access to the medium allows, the STA's buffers start filling up. At some point in time, the situation becomes so dire that the STA needs to access the medium—not to send that data that has now been waiting for a while, but instead to tell the AP that its buffers are very full. During this exchange, the traffic in the buffer continues to wait even longer. One major reason for creating the SCS QC process in 802.11be was to make this process more proactive, by allowing the STA to tell the AP in advance about the volume of traffic expected for the next few seconds.

In essence, the issue comes from the AP being unaware of the STA's internal states. If the AP had a better view of the STA's experience in the BSS, then the AP could anticipate the STA's needs and fulfill them before the STA starts suffering. Elements that are locally experienced, such as those indicating whether transmissions are successful (if a STA sends a frame that the AP fails to acknowledge until the sixth attempt, the AP might not even know that the first five attempts occurred) or how often the channel is busy from the STA's perspective (especially when it needs to send some traffic) could be very useful for the AP infrastructure to watch the experience degradation of individual STAs and proactively allocate more airtime where needed.[12]

Unfortunately, this problem is real but touches the delicate field of the STA automatically sharing information that may be used to infer the activity of the user of the STA. There are obvious benefits, but also risks in the field of privacy, if not implemented safely. In the end, the fear of privacy violations proved too dire for this direction to be explored in depth. However, the issues persist, and the dialog will likely continue for the next Wi-Fi generation.

12 See contribution 802.11-13/0849.

Wi-Fi 7 Expected Challenges

Wi-Fi 7 enables an extensive set of capabilities. In the chapters that follow, you will read more details about how each feature is expected to operate. However, many of them are only as efficient as their implementations, and you will find that many operational details are left to the implementers. As these features were designed and discussed in 802.11be, possible challenges were raised that you might want to be aware of to better appreciate their boundaries.

Link Aggregation/Diversity

The multi-link operation (MLO) construct was added to Wi-Fi 7 to standardize the behavior of STAs with two or more Wi-Fi radios that are capable of simultaneous connection/association to the WLAN. The primary advantage of this approach, in contrast to legacy approaches, is that the client device (STA) appears to have only a single network interface and thus needs only a single Layer 2 (Ethernet) network connection and associated IP address (via a virtual MAC layer abstraction).

The potential benefits are numerous, including more throughput from simultaneous use of two Wi-Fi radios (e.g., 5 GHz + 6 GHz, 6 GHz + 6 GHz, 2.4 GHz + 5 GHz). Practically, a STA device may integrate one or more radios. Early multi-band devices (e.g., 2.4 GHz and 5 GHz) incorporated two radios, one for each band. Later systems started collapsing the radios into a single antenna that was shared by both radios, and subsequently into a single radio. A smart algorithm would switch the radio from one band to the other. This design made all the more sense given that a STA is subject to self-interference, like any other radio device (including the AP). Now, if the STA implements MLO, it can transmit or receive over two or more bands. As you will see in Chapter 5, "EHT Physical Layer Enhancements," and Chapter 6, "EHT MAC Enhancements for Multi-Link Operations," the main question is centered on the idea of simultaneity: Should the STA send and/or receive on two bands simultaneously? This would mean that the STA would send a signal (possibly at 20 dBm) while receiving, likely on another antenna and radio positioned millimeters away from the 20 dBm transmitter, a –60 or –70 dBm signal. Is this even possible? Maybe, depending on the implementation. Thus, the 802.11be designers ended up envisioning three types of STAs:

- Those that would have a single radio/antenna, and could transmit or receive on only one band at a time (switching from one band to the next; i.e., Multi-Link Single Radio [MLSR])

- Those that would have two or more radios, but that would be too close to allow simultaneous operations (enhanced MLSR)

- Those that would be really simultaneous (simultaneous Tx/RX, STR)

The performances that you can expect from an MLO STA will depend on its hardware, and its ability to support one mode or the other.

Another important practical component of Wi-Fi 7 MLO is the ability for the AP to advertise which links are available to the STAs. The default mapping of "any TID to any link" allows the STA and AP to use any link for any TID. While this is very flexible, it can lead to unintended consequences:

- **Faster CSMA collapse:** The AP/STA can now attempt a TXOP channel access per link, as opposed to one link, effectively increasing the overall collision probability.

- **Link imbalance:** If all or many STAs use the same preferred link (e.g., 6 GHz for its expected higher bandwidth, or 2.4 GHz for its expected higher RSSI), then this link can easily be congested while other links may be under-utilized.

- **Degraded QoS:** As a side effect of the first two consequences, the QoS (especially latency) of critical traffic can be detrimentally affected.

In managed Wi-Fi networks, the AP can use basic load-balancing features to mitigate these issues. In particular, a policy may be put in place that offers STAs with critical TIDs (e.g., collaborative tools, video, AR/VR headsets, IIoT devices) all available links, including one link with high determinism/reliability (e.g., 6 GHz), and limits STAs with only noncritical TID traffic to links with sufficient throughput (e.g., 2.4 and 5 GHz). Wi-Fi 7 also provides the AP with the capability to repurpose briefly or permanently one of its radios (e.g., to permit a software upgrade) without disrupting STA connectivity on the remaining links. These capabilities are achieved through the advertised TID-to-link mapping and MLD reconfiguration features. Here again, the performance of the system will highly depend on the way MLO is implemented by the AP vendor, and how smart the non-AP STAs are, in choosing their preferred links.

SLA Management

Recall that the Stream Classification Service (SCS) allows a STA (client) to request a QoS treatment for specific flows sent from the AP. For example, the STA could request that all video traffic (marked as IP DSCP 34) from a certain web server be classified as video (TID = 5, AC_VI) when transmitted from the AP. In Wi-Fi 7/802.11be, this SCS capability is extended to specify a service level agreement (SLA) between the STA and AP. The STA and AP agree upon the required traffic characteristics, described by the QoS Characteristics (QC) element, for traffic flows such that the AP is able to explicitly schedule resources accordingly.

You will learn more details about the mechanics of this enhanced SCS in Chapter 7, "EHT MAC Operations and key features." Two of the key parameters in the QC element are the minimum and maximum service intervals (Min SI/Max SI), which specify the time between expected channel access for a specific traffic flow for uplink or downlink. This flow-specific knowledge allows the AP to schedule its downlink flow, and also schedule the access for the STA via trigger-based OFDMA for the uplink flow. The 802.11be amendment does not specify how the STA should make these requests, how the AP should schedule these transmissions, or how it should orchestrate multiple requests from multiple STAs. The process is simple on paper, but can be tricky in real-life scenarios.

Consider the example of an XR headset. Conceptually, most of the traffic will go downstream. Practically, the headset also provides feedback to the server. The exchange is systematically bidirectional: The STA provides position and movement feedback to the server, and the server returns the matching new visuals. Therefore, a logical QoS structure is a bidirectional traffic profile. However, such a profile does not exist in Wi-Fi 7 as each flow (and therefore each direction) is specified independently, with its own SCS request. Instead, the STA must request two traffic profiles and carefully quantify the volume for the upstream and downstream traffic flow. It must also identify the time at which a flow is interleaved with the other so that the up and down transmissions follow each other, one responding to the other, as the application expects. A deviation or miscalculation could lead to misalignments in which the server does not receive the information it needs for the next visual, or the STA does not have space to send updated position information before the server's transmission turn comes. As the standard develops, STA vendors (and application developers) will learn the right values and the right exchange tempo, but the first iterations may be suboptimal.

Similarly, periodic flows (e.g., video-conferencing) require only a soft bound on delay (e.g., 5–50 ms), as the application can adapt to changing network conditions via coding (e.g., H.264). This application-based adaptation mechanism might be triggered by any condition, which means that the minimum data rate and burst size requested might change quite suddenly. If the threshold is set too low, the requested schedule will starve the application, forcing it to decrease the data rate. If it is set too high, the requested schedule will waste bandwidth, leading the AP to observe that many of the requested slots remain empty; this will force the AP to start starving the STA, without knowing the reasonable lower bound for where to stop. Here again, getting the schedule "just right" will probably require several generational iterations in the STAs' drivers.

Summary

The 802.11be project started in the first part of the 2010 decade, a time when video-conferencing and video calls were becoming mainstream, and when the first AR/VR applications were coming to market. Both signaled the need for higher bandwidth, but also higher efficiency, a reduced loss rate, and a better controlled delay bound. In short, capabilities were needed to remove these occasional perfect storms where sudden congestion events destroy the user experience, even on the best-designed Wi-Fi network. The 802.11be group studied multiple possible directions, and retained several features for higher efficiency:

- The ability to connect a STA to more than one AP radio (with MLO)
- An enhanced QoS exchange procedure (with SCS and the QC element) through which the STA and the AP can agree on the exact characteristics of the flow to prioritize
- A mechanism (R-TWT) to prevent slow STAs from transmitting on channels when high QoS efficiency is anticipated

The 802.11be group also defined faster modulations, wider channels in 6 GHz, and the ability to ignore subchannels of a wide channel affected by narrow interferences.

Not all ideas that were explored ended up being retained. Notably, multi-AP MLO was removed, as was the idea of full duplex operations, which would enable the AP to know the STA state and better anticipate its needs. A total of 16 spatial streams also fell out of the planned amendment. However, the ideas that were selected are expected to bring remarkable improvements in the user experience. Of course, the difficulty is not in creating the general idea, but rather lies in settling the details. In the next chapters, you will learn how these ideas operate, from the PHY to the MAC layer, and what they imply in terms of WLAN design.

Chapter 5

EHT Physical Layer Enhancements

physical layer (PHY) protocol data unit: [PPDU]
The unit of data exchanged between PHY entities to provide the PHY data service.

802.11-2020, Clause 3

The 802.11be amendment for Extremely High Throughput (EHT) STAs introduces several key elements intended to improve operations for highly demanding applications. Some of these improvements are implemented at the MAC layer, and address all elements of coordination and scheduling. These improvements also rely on innovations at the Physical layer (PHY). This chapter focuses on the 802.11be Physical layer enhancements, with particular attention being paid to the new features introduced in 802.11be: 320 MHz, multi-link operation, large-size and small-size multi-RUs, and 4096-QAM.

Wi-Fi7 in a Nutshell

This section highlights the major features of the Wi-Fi 7 PHY. If you are new to the last few generations of the 802.11 PHY, you can revisit this section after reading the other sections of this chapter in depth, since it provides the key points.

Revolution

Wi-Fi 7 brings five main PHY features, and a host of minor enhancements[1]:

- 320 MHz bandwidth (at 6 GHz) (optional)

- Multi-link operation (MLO) (see Chapter 6, "EHT MAC Enhancements for Multi-Link Operations," for details on optional/mandatory enhancements)

- Large multi-RUs (MRUs) and preamble puncturing (mandatory)

---

1 See P802.11be D5.0, clause 36.1.1.

- Small-sized MRUs (mandatory)

- 4096-QAM (also known as 4K-QAM) (optional)

The MLO feature requires changes across both MAC and PHY, and the MAC aspects are described in detail in Chapter 6.

In addition, some features that never took off or whose adoption has fallen off in recent Wi-Fi generations are not continued in 802.11be. These abandoned features include the following:

- A dedicated Single-User (SU) PPDU format, which is replaced by the MU PPDU format for simplicity

- A PPDU format optimized for extended range, since there is little need to enhance the Wi-Fi 6 extended range PHY format

- 80 + 80 MHz, which (arguably) is being upgraded to MLO

- Space-Time Block Coding (STBC), which is more expensive to implement given the larger FFT size introduced in Wi-Fi 6

- Dual-carrier modulation (DCM) for constellations and forward error correction (FEC) code rates above the BPSK rate – ½, to avoid data-rate duplication

The following features are defined in 802.11be but (at time of this book's writing) do not seem to be gaining much traction in the market:

- MCS 14 and MCS 15 (longer-range modes)

- A single PPDU combining both OFDMA and MU-MIMO

Evolution

Ever since the wise choices made in the 802.11a amendment, all subsequent mainstream Wi-Fi generations have evolved and expanded rather than being started over. Accordingly, the following characteristics are inherited and maintained by the Wi-Fi 7 PHY[2]:

- Burst transmission, where each burst of energy is termed a PHY protocol data unit (PPDU), sometimes known as a transmission or, occasionally, a packet.

- Two types of FEC codes: binary convolution coding (BCC) and low-density parity checking (LDPC) coding (see the "Construction of the Data Field" section later in this chapter). In 802.11be, LDPC is the superior technique and is the only choice if any of the following events occur: the RU is 484 tones or wider, the number of spatial streams is 5 or more, or the MCS is 10, 11, 12, 13, or 14.

2 See P802.11be D5.0, clause 36.1.1.

- FEC code rates of ½, ⅔, ¾, and ⅚.

- Multiple-phase shift keying (PSK) and quadrature-amplitude (QAM) constellations: BPSK, QPSK, 16-QAM, 64-QAM, 256-QAM, and 1024-QAM (also known as 1K-QAM). In 802.11be, MCSs up to 256-QAM are mandatory (or up to 64-QAM for 20 MHz-only clients).

- Spatial multiplexing, with up to 8 spatial streams defined. To allow for low-end APs such as mobile APs, support for 2 or more spatial streams is optional in 802.11be.

- 20, 40, 80, and 160 MHz bandwidths.

- MU-MIMO across the entire PPDU bandwidth, with up to eight users.

- OFDMA with one user per resource unit (RU), where the total number of tones (equal to data + pilot tones) for the defined RUs are 26 (24 + 2), 52 (48 + 4), 106 (102 + 4), 242 (234 + 8), 484 (468 + 16), 996 (980 + 16), and 2×996 or 1992 (1960 + 32) subcarriers.

Purposes and Opportunities for the Physical Layer

Before diving into the PHY features of Wi-Fi 7 in detail, it is useful to take a step back. In this section, we'll provide a short review of the PHY design goals.

In a communications system, the core purpose of the PHY is to deliver a very high data rate to the upper layers (most obviously the Data Link layer, including the MAC layer, the Networking layer, and ultimately the Application layer). The secondary goal is to minimize inefficiencies, such as those that arise when overhead consumes an appreciable fraction of the resources used to deliver data. A third goal is to reduce sensitivity to interference from neighboring wireless systems, which might be 802.11 or some other unlicensed technology. A fourth goal is to provide a low latency capability to the upper layers.

As you saw in Chapter 4, "The Main Ideas in 802.11be and Wi-Fi 7," the wireless PHY data rate is the product of the three parameters—data bits per constellation point (via denser constellations), bandwidth (more constellation points sent in the same amount of time), and number of spatial streams via additional antennas (more highway lanes for the constellation points). Thus, higher data rates are achieved by increasing one or more of these parameters.

However, increasing any of these parameters comes with various trade-offs. The use of denser constellations adds engineering challenges, such as more difficult analog RF design and a somewhat higher power draw. There is also the fundamental physics constraint whereby the constellation points in denser constellations are inherently closer together and more sensitive to noise. These denser modes are limited to shorter-range operation, where the useful signal is much stronger than the noise. The gains from increased constellation density are steadily decreasing: Going from 64-QAM to 256-QAM provides a $\log_2(256)/\log_2(64) = 1.33 \times$ data rate increase, but then 1024-QAM and 4096-QAM provide diminishing data rate gains of $10/8 = 1.25 \times$ data rate and $12/10 = 1.2 \times$ data rate, respectively. At a short range, where free space propagation is a reasonable model, each of these jumps in data rate comes with a halving of the range.

Wider bandwidth operation provides a bigger payoff. A doubling of bandwidth provides an unalloyed doubling of data rate. Once again, there is the challenge of more difficult analog RF design. In addition, the conversion rate of the analog-to-digital converters (ADCs) and digital-to-analog converters (DACs) and the amount of digital signal processing in the baseband doubles, creating a somewhat higher power draw. Figure 5-1 shows the baseband in the context of an entire ratio.

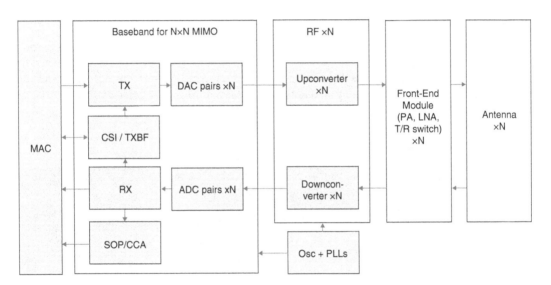

Acronym Key:

MAC (Medium Access Controller)

TX (Transmitter)

CSI (Channel State Information)

TXBF (Transmit Beamforming)

RX (Receiver)

SOP (Start of PPDU)

CCA (Clear Channel Assessment)

Osc (Oscillator)

PLL (Phase Locked Loop)

PA (Power Amplifier)

LNA (Low Noise Amplifier)

T/R (Transmit/Receive)

FIGURE 5-1   Major Components of the PHY of One Radio

The physics cost of the doubled data rate is a relatively modest 29% reduction in range: Given the same transmit energy, each constellation point is sent with only half as much energy as it would on a channel half as wide, and has a reach scaled by the square root of a half (approximately 0.71).

Finally, another possibility is increasing the number of spatial streams. This increase comes with clear costs in terms of extra RF hardware, such as antennas and connectors, extra transceiver paths in the RF, and nonlinearly more complicated baseband signal processing. For instance, an 8×8 MIMO equalizer requires 8 times more compute power than a 4×4 MIMO equalizer. Additionally, to increase the number of spatial streams, each antenna needs to be adequately separated from its neighbors, and a rich scattering environment is needed around and between the transmitter and receiver to afford each spatial stream a relative independence from the other spatial streams. A perverse side effect of more antennas is the need for more training, which expands the preamble length.[3] Of course, extra antennas are valuable in any event, as they can be repurposed for improved reliability during the receive operation and more effective beamforming during the transmit operation.

Efficiency is another key means to increase the data rate. PHY overheads come from the preamble, the cyclic extension, and subcarriers that don't carry data. Figure 5-2 summarizes these overheads. In the figure, green indicates data-bearing, and red/pink indicates overheads; the figure assumes EHTTB PPDU format, EHTLTF $\times$ 2 with two OFDM symbols, 0.8 µs GI, 15 OFDM symbols of data, 16 µs PE, and 16µs SIFS.

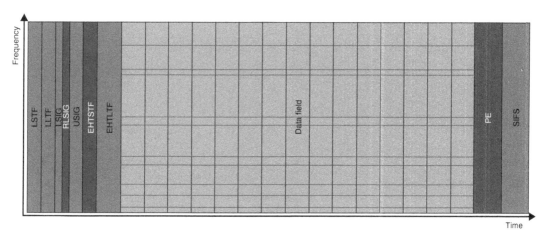

FIGURE 5-2   Summary of Overheads in the PHY PPDU

As data rates increase, it is important that the total preamble and post-amble duration continue to be dwarfed by the payload duration. This vital observation motivated key 802.11 features such as the following:

- Time-axis aggregation of multiple packets to a user: AMSDU and AMPDU aggregation in Wi-Fi 4.[4]

---

3 See P802.11be D5.0, clause 36.3.12.10.
4 See P802.11-REVme D5.0, clauses 9.3.2.2 and 9.7.

- Spatial-axis aggregation of multiple users' packets: DL-MU-MIMO added in Wi-Fi 5[5] and UL-MU-MIMO added in Wi-Fi 6.[6] MU-MIMO also helps when the AP has more antennas than clients, because the AP can use its antenna resources to send up to its maximum number of spatial streams to multiple users in parallel.

- Frequency-axis aggregation of multiple users' packets: both UL and DL OFDMA in Wi-Fi 6.[7]

> **Note**
>
> Notice the use of "user" rather than "STA." The PHY is not aware of whether a payload that the PHY transmits is for one STA (unicast) or more STAs (groupcast), so the term "user" is employed to accommodate both scenarios.

In terms of subcarrier design, Wi-Fi 6 brought a major redesign, from 64 subcarriers (or tones) per 20 MHz to 256 subcarriers per 20 MHz.[8] When there are more subcarriers, the relative cost of a fixed number of pilot tones diminishes. Additionally, because each subcarrier is narrower in the frequency domain, its side lobes drop faster and data subcarriers can be packed closer to the frequency edge of transmissions without degrading resilience to other transmissions in adjacent channels.

A somewhat indirect path to greater efficiency is through more effective FEC coding that can achieve a given level of reliability but with reduced coding overhead (e.g., the use of code rate ⅚ in place of code rate ¾). The exemplar here is LDPC coding, which was first introduced in Wi-Fi 4,[9] and is now mainstreamed in mid- and high-end devices.

PPDU Formats and Their Fields

The 802.11be Extremely High Throughput (EHT) PPDU has two formats: multiuser (MU) and trigger-based (TB). Both PHY formats closely follow their respective 802.11ax/HE MU and TB PPDU formats, as shown in Figure 5-3.

---

5 See P802.11-REVme D5.0, clause 21, and especially clauses 21.2.2, 21.3.3 and 21.3.11.
6 See P802.11-REVme D5.0, clause 27, and especially clause 27.3.3.
7 See P802.11-REVme D5.0, clause 27, and especially clause 27.3.2.
8 See P802.11-REVme D5.0, clauses 21.3.6 and 27.3.9.
9 See P802.11-REVme D5.0, clause 19.3.11.7.

HE MU PPDU format

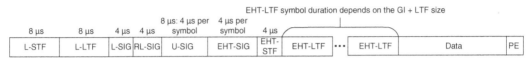

ETH MU PPDU format

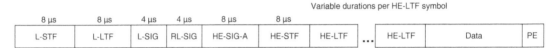

HE TB PPDU format

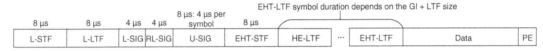

ETH TB PPDU format

FIGURE 5-3   Multiuser (MU) and Trigger-Based (TB) PHY Formats for EHT and HE[10]

The EHT TB format is used for uplink transmissions specifically solicited by the AP, and the EHT MU format is used for everything else, including both uplink and downlink single-user transmissions.[11] Each PPDU is made up of the following fields:

- **Legacy Short Training Field (LSTF)[12]:** for start-of-PPDU detection, automatic gain control (AGC), and coarse carrier recovery. Each 20 MHz of this field is identical to the 802.11a STF field, so the LSTF field is a frequency-domain duplication of the STF field across the whole PPDU bandwidth.

- **Legacy Long Training Field (LLTF)[13]:** for fine carrier recovery, timing recovery (together with the LSTF), and channel estimation toward equalizing the subsequent SIG fields of the preamble. Each 20 MHz of this field is identical to the 802.11a LTF field, so the LLTF field is a frequency-domain duplication of the LTF field across the whole PPDU bandwidth.

---

10 See P802.11be D5.0, clause 36.3.4.
11 See P802.11be D5.0, clauses 36.3.12.7.2 and 35.11.1.3, and then P802.11-REVme D5.0, clause 26.11.2.
12 See P802.11be D5.0, clause 36.3.12.3.
13 See P802.11be D5.0, clause 36.3.12.4.

- **Legacy SIG (LSIG) field** and **Repeated LSIG (RLSIG) field**[14]: the RLSIG is a twin of the LSIG field and they are used together to identify that a PPDU is an EHT PPDU (i.e., 802.11be). The fields also indicate whether the PPDU is in MU PPDU format or TB PPDU format, and report the length of the PPDU, in units of 4 µs. Each 20 MHz of the LSIG and RLSIG is identical to the 802.11a SIGNAL field, except that known training is inserted at subcarriers ±27 and ±28 as a belated augmentation of the training in the LLTF, which spans subcarriers –26 to +26 (excluding DC) per 20 MHz. In this way, the USIG and EHT SIG fields can carry 52 data-bearing subcarriers per 20 MHz instead of just 48. Like the previous subfields, the LSIG and RLSIG fields are frequency-domain duplicated across the whole PPDU bandwidth. Figure 5-4 shows how a receiver can distinguish the various 802.11 OFDM-based PHY formats.

- **Universal SIG (USIG)**[15]: contains information for third-party STAs and common information for the rest of the PPDU. A third-party STA does not transmit the PPDU, nor is it an intended receiver of the PPDU's payload, but it still needs to act upon the PPDU's reception (e.g., recognize that a PPDU is making the channel busy and other transmissions should be deferred). The USIG field is broadly similar to the HT-SIGA, VHT-SIGA, and HE-SIG fields defined in Wi-Fi 4, 5, and 6, respectively. One goal for the USIG field is to stabilize the design of this SIG field, making it future-proofed for subsequent Wi-Fi generations. Some bits of the field are used in 802.11be, whereas others are left to be defined by future generations. The USIG is frequency-domain duplicated across the whole PPDU bandwidth.

- **EHT SIG field**[16]: found in MU PPDUs, but not TB PPDUs. This field has the most complicated definition, because it is responsible for compactly defining the remaining format of the PPDU: the use of OFDMA and MU-MIMO, which users' content is included, which frequency and spatial resources are allocated to each user, and each user's MCS. In TB PPDUs, the EHT SIG field is not needed because the same information was sent immediately beforehand by the AP in its Trigger frames. Note that 20 MHz PPDUs have a single content channel, whereas PPDUs with bandwidth 40 MHz and wider support two parallel content channels per 80 MHz to increase the data rate: Each 80 MHz frequency subblock of the EHT-SIG comprises a contiguous pair of 20 MHz content channels, which are frequency-domain duplicated up to the 80 MHz bandwidth. The content channels are encoded either in a compact manner for non-OFDMA PPDUs or in a more generic manner for everything else. Each content channel starts with a Common field for U-SIG overflow (i.e., for signaling USIG overflow bits) and, if the PPDU uses OFDMA, the RU/MRU allocation and the number of User fields per RU/MRU. The Special User Info, if present, is next and carries extended common information. After that, User fields are sent, where each User field is 22 bits long and

---

14 See P802.11be D5.0, clauses 36.3.12.5 and 36.3.12.6.
15 See P802.11be D5.0, clause 36.3.12.7.
16 See P802.11be D5.0, clause 36.3.12.8.

contains the STA-ID of the recipient user (i.e., the LSBs of the STA's AID), the MCS, how many and which spatial streams are intended for the user, whether BCC or LDPC coding is used, and whether the RU/MRU is beamformed. Broadcast RUs are signaled via a STA-ID equal to 0 (or somewhat higher in the case of multiple BSSIDs) or 2047. CRC and Tail fields are regularly interspersed to increase the decoding reliability. Figure 5-5 shows the format of the EHT SIG field for a 160 MHz PPDU. The number of OFDM symbols of the EHT SIG field is signaled in the preceding U-SIG field.

- **EHT Short Training Field (EHT-STF)**[17]: For AGC refinement and still-finer carrier recovery and timing estimation.

- **EHT Long Training Field (EHT-LTF)**[18]: For channel estimation toward equalizing the Data field. The number of EHT-LTFs is signaled in the Common field of the EHT-SIG field. There are at least as many EHT-LTFs as there are spatial streams per RU.

- **Data field**[19]: Carries the payload, termed the PHY Service Data Unit(s) [PSDU(s)].

- **Packet Extension (PE)**[20]: Keeps the wireless medium appearing busy, yet carries no data and thus provides the receiver with extra time, beyond SIFS, to complete processing of the PPDU.

A Wi-Fi 7 device must also support Wi-Fi 6 and prior generations, so the PHY receiver must perform PHY format detection on each received PPDU, as shown in Figure 5-4. The acronyms and notation in this figure are Q-BPSK and QBPSK (quadrature BPSK), HTGF (HT greenfield PHY format), HTSIGAx (OFDM symbol x of the HT-SIG-A field), HEMU (HE multiuser PHY format), HEER (HE extended range PHY format), HESU (HE single-user PHY format), HETB (HE trigger-based PHY format), EHTMU (EHT multiuser PHY format), EHTTB (EHT trigger-based PHY format), HTMF (HT mixed-format PHY format), VHTSIGAx (OFDM symbol x of the VHT-SIG-A field), DataX (OFDM symbol X of the Data field), HESIGAx (OFDM symbol x of the HE-SIG-A field), VHTSTF (VHT Short Training Field), USIGAx (OFDM symbol x of the USIG field), [x] (bit x), [x:y] (bits x to y inclusive).

---

17 See P802.11be D5.0, clause 36.3.12.9.
18 See P802.11be D5.0, clause 36.3.12.10.
19 See P802.11be D5.0, clause 36.3.13.
20 See P802.11be D5.0, clause 36.3.14.

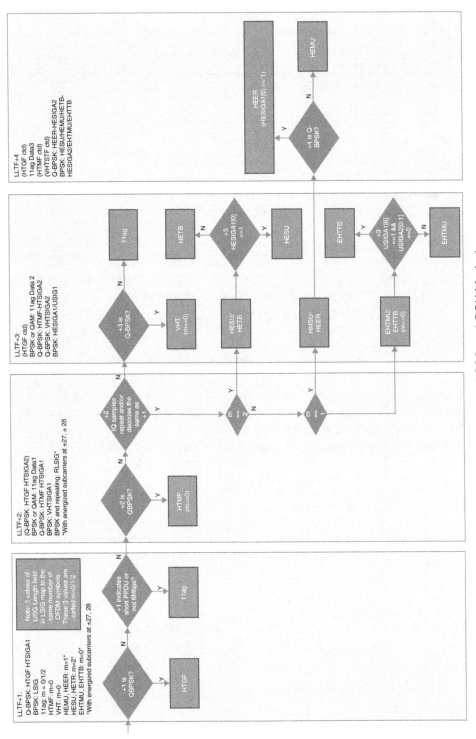

FIGURE 5-4   Summary of PHY Format Identification (Ignoring DSSS and CCK Modes)

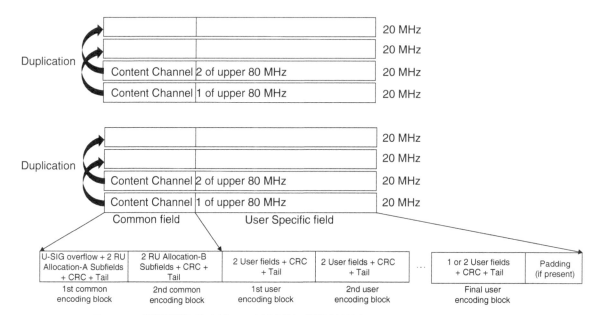

FIGURE 5-5   Format of EHTSIG Field for a 160 MHz PPDU Without Preamble Puncturing

Construction of the Data Field

This section describes the construction of the Data field. In large part, the SIG fields are merely special cases of the Data field. The important simplifications for the SIG fields are described in the next section. The Data field processing is designed in an evolutionary manner with respect to past Wi-Fi generations. Figure 5-6 shows the major processing blocks used to generate the Data field.

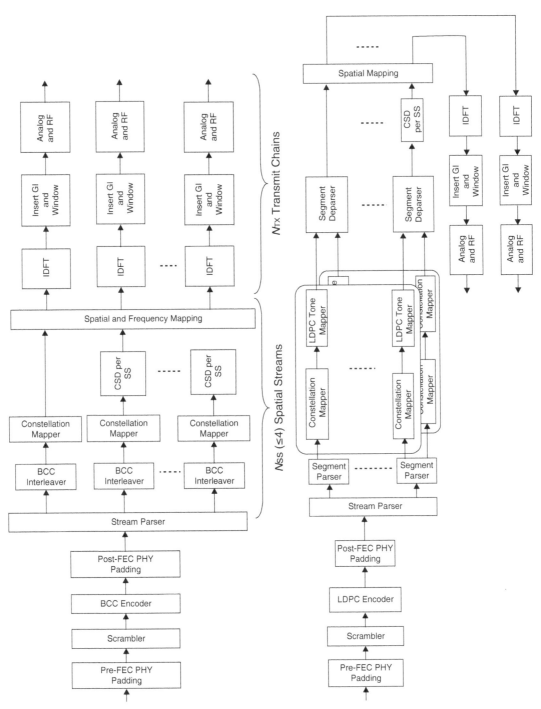

FIGURE 5-6 BCC and LDPC Processing[21]

For the simple case of no DCM and for one user (in total or one of many), the process is as follows:

1. For the Data field, the user's payload from the MAC (i.e., the PHY Service Data Unit [PSDU]) is prefixed by the SERVICE field and postfixed by 0–7 pre-FEC padding bits (and a 7-bit tail in the case of BCC) so that it exactly aligns with the space available for the user's data bits in the Data field. If more than 7 bits of padding is required, the padding is provided by the MAC.[22]

2. To reduce the effect of long sequences of zero or one bits in the PSDU causing a high peak-to-average in the time domain waveform (i.e., after the IFFT), the user's padded PSDU is scrambled by XORing the padded PSDU using a repeated length-2047 scrambling sequence. This is a new and longer scrambling sequence than the traditional length-127 sequence, because the number of data bits per OFDM symbol is now vastly larger: 4×980 [tones] $\times 12 \times \frac{5}{6}$ [data bits per constellation point] $\times 8$ [spatial streams] = 313,600 data bits.[23]

3. To encourage adoption of the powerful low-density parity check (LDPC) coding technique, only LDPC is supported for RUs wider than 242 tones, or with more than 4 spatial streams or the higher MCSs 10–13 (and for MCS 14). BCC or LDPC can be used for RUs up to and including 242 tones, 4 spatial streams, and MCS 9 (and also MCS 15).

The BCC encoding process uses the traditional 802.11 rate ½, $k = 7$ encoder, followed by a puncturer to reach code rates of ½, ⅔, ¾, and ⅚.[24]

The LDPC encoding process is quite complicated, and a good starting point to understand it in more depth is the 802.11n (HT) clause.[25] At a high level, first the size of the LDPC code-words and the number of LDPC codewords are determined. The number of LDPC codewords is chosen by design to exceed the number of coded bits made available in the user's Data field under some circumstances (and later some parity bits must be punctured to compensate). Next, extra padding bits (called shortening bits) are inserted as needed at the end of the data bits for each LDPC codeword, so as to achieve the expected number of data bits for the number of LDPC codewords. Then the core encoding operation is performed: The LDPC parity bits are calculated. The shortening bits are discarded, and the parity bits are appended to the data bits of each codeword. Finally, the number of coded bits must be matched to the number of coded bits made available in the user's Data field, either by repeating the data + parity bits within each codeword (if they are too few), or by having some parity bits be punctured from the end of each LDPC codeword (if there are too many). The shortening, repeating, and puncturing processes are applied more-or-less uniformly to each LDPC codeword. One important complication is that, if the proportion of punctured bits is too high (especially if the proportion of shortening bits is low)—for any user, if more than one—then in 802.11ax and 802.11be an extra "quarter"

---

22 See P802.11be D5.0, clauses 36.3.13.2, 36.3.13.3.2, and 36.3.13.3.5, plus P802.11-REVme D5.0, clauses 27.3.12.1, 27.3.12.2, and 10.12.6.
23 See P802.11be D5.0, clause 36.3.13.2.
24 See P802.11be D5.0, clause 36.3.13.3.2, plus 802.11-REVme D5.0, clause 27.3.12.5.1.
25 See P802.11-REVme D5.0, clause 19.3.11.7.5.

OFDM symbol (called a symbol segment) is added (for all users) to the Data field for extra parity bits to reduce the level of puncturing.[26]

4. To support MIMO, the coded data must be split into multiple spatial streams. For this purpose, the stream parser[27] splits the coded data bits into blocks of s bits, equal to the number of coded bits per constellation axis, then allocates each block sequentially to one of the user's spatial streams in turn. As some spatial streams are likely to be more reliable than others, this parsing arrangement is designed to distribute the LDPC parity bits across all the spatial streams while enabling certain hardware optimizations. The same stream parser is used with the 802.11 binary convolutional code (BCC), where the coded bits for each data bit are distributed across all the spatial streams, too.

5. In 802.11ac/Wi-Fi 5, as a design aspect of the noncontiguous 80 + 80 MHz mode, each 80 MHz frequency sub-block of the Data field was constructed relatively independently, for both 80 + 80 MHz and 160 MHz modes, using two 80 MHz frequency sub-blocks. The 20, 40, and 80 MHz PPDUs make up single-frequency sub-blocks. This preexisting design makes incorporating the new 320 MHz bandwidth a straightforward process, because the one or two frequency sub-blocks are now joined by a new alternative: four 80 MHz frequency sub-blocks. The block responsible for this processing is labeled the segment parser by tradition, due to the connection to 80 MHz segments in an 80 + 80 MHz PPDU, but is more properly understood to be a frequency sub-block parser.[28] The so-called segment parser splits the user's coded bits per spatial stream into substreams, one for each 80 MHz frequency sub-block. For users in an RU occupying less than or equal to 80 MHz (such as all users encoded via BCC), the segment parser is a pass-through. Otherwise, for users in an RU occupying less than or equal to 160 MHz (including the former 80 + 80 MHz mode), the segment parser outputs two substreams per spatial stream. Beyond both of those, for users in an RU occupying less than or equal to 320 MHz, the segment parser outputs *four* substreams per spatial stream.

6. For BCC-encoded users only, a block interleaver[29] operates on the coded bits of each spatial stream of each OFDM symbol. It spreads out the coded bits for each data bit widely, thereby capturing the multipath diversity gains.

7. The constellation mapper[30] converts the coded bits from the previous stage to Gray-coded constellation points—namely, BPSK, QPSK, 16QAM, 64QAM, 256QAM, 1K-QAM, or (now with 802.11be) 4K-QAM (see Chapter 4 and the later discussion in this chapter). The constellation mapper is a simple memoryless block.

8. For 802.11n/Wi-Fi 4, interleaving for LDPC, unlike for BCC, was not included because the parity bits in an LDPC codeword are already calculated over widely spread data bits and the

26 See P802.11be D5.0, clause 36.3.13.3.4.
27 See P802.11be D5.0, clause 36.3.13.4.
28 See P802.11be D5.0, clause 36.3.13.5.
29 See P802.11be D5.0, clause 36.3.13.6.
30 See P802.11be D5.0, clause 36.3.13.7.

multipath diversity gains over 20 or 40 MHz were not appreciable. However, as the constellation sizes increases, a given LDPC codeword, without interleaving, is confined to a narrower range of subcarriers and becomes more susceptible to multipath fading. Moreover, the wider 80 MHz frequency sub-block provides greater opportunities for frequency diversity. For this reason, the LDPC Tone Mapper block[31] was introduced in 802.11ax/Wi-Fi 6. It also behaves like a block interleaver for each OFDM symbol, but operates on constellation points rather than individual bits. By doing so, it spreads the constellation points of an LDPC codeword widely in each frequency sub-block to capture the available frequency diversity. Wi-Fi 7 retains this new performance-enhancing capability.

9. Next, the constellation points from the 1, 2, or 4 frequency sub-blocks of each spatial stream are merged by the segment deparser[32] to create another OFDM symbol's worth of constellation points. This is a simple frequency-domain concatenation operation.

10. Pilot tones[33] are then inserted. These single spatial streams of known BPSK constellation points are inserted at a small number of known subcarrier indices. They are used for tracking phase noise, residual carrier frequency offsets (and hence timing drift), and amplitude changes.

11. To avoid unintentional beamforming in the absence of genuine beamforming, different delays are applied cyclically (i.e., per OFDM symbol) to the second and higher spatial streams by the cyclic shift diversity (CSD) per Spatial Stream (SS) block.[34]

12. Spatial mapping[35] converts the number of spatial streams to the number of available antennas. In spatial mapping, beamforming matrices are applied if available. For example, if there is a single user in an RU and the MAC determines it needs to send the user two spatial streams, yet the PHY supports four antennas, then the beamforming matrices are 4×2 for the 2×1 vector of spatial streams. If beamforming matrices are not available, then various alternatives are available—for instance, sending a spatial stream out of two or more antennas (hence the previous step) or transmitting the spatial streams directly out of that number of antennas, and leaving the other antennas unenergized. For the preceding example, the spatial mapping matrices are then

$$\begin{bmatrix} 1 & 0 \\ 0 & 1 \\ 1 & 0 \\ 0 & 1 \end{bmatrix} / \sqrt{2} \text{ and } \begin{bmatrix} 1 & 0 \\ 0 & 1 \\ 0 & 0 \\ 0 & 0 \end{bmatrix}, \text{ respectively}$$

Each parsed constellation point per spatial stream per OFDM symbol is intended for a different subcarrier, so the processed constellation points are really defined in the frequency domain. For this reason, every OFDM symbol, for every spatial stream, the Inverse Discrete Fourier

---

31 See P802.11be D5.0, clause 36.3.13.8.
32 See P802.11be D5.0, clause 36.3.13.9.
33 See P802.11be D5.0, clause 36.3.13.11.
34 See P802.11be D5.0, clauses 36.3.13.12 and 36.3.12.2.2.
35 See P802.11be D5.0, clause 36.3.13.12, plus P802.11-REVme D5.0, clause 19.3.11.11.2.

Transform (IDFT) block converts a vector of parsed constellation points into time-domain samples.[36] For instance, for an RU spanning 320 MHz, given a subcarrier spacing of 20 MHz/256 = 78.125 kHz (used for Wi-Fi 6 and Wi-Fi 7), the vector comprises 320/0.078125 = 4096 entries. The IDFT block is typically implemented as an inverse fast Fourier transform (IFFT) algorithm. In addition, to simplify time-domain filtering, the IFFT is typically over-sized by a factor of between 2 and 4 (inclusive), with the higher frequency subcarriers set to zero. In the time domain, this leads to oversampling by the same factor.

13. To simplify the receiver implementation, the end of each OFDM symbol is repeated at the start as a cyclic prefix (called somewhat incorrectly a guard interval [GI]). Then, consecutive OFDM symbols are overlapped, windowed, and added. The windowing assists with suppression of frequency-domain side lobes.[37]

Major Wi-Fi7 PHY Innovations

In this section we describe the PHY features that offer a marked benefit for Wi-Fi 7 devices: 320 MHz (for increased speed), multi-link operation (which, according to implementation, can provide one or more benefits—increased speed, lower latency, or increased reliability), large-size multi-RUs and preamble puncturing (for interference avoidance), small-size multi-RUs (for scheduling efficiency), and 4096-QAM (for increased speed at short range).

320 MHz Bandwidth

320 MHz is the most high-profile feature of 802.11be, as this new bandwidth doubles the peak data rate. The 802.11be amendment defines 320 MHz channel bandwidths judiciously for 6 GHz only, based on experience (see the sidebar).

Why No 320 MHz at 5 GHz?

Considering the FCC as an example regulatory domain, 83.5 MHz of spectrum was allocated at 2.4 GHz. With some spectrum excised for guard bands, 802.11 defined three non-overlapping 20 MHz channels: channels 1, 6, and 11. For classic Wi-Fi (at 2.4, 5, and 6 GHz) a channel number increment represents 5 MHz, so these channel center frequencies are separated by 25 MHz. The 802.11n amendment also introduced 40 MHz wide channels, where the early preamble fields are a contiguous, frequency-domain duplication of the 20 MHz preamble. This design was well matched to 5 GHz but introduced two issues at 2.4 GHz:

■ The 40 MHz PPDU design does not nicely align with channels 1, 6, and 11. For instance, the two 20 MHz preamble copies in a 40 MHz PPDU can be centered at channels 1 and 5, or 2 and 6, or 6 and 10, or 7 and 11, but none of these pairs line up with channels 1 and 6.

---

36 See P802.11be D5.0, clauses 36.3.13.12, 36.3.10, and 36.3.11.
37 See P802.11be D5.0, clauses 36.3.13.12, 36.3.10, and 36.3.11, plus P802.11-REVme D5.0, clause 17.3.2.5.

Thus 40 MHz systems cannot coexist well with traditional 20 MHz Wi-Fi systems operating at channels 1, 6, and 11.

- If one Wi-Fi system occupies 40 MHz at 2.4 GHz, nearby Wi-Fi systems cannot have their own separate 40 MHz channel, raising fairness questions.

- Consuming around half the bandwidth of a band makes it harder to coexist with other non-Wi-Fi systems operating in the same band, such as Bluetooth.

Accordingly, the 40 MHz bandwidth in 2.4 GHz is little used, especially in enterprise environments. A major take-away from this experience for the industry has been that at least three channels per band, plus alignment of center channels, are required in practice.

At 5 GHz, similar issues come into play when the bandwidth gets much higher than 160 MHz. The FCC allocates 200 MHz at 5150–5350 MHz and 425 MHz at 5470–5895 GHz. However, the second spectrum block was allocated in multiple tranches, and the 802.11 channel assignment for each tranche was individually optimized (mainly) for 20 MHz channel bandwidths. This process led to discontinuous channel numbering. Specifically, there is a 25 MHz gap between the uppermost channel center defined for the 255 MHz of spectrum at 5470–5725 MHz and the lowermost channel center defined for the 170 MHz of spectrum at 5725–5895 MHz. With these limitations in mind, the 802.11ac amendment defined a trio of 160 MHz channels[38] (channel 50 centered at 5250 MHz within 5150–5350 MHz, channel 114 centered at 5570 MHz within 5470–5725 MHz, and channel 163 centered at 5815 MHz within 5725–5895 MHz), where none of these are adjacent (i.e., separated by 160 MHz or a channel number of 160 MHz/5 MHz = 32) and suitable for aggregation. Therefore, the 160 MHz bandwidth channels at 5 GHz cannot be bundled together to form 320 MHz–wide channels. Even if this bundling were possible, there is insufficient 5 GHz spectrum allocated to support three 320 MHz channels (which again is the accepted good practice).

Consequently, only 6 GHz is suitable for 320 MHz:

- It has enough spectrum for three non-overlapping channels, each of 320 MHz bandwidth.

- All legacy channel numbers within the spectrum are separated by 20 MHz.

The 320 MHz bandwidth is expected to be used mainly in the home. For single-family dwellings, the number of neighboring houses is relatively small, and the attenuation from two external walls is likely to create strong RF separation between devices in different houses. Thus, each house is likely to get relatively clean access to a single 320 MHz channel, where a typical use case is to deliver streaming video traffic to multiple devices via 320 MHz with orthogonal frequency division multiple access (OFDMA).[39]

For enterprise venues, the requirement for robust coverage means strong overlap between adjacent APs. In turn, APs are likely to be operating at bandwidths lower than 320 MHz so that adjacent APs are not co-channel. Even so, the 320 MHz bandwidth is a very important and valuable feature, because it

---

38 See P802.11-REVme D5.0, Annex E.1.
39 See Chapter 3 for an introduction to OFDMA, in Wi-Fi 6.

accelerates the uptake of the new and vastly important 6 GHz spectrum. As a commercial imperative, a leading-edge product must support the highest data rate available, at least until the costs become unreasonable. The inclusion of more antennas quickly adds to the silicon area and volume requirements, whereas the impact of increased bandwidth is much more modest. Thus, any client system or silicon vendor in the smartphone or laptop market needs to support 320 MHz *and thence 6 GHz*. In a very real way, 320 MHz is the vehicle by which 6 GHz becomes a mainstream Wi-Fi band. Certainly, this same tactic was used to good effect by 802.11ac/Wi-Fi 5 to drive broad adoption of 5 GHz, as that was the only band at the time that supported 80 and 160 MHz bandwidths.

In the early days of Wi-Fi, wider bandwidths (also known as channel bonding) drove increased spectral efficiency. At that time, guard tones were needed only at the edge of the wider bandwidth, and the interior guard tones could be converted to data tones. This process provided major improvements when Wi-Fi evolved from 20 MHz to 40 to 80 MHz and from 64 to 256 tones per 20 MHz. However, the potential gains beyond this level are limited to a few percent in theory, as shown in Table 5-1. Accordingly, 320 MHz is constructed with hardware reusability as a major design goal, and it comprises four 80 MHz frequency sub-blocks, each densely populated in accordance with the Wi-Fi 6 design.

TABLE 5-1   Evolution of the Percentage of Data Tones with Wider PPDU Bandwidths

Wi-Fi Generation	Bandwidth (MHz)	Number of Data Tones	Number of IFFT Tones	Percentage of Data Tones
802.11a/g	20	48	64	75
Wi-Fi 4	40	108	128	84
Wi-Fi 5	80	234	256	91
Wi-Fi 6	80	980	1024	96
Wi-Fi 7	80	980	1024	96
Wi-Fi 5	160	468	512	91
Wi-Fi 6	160	1960	2048	96
Wi-Fi 7	160	1960	2048	96
Wi-Fi 7	320	3920	4096	96

A key aspect of the wider bandwidths is that often they are used for transmission to multiple users via OFDMA, instead of a single user. 802.11be defines many RUs (and MRUs; see the "Large-Size MRUs and Preamble Puncturing" section and the "Small-Size Multi-RUs" section that follow), as shown in Figure 5-7. A 320 MHz PPDU can be assigned to a single user with a 4 × 996 tone RU, or two users each with a 2 × 996 tone RU, or another configuration taken from the vast array of other allowed combinations according to the available traffic. For example, a 320 MHz PPDU can carry six 26-tone RUs, three 52-tone RUs, two 106-tone RUs, and two 242-tone RUs in one 80 MHz frequency sub-block, plus two 484-tone RUs in a second 80 MHz frequency sub-block and one 2 × 996-tone RU in the remaining two 80 MHz frequency sub-blocks. Enterprise AP implementations typically support dozens of users in a single PPDU.

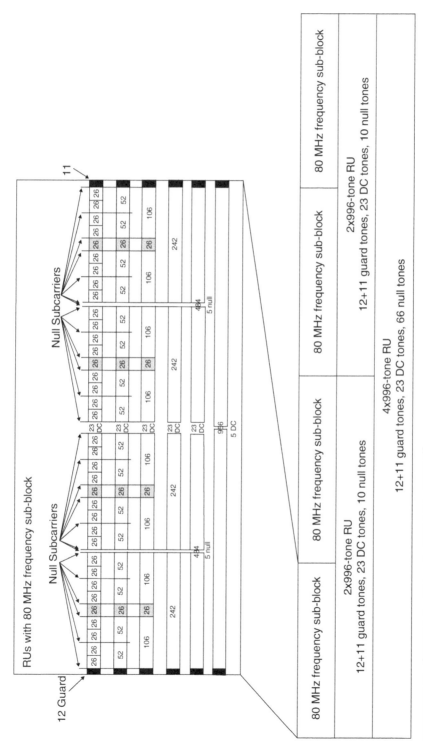

FIGURE 5-7   OFDMA for RUs of a 320 MHz PPDU[40]

40 P802.11be D5.0, clause 36.3.2.1.

To support the traditional MAC protection modes such as RTS + CTS and CTS-to-self, a non-HT duplicate mode is defined for 320 MHz.[41] This mode replicates 16 times the 20 MHz 802.11a PHY format. Each duplicate is subject to semi-controlled rotations to reduce the peak-to-average power ratio.[42] Given that the original 802.11a PHY format had no knowledge of channel bonding and the 2-bit backward-compatible bandwidth signaling mechanism introduced in Wi-Fi 5 supported signaling 20/40/80/160 MHz only,[43] an updated signaling scheme was required for 320 MHz. As 320 MHz is used exclusively in 6 GHz spectrum, certain legacy issues do not arise. Thus, it is possible to employ bit 7 of the Service field as an extra most-significant bit of the bandwidth signaling field, thereby converting it from a 2-bit field to a 3-bit field, where the newly defined value 4 indicates 320 MHz (and bit 7 of the Service field is set to 1).[44]

Multi-Link Operation

After the 320 MHz feature and the way it motivates support for 6 GHz communications, the next most important feature in Wi-Fi 7 is multi-link operation. In part, multi-link operation can be considered a standardized way to better harness the available hardware sources at both the AP device and the client device:

- The AP device typically contains multiple concurrent "IEEE APs" via radios for 2.4, 5, and 6 GHz.

- High-end AP devices typically provide some further radio flexibility or capability. For instance, the radio nominally operating at 2.4 GHz could instead operate at 5 GHz, or the AP device might include an additional radio at 5 GHz.

- A multiband client device includes transceiver paths that can switch from one band and channel to another band and channel.

- High-end client devices typically contain two concurrent "IEEE non-AP STAs," one nominally for Internet connectivity and a second nominally for peer-to-peer (P2P) connectivity. When and while there is no P2P connectivity, the second STA might also connect to the infrastructure, but—due to various networking constraints—do so via a different IP address (and thereby inherit various complexities and limitations).

The multi-link operation uses these capabilities to enable concurrent communications on multiple radios between an AP and a client—or at least some better approximation of this than is available via previous Wi-Fi generations.

The ever-present complication with concurrent operation is that a device's strong transmitted signals on one radio can interfere with the reception of weak (yet intended) signals on another radio. Similar

---

41 See P802.11be D5.0, clause 36.3.15.
42 The parameter $\gamma_{k,320}$ is defined in equation (36-12) in P802.11be D5.0, clause 36.3.11.4.
43 See P802.11-REVme D5.0, clauses 17.2.2, 17.2.3, and 17.3.5.5.
44 See P802.11be D5.0, clauses 17.2.2, 17.2.3, 17.3.5.2, and 17.3.5.5.

to the full-duplex problem described in Chapter 4, the transmitted signals are 10 billion times stronger than the maximum interference level allowed to meet the requirements for 802.11 reception, and are 10 trillion times stronger than the ideal interference level. This calculation is a function of the transmit power (e.g., 15 to 23 dBm) and the noise level. Noise level is a somewhat nebulous concept:

- At one extreme, 802.11 receivers are only required to achieve MCS 0 operation at a received signal level of −82 dBm. Assuming operation at a signal-to-interference-and-noise ratio (SINR) of 4 dB, then the noise + interference floor can be as high as −86 dBm.

- At the other extreme, most engineers would presume that an interference level of −107 dBm per 20 MHz (or 6 dB below the thermal noise floor of −101 dBm per 20 MHz) is almost completely negligible, given that all OFDM MCSs operate at a positive SNR.

Thus, in decibels, the transmit signals need to be attenuated by at least 101 dB within the reception channel to achieve satisfactory reception performance. There will be diminishing returns, however, if the attenuation can be increased toward 130 dB.

The first defense is the transmit spectral mask, where 802.11 requires at least 40 dB of suppression by the alternate adjacent channel. Most implementations achieve much greater suppression of the skirts of a 2.4 GHz signal at farther carrier frequencies, such as the 5 GHz band, and vice versa. Beyond this mask, other suppression techniques are applied: separate antennas with reasonable distance separation or polarization separation, frequency-dependent antenna inefficiency, and/or RF filters. However, these other techniques are especially challenging for small-form-factor and low-cost clients. For this reason, traditionally concurrent operation has been mostly confined to APs.

It is because of these constraints that the 80 + 80 MHz mode defined in Wi-Fi 5 and Wi-Fi 6 (yet little used) was tightly constrained: Either the STA (AP or non-AP) was transmitting on both 80 MHz frequency segments or was receiving on both 80 MHz frequency segments. In essence, the STA never had to receive on one frequency segment while transmitting on the other frequency segment. This made the 80 + 80 MHz mode implementable, but did limit its benefits in congested channels. That is, any activity on the primary 80 MHz segment prevented both primary 80 MHz and 80 + 80 MHz transmissions.

The multi-link operation feature takes a new and more valuable approach: It allows for independent contention and potentially independent transmission on each radio.

Before proceeding further, some specific terms need to be defined. Recall that some AP devices (i.e., the physical boxes that implement the IEEE APs) can include two radios on the same band—for example, channel 1 at 2.4 GHz, channel 36 at 5 GHz, and a third radio at channel 149 at 5 GHz. Furthermore, many AP devices support multiple BSSs on the same radio. Accordingly, the terminology needs to account for these extended implementations and there is a need for a narrower term than "band" or "radio." The term chosen was "link" and, in turn, "multi-link device" (MLD).

For illustrative purposes, consider an AP MLD with four affiliated IEEE APs, operating on 2.4, 5 GHz low, 5 GHz high, and 6 GHz, and a non-AP MLD with four affiliated non-AP STAs. When the non-AP

MLD associates to the AP MLD, the non-AP MLD can set up one, two, three, or (in theory) even four links with the AP MLD, where each link connects one affiliated STA of the non-AP MLD with one affiliated AP of the AP MLD.

There are multiple allowed client implementations of multi-link operation, as shown in Figure 5-8.

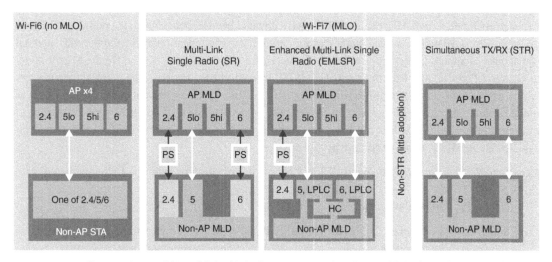

FIGURE 5-8   Comparison of Non-Multi-Link Operation and Different Multi-Link Operation Modes (LPLC = low power and low capability, HC = high capability)

The most ideal (and conceptually simple) non-AP MLD implementation is simultaneous transmit/receive (STR). All AP MLD implementations are required to be STR. In STR, there is sufficient isolation between the links that they operate essentially independently; any link can be transmitting or receiving at a given time while other links are also transmitting or receiving at that same time. Thus, an STR client requires enough baseband processing and transceivers so that all of its radios can operate at the same time. Such high-end clients are expected to represent a smaller proportion of the overall client market. A wireless deployment of AP devices that use wireless mesh for backhaul is another likely scenario for this technology.

The benefits of multi-link operation with STR clients include the following outcomes:

- Higher peak throughput, from aggregating the throughput achievable via each link.

- Lower latency in the presence of congestion, because data can be sent over the first link that is clear and where backoff has completed.

- Higher reliability in the presence of congestion, because retries can be sent on other channels with reduced correlation (and oftentimes less congestion).

- Great client flexibility to accommodate in-device coexistence, such as co-located Bluetooth or cellular radios. For instance, while a Bluetooth radio is active at 2.4 GHz, Wi-Fi communications can occur without interruption at 5 or 6 GHz.

At the other extreme, especially for highly constrained devices such as in the IoT market, no additional client hardware can be tolerated. Given these devices can tune to 5 GHz (and even 6 GHz), they can associate to an AP MLD and set up links with many or all of the affiliated APs of the AP MLD. These clients are called multi-link single radio (MLSR). Critically, a non-AP MLD in MLSR mode reports itself as in power save mode to all but one affiliated AP, at most. At any given time, the client performs all communications through a single affiliated AP using its one baseband and with all its transceivers tuned to that AP's channel. If the AP's channel becomes congested, the non-AP MLD can go into power save mode on that channel, and then initiate communications with another affiliated AP (on another channel). Essentially only a single successful transmission is needed for the MLSR client to change APs (affiliated with the same AP MLD), instead of the previous scheme of authentication, secure association, and negotiation of agreements (e.g., Block Ack agreements). Because APs never initiate communications with a client in power save mode, the other affiliated APs never attempt to transmit to the client while it is not listening. When traffic and congestion are light, STR clients can also operate in MLSR mode to achieve power savings.

Relative to Wi-Fi 5 and Wi-Fi 6 clients, Wi-Fi 7 clients in MLSR mode offer the following benefits:

- No change in peak throughput

- Somewhat lower latency and higher reliability in the presence of congestion, because there is a lower cost for the client to change to a lower congestion channel (and then congestion avoidance can occur more frequently)

- Great client flexibility to accommodate in-device coexistence, such as co-located Bluetooth or cellular radios

In the middle of these extremes, three other modes are defined[45] that also avoid the need for antenna separation or RF filtering:

- **Non-STR (NSTR):** This mode is introduced merely to note that it is a somewhat complicated mode and is expected to see little adoption.

- **Enhanced MLMR (EMLMR) mode:** A non-AP MLD in EMLMR mode on N links can be implemented as N full-capability radios (each supporting one or more spatial streams), where the transceiver resources can be fluidly moved from one radio to another after a short delay. This mode is not Wi-Fi certified.

- **Enhanced MLSR (EMLSR) mode:** A non-AP MLD in EMLSR mode on N links can be implemented as N low-power, low-capability radios plus a single high-capability radio. In some implementations, the single high-capability radio can take the place of one low-power radio. The low-power radios need to be capable of supporting only one spatial stream, 20 MHz, and low MCS (e.g., up to 16-QAM) operation.

---

45 See Chapter 6 for more details on multi-link operation modes and their operations.

For client-initiated TXOPs in EMLSR operation, the low-power radios are responsible for monitoring physical carrier sense (i.e., clear channel assessment) and virtual carrier sense by decoding (MU)RTS and CTS and related frames. The non-AP MLD's affiliated MACs can then perform the backoff procedure on each link in parallel. Then, when the non-AP MLD is about to be able to start a TXOP on one link, the high-capability radio can perform CCA across the entire PPDU bandwidth. If it is clear, it can begin transmission using multiple spatial streams, a wide bandwidth, and/or a high MCS.

For AP-initiated TXOPs in EMLSR operation, one of the affiliated APs must first warn the EMLSR non-AP MLD that the client's high-capability transmitter or receiver is required by sending the non-AP MLD an initial control frame (ICF) padded to a minimum length. Upon receiving the ICF, the non-AP MLD in EMLSR mode transmits a control response frame via the low-complexity radio and uses this time to wake up the non-AP MLD's high-capability transmitter (if triggered) or receiver (otherwise) and tune it to the affiliated AP's channel. At the end of the TXOP on one channel, after a timeout, the high-capability radio is again available for operation on any of the N links.

Relative to Wi-Fi 5 and Wi-Fi 6 clients, Wi-Fi 7 clients in EMLSR mode achieve the following benefits:

- No change in peak throughput

- Lower latency and higher reliability in the presence of congestion, in much the same way as an STR client

- Better client power savings, because only a client's low-power receivers need to be powered on

- Great client flexibility to accommodate in-device coexistence

Table 5-2 summarizes these main multi-link operation modes.

TABLE 5-2   Multi-Link Operation Opportunities Given Different Client Hardware Capabilities

Client HW Resources	AP HW Resources	Multi-Link Operation
Single non-AP STA with multi-band tuner	Simultaneous multi-link	MLSR
Single non-AP STA with multi-band tuner plus at least one low-power, low-complexity radio		EMLSR, MLSR
Multiple non-AP STAs		STR, EMLSR, MLSR

In the large proportion of the world where 6 GHz spectrum is allocated for Wi-Fi, a typical AP device or STR client supports 2.4 + 5 GHz + 6 GHz, with RF filters required between the 5 and 6 GHz bands. Unfortunately, China does not currently allocate any 6 GHz spectrum for Wi-Fi, so natural AP device designs in that country are 2.4 + 5 GHz or 2.4 + low 5 GHz + high 5 GHz. STR typically requires RF filters between the low 5 GHz and high 5 GHz sub-bands.

Large-Size MRUs and Preamble Puncturing

Wider channel bandwidths and 6 GHz spectrum in the United States, Canada, and multiple other regulatory domains are challenged by one issue: other wireless systems. The large-size MRU and preamble puncturing features are designed to mitigate this issue.

As described in Chapter 3, "Building on the Wi-Fi 6 Revolution," use of higher power in the UNII-5 and UNII-7 sub-bands of 6 GHz is authorized by an automated frequency control system that is designed to protect the operation of primary services (i.e., incumbents). Their signals are typically narrowband (e.g., 10 MHz wide), so oftentimes the AFC provides high-power access to a "notched" spectrum.

Similarly, other unlicensed systems such as New Radio Unlicensed (NRU) cellular can raise similar challenges as the incumbents. If these systems transmit for more than, say, 4–8 ms at a time, or if they transmit before checking that no other device is already transmitting, then their coexistence behavior is poor. Even a system with good coexistence but a high offered load and, therefore a high-duty cycle, may create complications. If such systems are relatively narrowband (e.g., less than 40–80 MHz), then in many scenarios it may be better for an AP MLD to simply avoid the spectrum on which they generate interference, by transmitting below, above, or on both sides of them. Such behavior is enabled by PPDU puncturing. 802.11ax defined certain preamble puncturing modes for use with OFDMA, but EHT goes well beyond that:

- Along with preamble puncturing for OFDMA, EHT provides for single-user and non-OFDMA MU-MIMO preamble puncturing and multi-RUs with certain edge and interior spectral gaps.

- EHT defines sounding for punctured channels.

These modes are defined because Wi-Fi 7 BSSs, operating at a wider bandwidth including 320 MHz, are more likely to encounter a scenario in which the best available spectrum overlaps an incumbent or a poorly coexisting system or a high-duty cycle system.

Recall that a PPDU can be split into two parts: the early part of the preamble, which comprises, almost exclusively, duplicated 20 MHz signals (the pre-HE/EHT modulated fields), and then the HE/EHT STF, HE/EHT LTF, Data field, and PE (the HE/EHT modulated fields). Providing EHT PPDUs with the ability to transmit around other wireless systems must address both PPDU parts, as well as the problem of sounding feedback. The solution is *puncturing*:

1. For the pre-EHT modulated fields,[46] it is most natural, and also relatively straightforward, to puncture transmissions on one or more individual 20 MHz subchannels. This technique works naturally for the LSTF, LTF, LSIG, RLSIG, and USIG fields. It has several benefits, not least that third-party STAs are still able to receive the LSIG field and defer for the duration of the PPDU. As a corollary, such puncturing does not make sense for a 20 MHz PPDU (else nothing would be sent), or for a 40 MHz PPDU (else it would be a 20 MHz PPDU or a PPDU not overlapping the primary channel). Thus, puncturing is limited to 80 MHz and wider PPDUs.

---

46 See P802.11be D5.0, clause 36.3.12.11.

2. For the EHTSIG field, there are two content channels on alternate 20 MHz channels within each 80 MHz frequency sub-block: one content channel on the odd-numbered 20 MHz sub-channels and one content channel on the even-numbered 20 MHz subchannels.[47] Consequently, it is important *not* to puncture transmission on both even-numbered or both odd-numbered sub-channels, unless there are no users at all on the entire 80 MHz frequency sub-block. Table 5-3 lists the preamble puncturing options for each 80 MHz frequency sub-block.

TABLE 5-3   Preamble Puncturing Options per 80 MHz Frequency Sub-block[48]

Puncturing Signaled	Preamble Puncturing Options
1111	No puncturing
0111, 1011, 1101, 1110	20 MHz out of 80 MHz punctured
0011, 1001, 1100	40 MHz out of 80 MHz punctured (with one odd-numbered and one even-numbered subchannel unpunctured)

3. For the EHT-modulated fields, it may be desirable for one user to use the entirety of the un-punctured channel, or at least a large swathe of it. That is, instead of using OFDMA and having some users transmit below the punctured spectrum and other users transmit above the punctured spectrum (which is all that Wi-Fi 6 offered), the 802.11be designers chose to join multiple RUs together, via extensions to the BCC interleaving block, the LDPC coded-bit calculations, and the segment parsing block. When formed as multiples of 20 MHz, these multi-RUs are called *large-size* multi-RUs. They enable transmission across 80, 160, or 320 MHz bandwidth channels *except* for one or two punctured portions of the bandwidth. The available options are rich yet somewhat more limited than the preamble puncturing options, as you can see in Table 5-4 (for non-OFDMA PPDUs) and Table 5-5 (for OFDMA PPDUs). In non-OFDMA PPDUs, one multi-RU is used per PPDU. In OFDMA PPDUs, the AP can mix and match between RUs and/or MRUs, and the punctured spectrum is achieved by selecting the RUs/MRUs such that no user is assigned an RU/MRU that overlaps the punctured spectrum.

TABLE 5-4   MRU Options in a Non-OFDMA PPDU[49]

PPDU Bandwidth (MHz)	MRU Options	Description of Puncturing
20	None	Full 20 MHz
40	None	Full 40 MHz
80	242-tone + 484-tone	60 MHz: any single 20 MHz subchannel punctured out of 80 MHz
160	484-tone + 996-tone	120 MHz: any single 40 MHz, on a 40 MHz boundary, punctured out of 160 MHz
	242-tone + 484-tone + 996-tone	140 MHz: any single 20 MHz subchannel punctured out of 160 MHz

47 See P802.11be D5.0, clause 36.3.12.8.6.
48 See P802.11be D5.0, clause 36.3.12.7.2.
49 See P802.11be D5.0, clause 36.3.2.2.3.1.

PPDU Bandwidth (MHz)	MRU Options	Description of Puncturing
320	484-tone + 2 × 996-tone	200 MHz: any single 80 MHz, on an 80 MHz boundary, plus any single 40 MHz, on a 40 MHz boundary, punctured out of 320 MHz
	3 × 996-tone	240 MHz: any single 80 MHz, on an 80 MHz boundary, punctured out of 320 MHz
	484-tone + 3 × 996-tone	280 MHz: any single 40 MHz, on a 40 MHz boundary, punctured out of 320 MHz

TABLE 5-5   MRU Options in an OFDMA PPDU[50]

Portions of the PPDU Bandwidth	MRU Options	Description of Puncturing
80 MHz aligned with the 80 MHz boundaries of an 80/160/320 MHz PPDU	242-tone + 484-tone	60 MHz: any single 20 MHz subchannel punctured out of the 80 MHz
160 MHz aligned with the 160 MHz boundaries of a 160/320 MHz PPDU	484-tone + 996-tone	120 MHz: any single 40 MHz, on a 40 MHz boundary, punctured out of the 160 MHz
320 MHz of a 320 MHz PPDU	484-tone + 2 × 996-tone	200 MHz: any single 80 MHz, on a 80 MHz boundary, plus any single 40 MHz, on a 40 MHz boundary, punctured out of the 320 MHz
	3 × 996-tone	240 MHz: any single 80 MHz, on an 80 MHz boundary, punctured out of the 320 MHz
	484-tone + 3 × 996-tone	280 MHz: any single 40 MHz, on a 40 MHz boundary, punctured out of the 320 MHz

Figures 5-9 through Figure 5-12[51] show examples of these large-size RUs and multi-RUs.

The sounding procedure leverages these capabilities:

- The NDP PPDU used for sounding can be punctured.[52]

- The feedback of the sounded channel can be limited to the punctured bandwidths shown in Table 5-4 and Figures 5-9 to 5-12.[53]

The large-size MRU and preamble puncturing technologies enable the preamble puncturing feature (also see the "Preamble Puncturing" section in Chapter 8). In addition, because the channelization defines overlapping 320-1 MHz and 320-2 MHz channel widths, it is *theoretically* possible to use puncturing to achieve a contiguous 240/200 MHz channel plan (Figure 5-13), although STAs are expected to have sub-par immunity to adjacent channel interference. This channelization is unique to 320 MHz,

50 See P802.11be D5.0, clause 36.3.2.2.3.2.
51 See P802.11be D5.0, clauses 36.3.2.2.3.1 and 36.3.2.2.3.2.
52 See P802.11be D5.0, clause 9.3.1.19.4.
53 See P802.11be D5.0, clauses 9.4.1.66 and 9.3.1.19.4.

however, and there is no ability to create a contiguous 60 or 120 MHz channel plan, as neither over-lapping 160-1 MHz and 160-2 nor overlapping 80-1 and 80-2 MHz channel widths are defined.

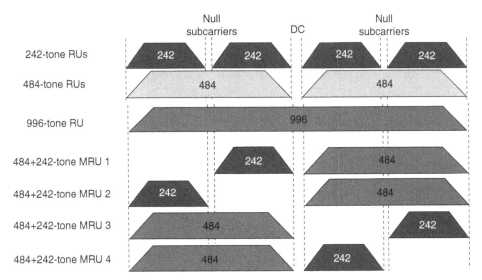

FIGURE 5-9 Allowed Large-Size RUs and MRUs in a Non-OFDMA or OFDMA 80 MHz EHT PPDU. Here, 242, 484, and 996 tones correspond to 20, 40, and 80 MHz, respectively, so 242 + 484 tones span 20 + 40 = 60 MHz.

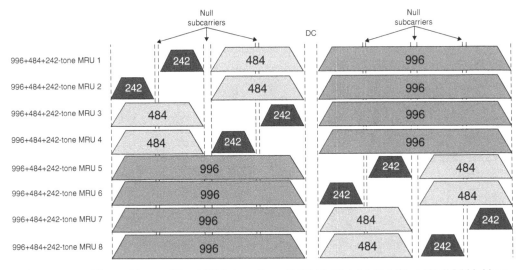

FIGURE 5-10 Allowed Large-Size MRUs in a Non-OFDMA (Only) 160 MHz EHT PPDU. Here, 242 + 484 + 996 tones span 20 + 40 + 80 = 140 MHz.

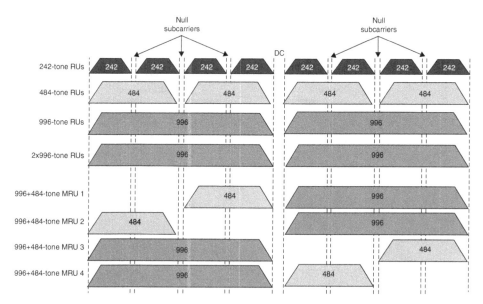

FIGURE 5-11 Allowed Large-Size RUs and MRUs in a Non-OFDMA or OFDMA 160 MHz EHT PPDU. Here, 484 + 996 tones span 40 + 80 = 120 MHz.

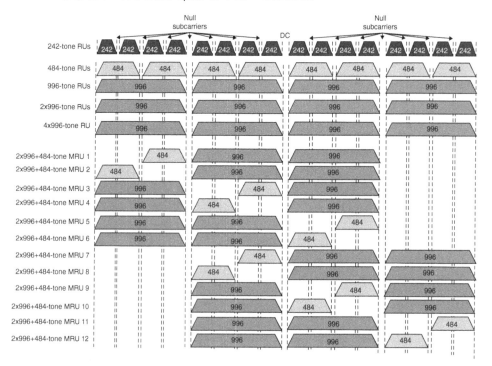

FIGURE 5-12 Allowed Large-Size RUs and MRUs in a Non-OFDMA or OFDMA 320 MHz EHT PPDU. Here, 484 + 2 × 996 tones span 40 + 2 × 80 = 200 MHz, 3 × 996 tones span 40 + 2 × 80 = 240 MHz, and 484 + 3 × 996 tones span 40 + 3 × 80 = 280 MHz.

Small-Size Multi-RUs

When forming an OFDMA PPDU, the Wi-Fi 6 AP scheduler has essentially a power-of-2 choice: assign a user to 24, 48 (2×), 102 (2.1×), 234 (2.3×), 468 (2×), 980 (2.1×), or 1960 (2×) data tones. If one user had a little more data than would fit in an RU, there was a big jump to the next RU size. This left the following choices:

1. A complicated fragmentation scheme, which received little implementation attention and so did not provide for interoperability

2. Defer the sending of some MPDUs, when there were multiple MPDUs to send, in the AMPDU in this PPDU, with increased delay for those MPDUs

3. Use the wider RU and append desultory padding (EOF padding subframes) in the AMPDU containing the transmitted MPDU(s)

4. Drop down the MCS somewhat, which provides increased robustness but no extra data transfer

None of these choices enabled the Wi-Fi 6 AP scheduler to interoperably and efficiently deliver as much data as possible. This limitation was recognized by the Wi-Fi 7 designers, who introduced the multi-RU feature as a solution. It provides the scheduler with additional intermediate RU/MRU sizes to work with, so the scheduler can allocate a "just right"-sized RU/MRU for the data of each user. We have already seen that a variety of large-size RUs and MRUs are available that can be used in OFDMA PPDUs.

The Wi-Fi 7 designers also included small-size RUs, which bond 26-tone and 52-tone RUs, or 26-tone and 106-tone RUs, into an MRU.[54] Small-size MRUs use the same technology as large-size MRUs.

In this way, the Wi-Fi 7 scheduler for an OFDMA PPDU has a broad and dense range of RU/MRUs to choose from: Options include 24, 48, 72, 102, 126, 234, 468, 702, 980, 1448, 1960, 2428, 2940, 3408, 3920, and 3920 data tones for each user.

Figure 5-14 shows the available small-size 26 + 52-tone RUs, while Figure 5-15 shows the available small-size 26 + 106-tone RUs. An important side benefit of the design of these small-size MRUs is that each central 26-tone RU (the one at the middle of nine in each 20 MHz subchannel and the one that surrounds the DC tones of the 20 MHz subchannel) can be bonded to an adjacent and wider RU, such that FEC coding across all the MRU's tones can dilute any DC-related receiver-side impairments arising from the central 26-tone RU.

4096-QAM

Since the first OFDM Physical layer (802.11a), 802.11 has evolved from supporting up to 64-QAM (i.e., the constellation comprises an 8×8 grid of constellation points) to 256-QAM (a 16×16 grid) to 1024-QAM (a 32×32 grid) in Wi-Fi 6, and to 4096-QAM (a 64×64 grid) in Wi-Fi 7. This evolution is depicted in Figure 5-16.

---

54 See P802.11be D5.0, clause 36.3.2.2.2.

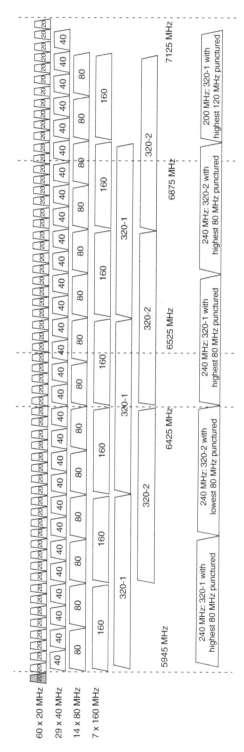

FIGURE 5-13   A Contiguous 240/200 MHz Channel Plan Using MRUs

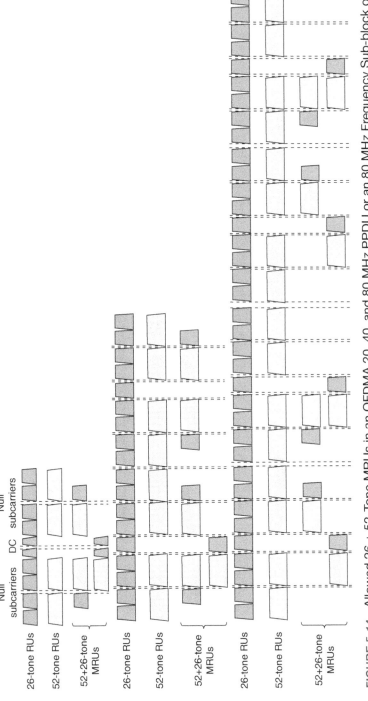

FIGURE 5-14 Allowed 26 + 52-Tone MRUs in an OFDMA 20, 40, and 80 MHz PPDU or an 80 MHz Frequency Sub-block of a 160/320 MHz PPDU, Plus the 26-Tone RUs for Context

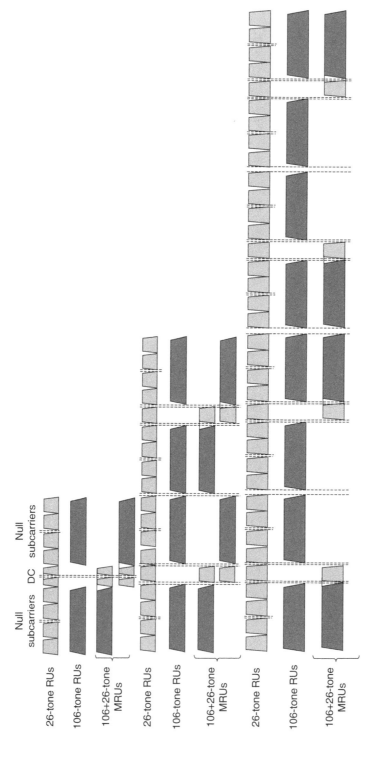

FIGURE 5-15 Allowed 26 + 106-Tone MRUs in an OFDMA 20, 40, and 80 MHz Frequency Sub-block of a 160/320 MHz PPDU, Plus the 26-Tone RUs for Context

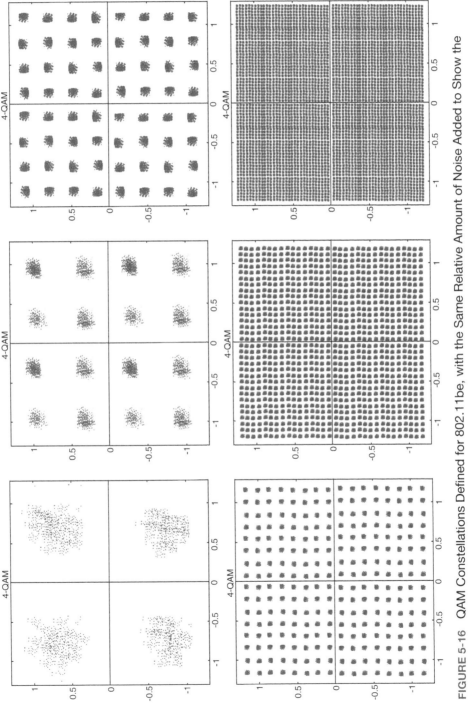

FIGURE 5-16   QAM Constellations Defined for 802.11be, with the Same Relative Amount of Noise Added to Show the Reduced Tolerance of the Denser Constellations to Noise

Two code rates are defined for 4K-QAM: rate ⅔ and ⅚. This leads to the numbers of data bits per (grouped) constellation point for the different MCSs shown in Table 5-6.

TABLE 5-6   Available MCSs in EHT[55]

MCS	Constellation and Code Rate	Data Bits per Grouped* Constellation Point(s)
0	BPSK-r1/2	0.5
1	QPSK-r1/2	1
2	QPSK-r3/4	1.5
3	16-QAM-r1/2	2
4	16-QAM-r3/4	3
5	64-QAM-r2/3	4
6	64-QAM-r3/4	4.5
7	64-QAM-r5/6	5
8	256-QAM-r3/4	6
9	256-QAM-r5/6	6.667
10	1024-QAM-r3/4	7.5
11	1024-QAM-r5/6	8.333
12	4096-QAM-r3/4	9
13	4096-QAM-r5/6	10
14 (-2)[#]	BPSK-r1/2 with DCM and duplication	0.119
15 (-1)[#]	BPSK-r1/2 with DCM	0.25

*DCM "groups" two constellation points; DCM with duplication "groups" four constellation points.

[#]MCS 14 and MCS 15 are longer-range modes and can be considered as MCS 2 and MCS 1, respectively; but they are not being certified by the Wi-Fi Alliance.

The number of data bits per constellation point are reasonably well spaced. This is a desirable property for rate selection. As a rule of thumb especially at higher constellation densities, 1 extra data bit requires a 3 dB better SINR, so the rate selection algorithm always has a rate available that achieves close to the maximum data rate.

Friis Equation

The Friis equation allows us to determine received signal level in free space conditions. The contemporary rendition of the Friis equation is

$$\text{Received Power [dBm]} = \text{Transmitted Power [dBm]} + \text{Transmitter Antenna Gain [dBi]} + \text{Receiver Antenna Gain [dBi]} + 20 \times \log10(c/(4\pi f_c d))$$

where c is the speed of light in units of meters per second, f_c is the carrier frequency in units of Hz, and d is the distance in units of meters.

---

55 See P802.11be D5.0, clauses 36.3.12.8.5 and 36.5.

For MCS 13, the standard requires that STAs achieve a 10% PER at a received signal level of between –46 dBm (20 MHz) and –34 dBm (320 MHz).[56] Because the highest MCSs require a very clean signal, conducted power is typically lowered compared to other MCSs. So, we assume 10 to 18 dBm conducted power, 4 dBi antenna gain at the AP, and 0 dBi antenna gain at the client, where the Friis equation reports 48.7 dB distance-based attenuation at 1 meter (for the mid 6 GHz band). Then, the pathloss budget for 20 MHz to 320 MHz is from 19.3 to –0.7 dB. Having 2 or 4 times more receiving antennas than transmitted spatial streams (e.g., 2 or 1 spatial streams received by 4 antennas, respectively) provides a 3 dB or 6 dB link budget improvement, respectively, and is often vital to achieve a useful range with MCS 13.

Assuming free space propagation, which is reasonable given the likelihood of line-of-sight propagation at these distances, the nominal range is as short as 0.9 meter (for 10 dBm, 320 MHz, 1 spatial stream, and 1 receiving antenna), or 2.3 meters (now with 18 dBm), or 4.6 meters (now with both 18 dBm and 4 receiving antennas), or even 9.2 meters (now with 18 dBm, 4 receiving antennas, and a 6 dB better noise figure than is presumed in the IEEE sensitivity requirements). Clearly, better implementations can convert MCS 13 from barely usable to usable throughout most of a 300 m^2 BSS.

However, the IEEE-defined –34 dBm implies that a very high SINR is available and there is no traffic at nearby BSSs. A moderate 18 dBm EIRP 20 MHz interferer must be at least 98 meters away before its interference drops below –90 dBm and is reasonably negligible, assuming free space pathloss for the first 5 meters, then a pathloss exponent of 3.5 thereafter. Accordingly, MCS 13 is unlikely to be selected for wider bandwidths, either because of a lack of RSSI or because the MCS was attempted but failed owing to a collision with a neighboring wireless system.

Minor Wi-Fi 7 PHY Innovations

In this section, we describe additional changes made to the 802.11be PHY to support the major Wi-Fi PHY innovations described in the previous section. These changes include sounding and beamforming (upgraded for 320 MHz, preamble puncturing, and OFDMA), extra EHT-LTFs (for enhanced receiver performance), and the packet extension (extended to provide more receiver processing time for 320 MHz, MCS 12, and MCS 13).

Sounding and Beamforming

Explicit sounding and beamforming were successfully adopted with Wi-Fi 5. This feature is valuable for increasing the data rate at a given range, and accordingly has continued to be updated in subsequent Wi-Fi generations. In terms of the PHY activities, explicit sounding is a three-step process:

1. A beamformer with multiple antennas (e.g., 4 or 8), such as an AP, transmits a special null data packet (i.e., PPDU) so that the recipient beamformee can measure the full wireless channel, including impairments from all the beamformer's antennas (N_{TX}) to all the beamformee's

---

56 See P802.11be D5.0, clause 36.3.21.2.

antennas (N_{RX}), at every nth subcarrier (where n equals 4 or 16 for EHT). The measurements are 3D—that is, a linear array (over subcarriers) of $N_{RX} \times N_{TX}$ matrices of complex numbers—and are called channel state information (CSI).

2. The beamformee compresses the CSI. Compression is a computationally intensive process. It nominally involves a singular value decomposition for each subcarrier's $N_{RX} \times N_{TX}$ CSI matrix to obtain the singular values and the right singular vectors (i.e., the V matrix). The singular values are converted to quantized channel quality indicators (CQI). Given rotations are applied to the V matrix to extract so-called phi and psi angles. Phi and psi are quantized, for compression, and different quality levels (bitwidths) are defined.

3. The quantized phi and psi values and/or the CQI values are sent back to the beamformer.

Finally, the beamformer uses the explicit sounding feedback to calculate beamforming matrices (and the CQI feedback is used to help select the MCS) for subsequent transmission to the beamformee.

Wi-Fi 7 extends support for beamforming, via the following mechanisms:

■ Sounding for 320 MHz bandwidths.

■ Sounding for partial bandwidths for (1) any of the RUs that are at least as wide as 20 MHz and (2) any of the large-size MRUs—that is, at a 20 MHz resolution. Partial bandwidth sounding is well matched to the needs of the preamble puncturing feature and can considerably lower the overheads of beamforming for OFDMA.

■ Downlink beamforming for OFDMA, as RUs subject to beamforming are so indicated in the corresponding User fields in the EHT-SIG field.

Extra EHT-LTFs

MIMO equalization at the receiver requires an estimate of the wireless channel (CSI) between transmitter and receiver. For the Data field, the estimate is made (at least initially) from the EHT-LTF field. There is additive noise on the OFDM symbols during the Data field but also on the EHT-LTF OFDM symbol. In a basic implementation for a one spatial stream PPDU, this adds an extra 3 dB of noise. Better implementations take advantage of the correlation of the CSI in the frequency domain to perform frequency-domain smoothing, in an effort to reduce the noise on the estimated CSI and lower its impact to well below 3 dB.

However, such channel smoothing is harder with beamforming (since the channel may be less smooth). It is also more difficult in expansive environments (e.g., outdoors, distribution centers, and hangars), where the multipath delay spread can be relatively long such that the CSI's frequency domain correlation is much lower. Worse, some forms of MIMO equalization—especially Zero Forcing (ZF), but to a lesser extent, Minimum Mean Square Error (MMSE)—can amplify noise from the CSI estimate.

For all these reasons, it is desirable to provide a mechanism that offers improved channel estimation. The feature defined in Wi-Fi 7 to meet this need is called extra LTFs. Instead of transmitting essentially the same number of EHT-LTF OFDM symbols as spatial streams, the extra LTF feature enables the transmitter to send more—for up to a maximum of 8 EHT-LTF OFDM symbols. The receiver can use the extra symbols for the following purposes:

- Extra averaging to improve the CSI estimate (akin to the processing of the LLTF field).

- Improved estimation of the signal to interference plus noise ratio (SINR) of the trained subcarriers. This improved SINR estimate can be used for better MIMO equalization and decoding, and can provide increased robustness to narrowband interferers.

- Spatial suppression of ongoing interference, in the most advanced implementations.

Packet Extension

The Packet (i.e., PPDU) Extension (PE) field is a necessary evil. The PE field carries no data, but provides the receiver with extra time, beyond SIFS, to complete processing of the end of the PPDU. The PE field is a raw waveform that is undefined except for the constraint that its power spectral density (PSD) must closely track the PSD of the Data field so that other wireless devices continue to perceive that the wireless medium is busy.

Because EHT introduces 320 MHz and MCS 12 and MCS 13, the receiver needs extra time to process the increased data load at the end of the PPDU. Accordingly, EHT newly defines a nominal packet padding of 20 μs, in addition to the existing values of 0, 8, and 16 μs.

Recall that the final OFDM symbol in the Data field has the usual length but is considered as four quarters (four symbol segments). One, two, three, or four of the symbol segments carry coded data, and the remaining symbol segments are filled with undefined post-FEC padding bits. A final OFDM symbol with just one quarter of the usual number of coded bits still takes the same time for the FFT and for MIMO equalization, but the time taken for decoding should be reduced by approximately a factor of 4. More broadly, given that the OFDM symbol duration is 16 μs or nearly so, each symbol segment containing post-FEC padding bits nominally reduces the required decoding time by 4 μs.

This characteristic allows the nominal packet padding to be achieved via a mix of post-FEC padding bits and the PE field. Considering a required nominal packet padding of 20 μs, if the final OFDM symbol carries one, two, three, or four symbol segments of coded data, then the PE field's actual duration is 8, 12, 16, or 20 μs, respectively.

PHY Capability Signaling

APs and clients declare their PHY capabilities as part of the EHT Capabilities element in several management frames. Figure 5-17 lists these PHY capability fields, organized into themes.

Most capabilities bits are Booleans (is the feature supported or not), and their meaning is self-explanatory. A few capability fields merit further explanation, as follows, where [letter] refers to the tag added at the end of the field or theme name in Figure 5-17:

- **[A]:** This field equals the maximum number of spatial streams sent by a single-user (SU) beamformer in an EHT sounding NDP, minus 1.

- **[B]:** Sounding feedback can be compressed via subcarrier grouping, described by the parameter Ng. Beamformees must support light compression to every Ng = 4th subcarrier and may support heavy compression to every Ng = 16th subcarrier (optional, with support indicated here).

- **[C]:** The phi and psi angles in the sounding feedback can be lightly compressed (mandatory) or heavily compressed (optional, with support indicated here).

- **[D]:** If we describe the compressed beamforming feedback matrix at a subcarrier as an $N_r \times N_c$ matrix, where N_r is the number of beamformer antennas and N_c is the number of spatial streams, then the Max N_c field is set to the maximum N_c supported by the beamformee for sounding feedback minus 1.

- **[E]:** Calculating the compressed feedback is computationally intensive, and some implementations may not be able to calculate the feedback and populate the feedback frame as fast as the 802.11be data rate. This field, when set to true, indicates that the beamformee cannot send compressed feedback at faster than 1500 Mbps.

- **[F]:** This field equals the maximum number of spatial streams that a STA can receive in an EHT sounding NDP or in a DL MU-MIMO transmission.

- **[G]:** This field enables a non-AP STA to indicate whether it has the same, or lowered, capability to transmit and/or receive MCS 10–13 on RU/MRUs with fewer than 242 tones.

- **[H]:** The expression "non-OFDMA UL-MU-MIMO" is a precise term for what is often just called UL-MU-MIMO. There is one RU/MRU in the PPDU (i.e., no OFDMA).

- **[I]:** For power savings, some clients might be currently operating in a 20 MHz mode, where receiving that portion of a wider PPDU can be challenging, and so clients may opt into supporting the related feature (or not). Closely related, certain low-capability STAs might not support a bandwidth wider than 20 MHz, and can opt into advanced features like sounding feedback, DL MU-MIMO, and MRUs.

- **[J]:** A STA's required nominal packet padding can be a complicated function of bandwidth, MCS, and number of spatial streams. It is expressed via the optional EHT PPE Thresholds field, or the nominal packet padding can be set to a single number across all these PPDU parameters. These fields allow the nominal packet padding to be signaled as a single value.

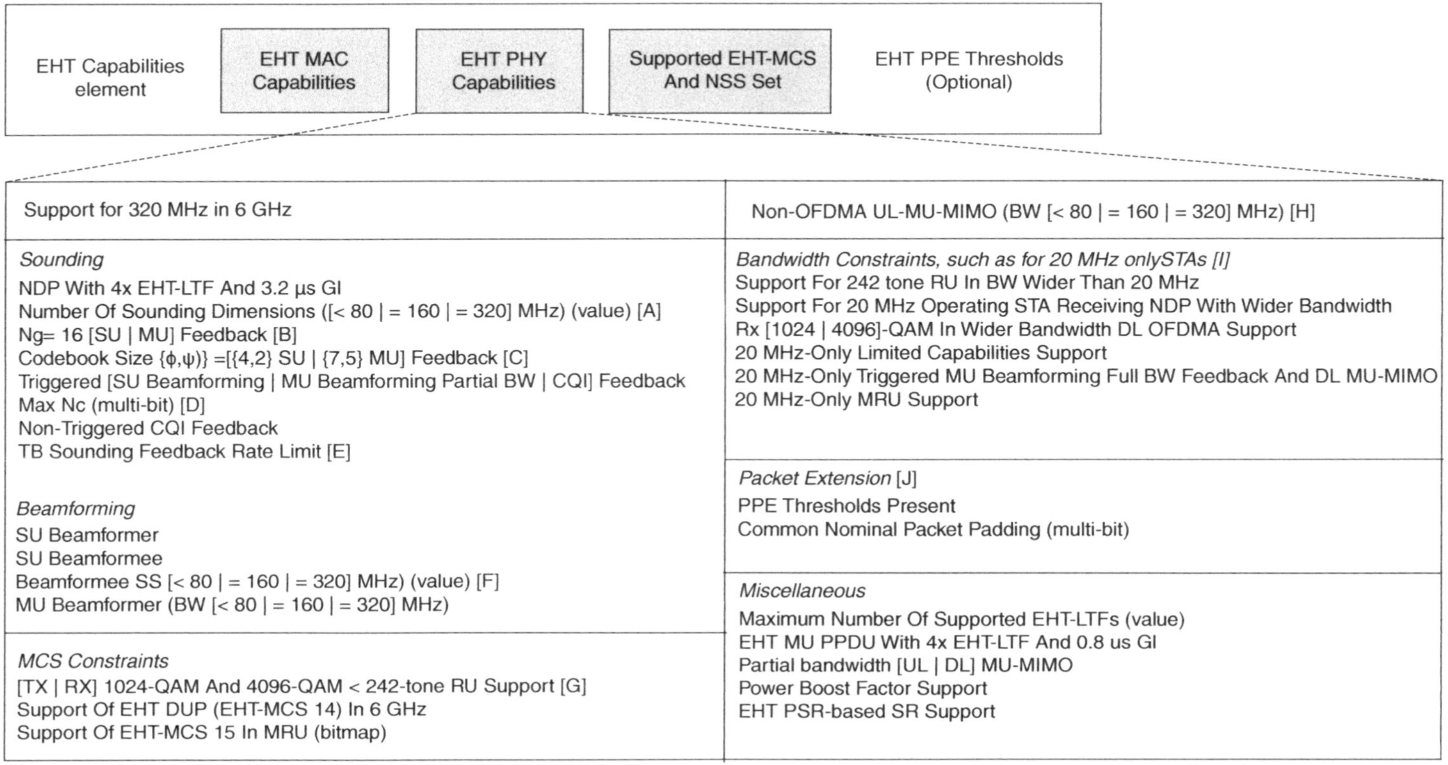

EHT Capabilities element	EHT MAC Capabilities	EHT PHY Capabilities	Supported EHT-MCS And NSS Set	EHT PPE Thresholds (Optional)

Support for 320 MHz in 6 GHz	Non-OFDMA UL-MU-MIMO (BW [< 80 | = 160 | = 320] MHz) [H]
Sounding NDP With 4x EHT-LTF And 3.2 μs GI Number Of Sounding Dimensions ([< 80 | = 160 | = 320] MHz) (value) [A] Ng= 16 [SU | MU] Feedback [B] Codebook Size {φ,ψ} =[{4,2} SU | {7,5} MU] Feedback [C] Triggered [SU Beamforming | MU Beamforming Partial BW | CQI] Feedback Max Nc (multi-bit) [D] Non-Triggered CQI Feedback TB Sounding Feedback Rate Limit [E]	*Bandwidth Constraints, such as for 20 MHz onlySTAs [I]* Support For 242 tone RU In BW Wider Than 20 MHz Support For 20 MHz Operating STA Receiving NDP With Wider Bandwidth Rx [1024 | 4096]-QAM In Wider Bandwidth DL OFDMA Support 20 MHz-Only Limited Capabilities Support 20 MHz-Only Triggered MU Beamforming Full BW Feedback And DL MU-MIMO 20 MHz-Only MRU Support
	Packet Extension [J] PPE Thresholds Present Common Nominal Packet Padding (multi-bit)
Beamforming SU Beamformer SU Beamformee Beamformee SS [< 80 | = 160 | = 320] MHz) (value) [F] MU Beamformer (BW [< 80 | = 160 | = 320] MHz)	*Miscellaneous* Maximum Number Of Supported EHT-LTFs (value) EHT MU PPDU With 4x EHT-LTF And 0.8 us GI Partial bandwidth [UL | DL] MU-MIMO Power Boost Factor Support EHT PSR-based SR Support
MCS Constraints [TX | RX] 1024-QAM And 4096-QAM < 242-tone RU Support [G] Support Of EHT DUP (EHT-MCS 14) In 6 GHz Support Of EHT-MCS 15 In MRU (bitmap)	

FIGURE 5-17   PHY Capabilities Signaled Within the EHT Capabilities Element

Summary

In this chapter, you have learned about the major revolutionary and evolutionary features of the 802.11be PHY layer. You have seen MU PPDU format and the TB PPDU format, and the fields therein. This chapter also traced through the major processing blocks used when generating the Data field of EHT PPDUs. Finally, you have reviewed the major PHY innovations in Wi-Fi 7:

- 320 MHz bandwidths for higher data rates
- Multi-link operation for increased throughput, reduced latency, increased reliability, and greater client flexibility to accommodate in-device coexistence, according to the level of MLO supported by the client
- Large-size multi-resource units and preamble puncturing for transmitting around other wireless systems
- Small-size multi-resource units for improved OFDMA scheduling efficiency
- 4096-QAM for higher data rates at short ranges

The interested reader is encouraged to obtain the latest drafts produced by the 802.11be (and 802.11-REVme) task groups to understand the fine-grained details of the 802.11be PHY.

Chapter 6

EHT MAC Enhancements for Multi-Link Operation

multi-link device: [MLD] A logical entity that is capable of supporting more than one affiliated station (STA) and can operate using one or more affiliated STAs, and that presents one medium access control (MAC) data service and a single MAC service access point (MAC SAP) to the logical link control (LLC) sublayer.

P802.11be, Draft 5.0, Clause 3

The 802.11be amendment, also called Extremely High Throughout (EHT), has introduced some fundamental new architecture enhancements and features at the Medium Access Control (MAC) layer. These include multi-link operation (MLO), SCS (stream classification service) with quality of service (QoS) characteristics, Restricted Target Wake Time (R-TWT), triggered TXOP sharing, larger Block Ack bitmap lengths (512 and 1024 bits in length), and EPCS (emergency preparedness communications service) priority access. Among these enhancements, the MLO is the most important feature added in 802.11be; it has led to a significant architecture redesign at the MAC layer with the introduction of the multi-link device (MLD) entity. MLO defines many new subfeatures and associated procedures between peer MLDs. Additionally, other features have been defined to enable dynamic MLO manageability. This chapter reviews key design aspects, functions, and procedures defined for MLO. Other EHT MAC layer enhancements are detailed in Chapter 7, "EHT MAC Operation and Key Features."

MLO Architecture

Before diving deep into the design aspects of specific features and procedures for MLO, it is useful to first highlight the motivation that led to the development of MLO (see Chapter 4, "The Main Ideas in 802.11be and Wi-Fi 7," for more details) and to explore the new MLD architecture defined in 802.11be. This section provides a brief overview of MAC layer evolution to support MLO and introduces the MLD architecture.

Evolution to MLO

AP devices typically support concurrent operation on multiple bands (e.g., 2.4, 5, and/or 6 GHz). Most STA devices (e.g., mobile phones, laptops) also typically have radios that support tuning and connecting to APs over any of these bands. Today, most devices support the 2.4 and 5 GHz bands, and 6 GHz support is becoming more common as well. Certain high-end devices could even support multiple radios, such as one for connectivity through the AP and another for peer-to-peer (P2P) connectivity. Such hardware support gives a STA the flexibility to connect to APs with different radio-band capabilities across different deployments and to select the best-performing radio for Wi-Fi connectivity.

Before MLO, an "AP device" would expose each of these radio bands as a separate AP. So, a typical Wi-Fi 6E AP device would host three separate APs—one on each of the 2.4, 5, and 6 GHz bands. The STA would pick one of these APs to associate, with the selection typically being based on signal strength (RSSI), congestion (such as the AP's advertised BSS load), and/or local device configuration (such as an internal scanning and preference algorithm developed by the device manufacturer). If the radio conditions deteriorated on that band, then the STA would scan to find another AP on the same or a different band (presumably providing better radio conditions) and establish connection with that AP. Given that the process of scanning and reassociation is disruptive to data connectivity, most clients exhibit sticky client behavior; that is, they stick to their current AP even when it is not providing the best Wi-Fi performance. For example, a STA would stick to 2.4 GHz band even when a high-performing 5 or 6 GHz band is available from the same AP device (see the "Client Roaming and Stickiness: Coverage Discontinuity" section in Chapter 4 for more explanation of the reasons for that stickiness). This behavior leads to increased congestion (on bands where most clients decide to stick) and suboptimal utilization of the network capacity, because other bands may not be utilized fully. The net result is sub-par Wi-Fi performance and undesirable end-user experiences.

With MLO, the 802.11be designers aimed to address these two key issues: (1) the need to associate again when STA is moving across radio bands (i.e., classic break-before-make behavior leading to connection disruption) and (2) the suboptimal use of the overall network bandwidth offered across multiple radio bands. The 802.11be amendment introduced the MLD architecture, which enables a STA to seamlessly and dynamically switch between radio bands offered by an AP device without needing to associate again, which facilitates best use of all the radio bands (according to STA capability) to achieve improved network utilization. This amendment enables a STA to simultaneously connect over multiple bands based on its capabilities and the multi-link (ML) modes supported (see the "MLO Modes" section later in this chapter). As described in Chapter 5, "EHT Physical Layer Enhancements," each of the radios (typically but not necessarily in different bands) offered by an AP device is referred to as a *link* in 802.11be—hence the term *multi-link device* (MLD).

The MLO feature allows seamless operation over multiple links between the AP and the client devices. It defines authentication, association, and establishment of multiple links between an AP MLD entity and a non-AP MLD entity (described in the next section), and enables simultaneous channel access and frame exchanges across multiple links (based on client device capability).

Multi-Link Devices

The new architectural entity defined for MLO is an MLD. An MLD is a logical entity that supports one or more 802.11 STAs, where each STA operates on a different link (either a different radio band or a different channel in the same band) within the MLD. These STAs, which are called affiliated STAs of an MLD, are co-located in the same physical device. The link corresponding to each affiliated STA is identified by a link ID. An MLD can be an AP MLD (in an AP device) with one or more affiliated APs, or a non-AP MLD (in a STA device) with one or more affiliated non-AP STAs. An AP MLD advertises links corresponding to each of its affiliated APs in the beacon and probe responses sent by its affiliated APs for enabling multi-link connectivity. All affiliated APs of an AP MLD are part of the same ESS and advertise the same SSID. In practice, an AP device might support multiple SSIDs, and then contain multiple virtualized AP MLDs.

A multi-link association is performed at the MLD level, where a non-AP MLD associates with an AP MLD, thereby establishing simultaneous association over all (or a subset) of the links offered by the AP MLD. Independent channel access is performed on each of the established links. The same security key material is used across all associated links between peer MLDs for exchange of individually addressed data frames. Each MLD link is assigned its own group keys for protected delivery of group-addressed frames. MLO-related security aspects are captured in the "MLO Security" section later in this chapter.

In an MLD (both the AP MLD and the non-AP MLD), the MAC layer functions are split across two sublayers[1]: an MLD Upper MAC sublayer, which provides common MLD upper MAC functions across all the links of the MLD, and an MLD Lower MAC sublayer, which provides link-specific MAC functions for each of the MLD links (as shown in Figure 6-1) for an MLD with two links (e.g., on 2.4 and 5 GHz). The MLD Upper MAC presents a single MAC data service that encompasses all links and exposes a single MAC Service Access Point (SAP), which is the MLD MAC SAP to the upper layer (the LLC sublayer of the Data Link layer as detailed in Chapter 1, "Wi-Fi Fundamentals"). The MLD upper and lower MAC functions are coordinated and managed by the STA Management Entity (SME) using MAC Layer Management Entity (MLME) primitives. The MLD Lower MAC for a given link interfaces with the PHY layer (via the PHY SAP) corresponding to that link to exchange A-MPDUs/ PSDUs over that link (see Chapter 5 for details of PHY layer functions).

Each affiliated STA (AP or a non-AP STA) in an MLD has its own STA MAC address assigned, denoted as the STA MAC addresses x and y in Figure 6-1. Another MLD MAC address is assigned to the MLD itself, denoted as the MLD MAC address m in Figure 6-1. A separate MLD MAC address is used to establish MLD-level authentication, association, and generation of MLD-specific security key material. An associated non-AP MLD is registered with the DS network using the non-AP MLD MAC address, independent of how many links are established between the non-AP MLD and the AP MLD. An AP MLD, with which a non-AP MLD associates, establishes at the Distribution System (DS) the mapping of that non-AP MLD to the AP MLD for the proper delivery of data frames targeted for that non-AP MLD.

---

1 See P802.11be D5.0, clause 4.9.6.

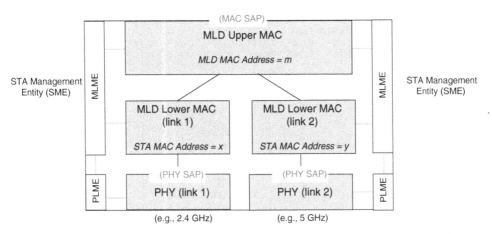

FIGURE 6-1   MLD MAC Architecture. SAP = Service Access Point, PLME = PHY Layer
Management Entity, MLME = MAC (Sub)Layer Management Entity.

> **Note**
>
> The MLD MAC address may differ from the affiliated STA MAC addresses. However, the
> architecture allows the MLD MAC address to be the same as one of the STA MAC addresses of
> that MLD.

On the AP MLD, the MLD Upper MAC exposes a single MAC SAP interface to the DS, as shown
in Figure 6-2. All the data frames destined for an associated non-AP MLD (identified by the non-AP
MLD MAC address) are delivered by the DS through the single (MLD) MAC SAP to that AP MLD.
The AP MLD then determines which of the associated links (one or more) it will use to deliver the data
frames to that non-AP MLD, and TID-To-Link mapping becomes relevant in this context (as will be
explained in the "Link Management" section later in this chapter).

In the AP MLD, the individual affiliated APs on each of the links continue to support association for
non-AP STAs that do not support MLO functions; such STAs are called non-MLD STAs. Each affiliated AP of an AP MLD also has a non-MLD Upper MAC sublayer to provide connectivity for such
non-MLD STAs, as shown in Figure 6-2. The non-MLD Upper MAC layer also provides link-specific
encryption/decryption of groupcast traffic and power save buffering of groupcast frames.

Each of the affiliated APs advertises a BSS that can provide connectivity for both the non-AP MLDs
and the legacy non-MLD STAs. Such architectural flexibility is important because most deployments
will have a mix of MLD and non-MLD client devices and the desired goal is to provide connectivity
across all such devices.

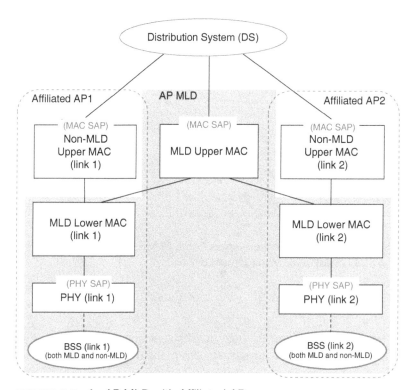

FIGURE 6-2   An AP MLD with Affiliated APs

A non-AP MLD with two affiliated STAs is shown in Figure 6-3. Each affiliated STA consists of an MLD Upper MAC layer, an MLD Lower MAC layer, and a PHY layer. The non-AP MLD architecture is designed such that either a client device can connect to an AP MLD and establish an MLD-level association to enable use of multiple links, or it can connect with a single AP (with a pre-802.11be, non-MLO, association) using one of the non-AP STAs over a specific link. For a non-MLO association, the MLD Upper MAC and MLD Lower MAC together provide MAC layer functionality and expose a MAC SAP to the upper layer. Such architectural flexibility is important for Wi-Fi 7 clients to be able to operate across both Wi-Fi 7 and other network deployments supporting Wi-Fi 6 and earlier generations of Wi-Fi.

The MLD Upper MAC sublayer provides MLD-level functions and services across the multiple links of the AP MLD. These include authentication, (re)association, management of security context (pairwise keys and group keys), and exchange of MLD-level management information (e.g., multi-link elements) via the MLD Lower MACs. On the transmission data path, the MLD Upper MAC provides encryption/decryption for individually addressed frames, selection of appropriate link(s) for (re)transmission of MPDUs, and, if needed, power-save buffering for MPDUs on the AP MLD. On the reception data path, the MLD Upper MAC supports merging of MPDUs across links and manages Block Ack scoreboarding across links for individually addressed frames. It also performs reordering and de-duplication of frames, then de-aggregation of A-MSDUs before MSDU delivery to the upper layer.

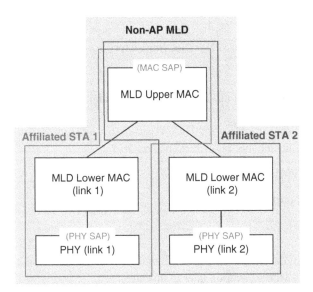

FIGURE 6-3   A Non-AP MLD with Affiliated STAs

The MLD Lower MAC sublayer provides link-specific functions including channel access (each link has its own EDCA parameters), link specific control frame exchanges (e.g., RTS/CTS, ACK, NDP) and per-link power-save management. On the transmission data path, the MLD Lower MAC sublayer enables delivery of MMPDUs (generated at the MLD Upper and Lower MAC sublayers) and MPDUs/A-MPDUs to the PHY layer as PSDUs. On the reception data path, the MLD Lower MAC filters frames targeted to the STA and manages Block Ack scoreboarding for individually addressed frames received over that link. Note that the overall Block Ack scoreboard management for Block Ack sessions established at the MLD level is done across the MLD Lower MAC and MLD Upper MAC sublayers, because MPDUs for a given Block Ack session can be delivered across multiple links. The "MLD Data Plane" section later in this chapter provides further details on the data frames delivery between peer MLDs.

MLO Functions Support

Support for the MLO feature is mandatory for both AP and client devices that aim to obtain the Wi-Fi Alliance's Wi-Fi 7 certification. The 802.11be amendment also mandates support for MLO for AP devices. For devices supporting the MLO feature, multiple MLO functions and related procedures are defined in 802.11be. In addition, the amendment specifies optional or mandatory support of these procedures for the AP and the STA side. The key MLO procedures and their mandatory/optional requirement in 802.11be amendment are as follows:

■ Multi-link discovery procedure (mandatory[2])

■ Multi-link (re-)setup procedure (mandatory)

2 A procedure indicated as mandatory without specific mention of an AP MLD or a non-AP MLD is mandatory for both the peer MLDs.

- Multi-link Block Ack support (mandatory)

- MLD-level sequence number (SN) and packet number (PN) space (mandatory)

- Multi-link power management (mandatory)

- Link management with default TID-To-Link mapping (TTLM) negotiation (mandatory)

- BSS parameter critical update procedure (mandatory)

- For an AP MLD, ML group-addressed frame delivery procedure (mandatory)

- For an AP MLD, serving a single radio non-AP MLD (mandatory)

- Multi-link reconfiguration (mandatory + optional[3])

- Link management (mandatory + optional)

These procedures are explained in more depth in this chapter. Requirements related to support for the different MLO modes are captured in the "MLO Modes" section later in this chapter.

MLD Discovery and Association

MLO involves the discovery of AP MLD(s) by the non-AP MLD and then establishing MLD-level authentication, association, and security context to allow exchange of data frames across multiple links. This section reviews these basic MLO functions.

MLD Discovery

Before a non-AP MLD can establish an association over multiple links with an AP MLD, it needs to discover the AP MLD and its affiliated APs. Recall from Chapter 1 that the 802.11 standard defines passive scanning (listening to beacons) and active scanning (Probe Request/Response frame exchange) for a STA to discover APs. MLO requires extensions to these 802.11 discovery procedures to accomplish discovery of information for the multiple links of an AP MLD. The following MLO-related discovery enhancements are defined on top of the existing AP discovery procedure:

- A new Basic Multi-Link element is defined.

- The RNR element is extended with MLD parameters.

- The Neighbor Report element is extended with MLD information.

- A new multi-link probe request/response feature is defined.

---

3 A procedure indicated as mandatory + optional has some aspects that are mandatory to support, whereas other aspects are optional to support.

The first three enhancements are targeted toward enabling discovery of some basic set of information for an AP MLD through the Beacon and Probe Response frames. The last enhancement (the multi-link probe request/response) is defined for a non-AP MLD to discover the complete set of information for an AP MLD, including complete profile information for all of its affiliated APs. The multi-link probe request/response makes use of the Basic Multi-Link element.

The Basic Multi-Link Element

For a non-AP MLD to discover an AP MLD, the relevant AP MLD information needs to be advertised in the BSS of each affiliated AP. Similarly, an AP MLD needs to learn the MLD capabilities of client devices. A new Basic Multi-Link element (Basic ML element) is defined in 802.11be to provide MLD-level information about peers MLDs, both for an AP MLD and for a non-AP MLD. The Basic ML element is a multi-link element with a type equal to Basic.[4] The Basic ML element includes (among other fields) information to identify the MLD using the MLD MAC Address field, and indicates the number of affiliated STAs for an AP in the Maximum Number Of Simultaneous Links field. A non-AP MLD also indicates its support for single-radio or multi-radio using the Maximum Number Of Simultaneous Links field. The Basic ML element also indicates various MLD-level capabilities in the EML Capabilities and the (Extended) MLD Capabilities and Operations fields, including indicating support for different MLO modes: EMLSR, EMLMR, STR, and NSTR.[5]

The Basic ML element can also include one or more Per-STA Profile subelements to provide link-specific information for each affiliated STA of an MLD, including the Link ID, the STA MAC Address, and a partial or complete list of information elements applicable for that affiliated STA in the STA Profile field. Figure 6-4 captures the Basic ML element showing some key fields. The Link ID for the affiliated AP transmitting the Basic ML element is provided in the Common Info field.

Each affiliated AP of an AP MLD (that is not a nontransmitted BSSID of a multiple BSSID set; see the sidebar) includes the Basic ML element in the Beacon and Probe Response frames it transmits. However, the Basic ML element in these frames does not include the Per-STA Profile subelements that provide a detailed profile for each of the affiliated APs. The main motivations for this design decision were to not expand the Beacon frame size and to minimize the overhead added to the Probe Response frames. The Basic ML element carried in the multi-link probe request/response frames provides detailed profiles of the affiliated APs, as explained in the "Multi-link Probe Request and Response" section later in this chapter. Additionally, the Basic ML element from an AP indicates whether there are critical BSS parameter updates for other affiliated APs of the AP MLD through the BSS Parameters Change Count field (also indicated for the current AP in the Common Info field; see Figure 6-4). This allows a non-AP MLD to determine when it needs to acquire Beacon frames from other affiliated APs to get the latest BSS parameters of those APs. This mechanism enables non-AP MLDs to avoid a periodic beacon scan of the channels of other affiliated APs, which is important to save power at the non-AP MLD (see the "BSS Parameters Critical Update" section).

4 See P802.11be D5.0, clause 9.4.2.312.2.
5 EMLSR: enhanced multi-link single radio; EMLMR: enhanced multi-link multi-radio; STR: simultaneous transmit and receive; and NSTR: non-simultaneous transmit and receive. These modes are described in the "MLO Modes" section in this chapter.

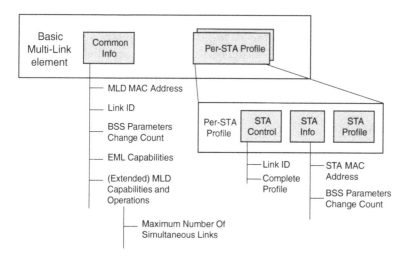

FIGURE 6-4   Basic Multi-Link Element (Showing Key Fields)

The Basic ML element is used by the non-AP MLD to indicate its MLD-level capabilities in management frames including the Probe Request, Authentication, and (Re)Association Request frames, as discussed later in the section on multi-link association.

Multiple BSSID and Transmitted/Nontransmitted BSSIDs

In many deployments, multiple SSIDs can be advertised across the AP devices. At each AP device in the deployment, each radio hosts the SSIDs via one BSS (each with its own BSSID) per SSID (e.g., SSIDs for home and guest users on 5 GHz). When the AP corresponding to each such BSS transmits its own Beacon and Probe Response frames, the process results in signif-icant network overhead. Wi-Fi 6/6E introduced a multiple BSSID feature, to reduce Beacon and Probe Response frames overhead for such deployments. The multiple BSSID feature enables advertising of multiple collocated BSSs (typically corresponding to different SSIDs) operating on the same channel in a single Beacon and Probe Response frame, thereby reducing network overhead.

A multiple BSSID set consists of a set of collocated BSSIDs operating on the same channel, where one of the BSSIDs is defined as a transmitted BSSID and other BSSIDs (one or more) are defined as nontransmitted BSSIDs. Only the AP of the transmitted BSSID transmits the Beacon and Probe Response frames. Each of these frames includes a Multiple BSSID element, which provides information for each of the nontransmitted BSSIDs in a separate Nontransmitted BSSID Profile element.

In MLO, given that all affiliated APs of an AP MLD belong to the same ESS (same SSID), only one of the APs of a multiple BSSID set is part of an AP MLD. If the affiliated AP of an AP MLD is a transmitted BSSID of a multiple BSSID set, then it advertises Beacon and Probe Response frames and includes the Basic ML element in those frames. If the affiliated AP of an AP MLD is a nontransmitted BSSID of a multiple BSSID set, then the AP corresponding to the transmitted BSSID of that multiple BSSID set advertises a Basic ML element for that AP MLD in a Nontransmitted BSSID Profile element in the Beacon and Probe Response frames.

RNR Extension with MLD Parameters

The Reduced Neighbor Report (RNR) element is used to advertise concise sets of information about neighboring APs in the Beacon and Probe Response frames, to enable non-AP STAs or non-AP MLDs to discover neighboring APs. For MLO, the RNR element has been extended to include MLD-specific parameters. Specifically, the TBTT Information field, which carries information about neighboring APs in the RNR, has been extended to include an MLD Parameters field, as shown in Figure 6-5. The MLD Parameters field indicates which APs are part of the same AP MLD using the AP MLD ID field and provides the Link ID assigned to the AP within its corresponding AP MLD. The MLD Parameters field also specifies other parameters related to the BSS parameters critical update (see the "BSS Parameters Critical Update" section) and indicates whether a link is disabled temporarily.

The AP MLD ID and the Link ID information discovered from the RNR can be used by a non-AP MLD when sending the multi-link probe request to discover detailed profile information for the affiliated APs of an AP MLD.

> **Note**
>
> The **AP MLD ID** is a short identifier (1 octet) for an AP MLD and is included (among other elements) in the Reduced Neighbor Report (RNR) element to indicate which reported neighboring APs are affiliated with the same AP MLD. It is unique across all AP MLDs reported by an AP in the RNR, but it is not a globally unique identifier for the AP MLD. In contrast, the **MLD MAC Address** (6 octets) is a globally unique identifier for an AP MLD that is included in the Basic ML element.

Neighbor Report Extension with MLD Information

The Neighbor Report element is defined in the 802.11 standard as providing information for neighboring APs. It can be included in management frames such as a Neighbor Report Response frame or a BSS Transition Management (BTM) Request frame to provide information and recommendations about neighboring APs to a non-AP STA. For MLO, the Neighbor Report element is extended to provide AP MLD information for reported APs. The Basic ML element is included for a reported AP in the Neighbor Report element; it indicates the MLD MAC address of the AP MLD with which the reported AP is affiliated. Using this information, a non-AP MLD can identify which reported APs are affiliated with the same AP MLD and use that information to trigger discovery for a desired AP MLD.

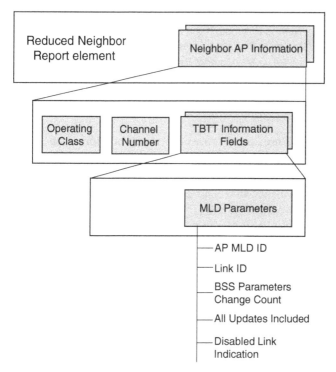

FIGURE 6-5   Reduced Neighbor Report Element

Multi-Link Probe Request and Response

The 802.11 standard defines the Probe Request/Response frame exchange for a non-AP STA for actively discovering information about nearby APs operating on a given channel. This operation is referred to as non-ML Probe Request/Response exchange in this chapter. For MLO, an extension to the 802.11-defined Probe Request/Response exchange is needed to enable a non-AP MLD to request information for the multiple affiliated APs of an AP MLD. This functionality is provided by the multi-link probe request/response procedure. The main goal is for a non-AP MLD to be able to discover the capabilities, parameters, and operational elements for the APs affiliated with an AP MLD. The 802.11be amendment defines a new Probe Request Multi-Link (ML) element that is used to request information about the affiliated APs of a given AP MLD. The Probe Request ML element is a multi-link element with a type equal to Probe Request.[6]

A multi-link probe request is a Probe Request frame that includes a Probe Request ML element. The targeted AP MLD for which the non-AP MLD is soliciting information is identified implicitly by the Address 1 or Address 3 field of the (ML) Probe Request frame, or explicitly by the AP MLD ID field in the Probe Request ML element included in the frame. If the AP addressed in the Address 1 or Address

6 See P802.11be D5.0, clause 9.4.2.312.3.

3 field is affiliated with the targeted AP MLD, then the AP MLD ID is set to 0; otherwise, the AP MLD ID indicates the targeted AP MLD for which the information is being solicited.

The multi-link probe request is addressed to an AP (either in both the Address 1 and Address 3 fields or only in the Address 3 field) from which the non-AP MLD has learned about the existence of the AP MLD (e.g., through the RNR element or the Basic ML element in the beacon). In the multi-link Probe Request, a non-AP MLD can solicit information either for all affiliated APs of the AP MLD or for a subset of the affiliated APs of the AP MLD. This is indicated by including one or more Per-STA Profile subelements in the Probe Request ML element indicating the Link IDs of the specific affiliated APs for which information is requested (see Figure 6-6). Additionally, the multi-link Probe Request gives the non-AP MLD the flexibility to either request the complete profile for an affiliated AP (which provides all applicable capabilities, parameters, and operational elements for the AP) or request partial profile for an affiliated AP by indicating the specific elements that are solicited (using a Request element and/or an Extended Request element included in the STA Profile field). When requesting complete profile for an affiliated AP, the Complete Profile Requested field is set to 1 in the STA Control field of the Per-STA Profile subelement for that affiliated AP. Note that a multi-link probe request is designed to request information for only one AP MLD.

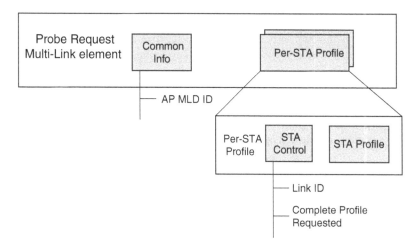

FIGURE 6-6 Probe Request Multi-Link Element

A multi-link Probe Response is a Probe Response frame that is transmitted in response to a multi-link Probe Request. It includes a Basic ML element that provides a complete or partial profile for each of the requested affiliated APs in a corresponding Per-STA Profile subelement, based on what was requested in the corresponding multi-link Probe Request. In the case of a multiple BSSID set (see the earlier sidebar), when a multi-link Probe Request is addressed to a nontransmitted BSSID and intended to discover information for the corresponding AP MLD, it is the transmitted BSSID of the same multiple BSSID set that transmits the corresponding multi-link Probe Response.

MLD Discovery Procedure

Now that we have an understanding of the various enhancements made to support MLD discovery in the 802.11be amendment, we will look at how the MLD discovery procedure happens. As described earlier, a non-AP MLD can learn basic information about an AP MLD through multiple means. For example, it can obtain this information from the Basic ML element and the RNR element received in the Beacon or (non-ML) Probe Response frames or from a Neighbor Report element received in one of these management frames.

The non-AP MLD can select one or more of the following methods to discover information about affiliated APs of an AP MLD:

- Passive scanning performed by each affiliated non-AP STA
- Active scanning performed by each affiliated non-AP STA
- Multi-link Probe Request/Response exchange on one of the links

Figure 6-7 shows these methods for AP MLD discovery by a non-AP MLD that has three affiliated non-AP STAs (on 2.4, 5, and 6 GHz) and is discovering an AP MLD that has three affiliated APs (on 2.4, 5, and 6 GHz). With passive scanning, each affiliated non-AP STA scans (i.e., sets its receiver to the channel and listens) the corresponding link of the AP MLD and receives Beacon frames, which provide complete information about that affiliated AP. As an example in Figure 6-7, the non-AP STA operating on 2.4 GHz performs passive scanning to receive beacons from the AP operating on the 2.4 GHz link. Similarly, the non-AP STAs operating on 5 GHz and 6 GHz would perform passive scanning to receive beacons from APs operating on those links. With active scanning, each of the non-AP STAs sends a (non-ML) Probe Request frame on the corresponding link to receive the (non-ML) Probe Response frame from the affiliated AP on that link, which provides the complete information about that affiliated AP. In Figure 6-7, the non-AP STAs operating on 2.4, 5, and 6 GHz each send a Probe Request frame to, and receive the Probe Response frame from, the APs operating on those links. Both the passive and active scanning methods of AP MLD discovery involves off-channel scanning to discover AP MLD information, which is disruptive to the client's current operation, especially for single-radio non-AP MLDs.

With the ML Probe Request/Response method, one of the non-AP STAs can send an ML Probe Request frame to receive information for all affiliated APs of an AP MLD without performing any off-channel scanning. For example, in Figure 6-7, the non-AP STA operating on 2.4 GHz sends an ML Probe Request frame, and AP1 responds with an ML Probe Response frame that provides information for all three affiliated APs of the AP MLD. This method of AP MLD discovery is preferred over the other two methods, because it does not require going off-channel for scanning and hence is the least disruptive to connectivity on the current link.

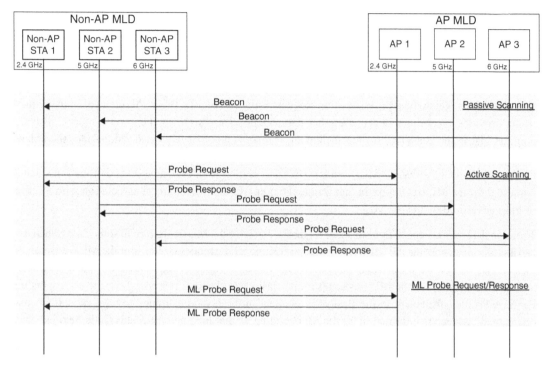

FIGURE 6-7 AP MLD Discovery Methods

Multi-Link Association

MLO enables a non-AP MLD to establish an association that spans multiple links of an AP MLD. This procedure of setting up multiple links between a non-AP MLD and an AP MLD is referred to as multi-link (ML) setup. An ML setup establishes association at the MLD level between the two peer MLDs (the non-AP MLD and the AP MLD).[7] An ML re-setup involves reestablishing an MLD-level association with another AP MLD or the same AP MLD (with a different set of links) within an ESS. The links that are established between a non-AP MLD and an AP MLD after successful ML (re-)setup are called the setup links.

Before the multi-link association operation, a non-AP MLD and the AP MLD perform MLD-level authentication by including a Basic ML element in the Authentication frame. The Basic ML element in the Authentication frame provides the MLD MAC address of the MLD from which the Authentication frame is sent.

For the ML (re-)setup, both the (Re)Association Request/Response frames are enhanced by adding the Basic ML element to perform the MLD-level association.[8] The Basic ML element provides the

7 See P802.11be D5.0, clause 35.3.5.1.
8 The ML re-setup is performed using Reassociation Request/Response frames.

Per-STA Profile information for the affiliated STAs of an MLD (either the AP MLD or the non-AP MLD) for multi-link association. This does not include the affiliated STA that is transmitting the (Re) Association Request/Response frame itself; its information is provided in the frame itself outside the Basic ML element. In the (Re)Association Request frame, the non-AP MLD requests one or more links for the ML (re-)setup by indicating those links in the Basic ML element. For each requested link other than the link on which the (Re)Association Request frame is sent, the Basic ML element in the request frame provides capabilities and operational parameters information for the corresponding affiliated non-AP STA in the Per-STA Profile subelement. If there is no other requested link besides the link on which the (Re)Association Request frame is sent, the Basic ML element does not include any Per-STA Profile subelement. A non-AP MLD sends a (Re)Association Request frame on one of the links that is requested for the ML (re-)setup; in that frame, the non-AP MLD can request association with a subset of links advertised by the AP MLD.

The AP MLD may accept all or a subset of requested links for the ML (re-)setup. In some cases, such as a capabilities mismatch, the AP MLD may end up rejecting all the requested links for the ML (re-)setup. In the (Re)Association Response frame, the AP MLD indicates the list of requested links that are accepted and that are rejected for the ML (re-)setup in the Basic ML element. For each link, the corresponding Per-STA Profile subelement in the Basic ML element in the response frame provides capabilities and operational parameters information for the AP operating on that link, and the Status Code field indicates the accepted or rejected status for that link for ML (re-)setup. The (Re)Association Response frame is transmitted by the same affiliated AP that received the (Re)Association Request frame. In addition, the (Re)Association Request and Response frames are exchanged on the same link where the Authentication frames are exchanged before the ML (re-)setup. An ML (re-)setup cannot be successful if the link over which the (Re)Association Request was sent cannot be accepted in the (Re)Association Response (e.g., for the parameters mismatch failure reason). So, after a successful ML (re-)setup, the current link where the ML (re-)setup was executed is always part of the setup links for the non-AP MLD.

During ML (re-)setup, an AP MLD assigns a single AID to a non-AP MLD. All the non-AP STAs affiliated with that non-AP MLD then use the same AID. After successful ML (re-)setup, the non-AP MLD is in power-save mode on all setup links other than the current link used for the ML (re-)setup. Additionally, the mapping between the non-AP MLD to the AP MLD is provided to the DS by the AP MLD. For each setup link that is successfully established, the corresponding non-AP STA is in the same associated state, as the non-AP MLD and the STA-level behavior based on the associated state continue to apply. However, no mapping between the affiliated non-AP STA and the corresponding affiliated AP is provided to the DS.

Note

The A1 address field (receiver address) and the A2 address field (transmitter address) in the individually addressed management and data frames sent over a link between peer MLDs are set to the corresponding STA MAC addresses for that link. The MLD MAC address is not used in the A1 and A2 address fields in such frames. Instead, the MLD MAC address is used in the key generation and encryption, as described in the next section on MLO security.

Figure 6-8 shows an example for the ML setup operation between an AP MLD advertising three links (APs 1, 2, and 3 on 2.4, 5, and 6 GHz, respectively) and a non-AP MLD with three affiliated non-AP STAs (non-AP STA 1, 2, and 3) that can operate on those links. After performing the authentication procedure on the 2.4 GHz link, non-AP STA 1 sends a (Re)Association Request frame on the same link to AP 1, requesting to set up association on three links: the 2.4 GHz, 5 GHz, and 6 GHz links. The AP MLD accepts the association request for all three links. After successful ML setup, the non-AP MLD and the AP MLD have established an association encompassing the three links (link 1, link 2, and link 3 in Figure 6-8).

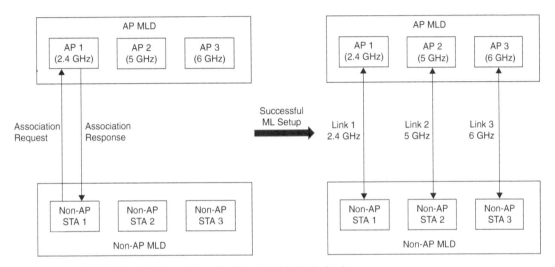

FIGURE 6-8 MLD-Level Association Setting Up Multiple Links

Inheritance in the Basic ML Element

When the Basic ML element is included in the (Re)Association Request and Response frames, it includes the complete profile information for the affiliated STAs of the reported AP MLD. The capabilities and operational parameters of the reported affiliated STAs in the Per-STA Profile subelements are likely to be similar to the reporting STA information sent in the frame outside the Basic ML element. To optimize the frame size, inheritance rules are applied that enable elements to be inherited from the reporting STA by the reported STAs that carry a complete profile in the Per-STA Profile subelement.[9]

In the complete profile for a reported STA in a Per-STA Profile subelement, an element that is specific to a reported STA is always included in the Per-STA Profile subelement and the included element's contents apply to the reported STA. An element that is carried in the frame outside the Basic ML element is considered inherited for a reported STA (and as a result the element's contents apply to the reported STA as well), under two conditions: (1) that element's identifier

9 See P802.11be D5.0, clause 35.3.3.5.

(i.e., element ID or element ID extension) is not included in the Non-Inheritance element carried in the Per-STA Profile (to explicitly exclude inheritance) and/or (2) that element is not explicitly excluded from being part of the Per-STA Profile subelement according to certain other rules.[10]

After the ML (re-)setup establishes one or more links between the non-AP MLD and the AP MLD, the 4-way handshake is executed between the peer MLDs to establish the PTK (Pairwise Transient Key) and exchange link-specific group keys. The MLO security aspects are explored in more detail in the "MLO Security" section of this chapter.

Figure 6-9 shows the end-to-end flow for the ML (re-)setup operation between a non-AP MLD and an AP MLD, including the ML authentication, the ML association establishing three links, the 802.1X/ EAP authentication (if applicable), and the 4-way handshake establishing security keys. Note that the 802.11be amendment requires that all the management frames for ML (re-)setup operations be exchanged on the same link. This means that the 4-way handshake must be executed on the same link where the (Re)Association Request/Response frames were exchanged. After the successful 4-way handshake, UL/DL data exchange is enabled on all links set up between the peer MLDs. Initially, all other links are in PS mode except the link where ML (re-)setup is completed.

Either of the peer MLDs in an MLD-level association can tear down all the setup links by sending a Disassociation frame through one of the affiliated STAs of the MLD. After disassociation, the non-AP MLD and all its affiliated non-AP STAs are in an unassociated state. Either of the peer MLDs can also send a Deauthentication frame that tears down all setup links and changes the MLD state to state 1 (unauthenticated and unassociated).

MLO Association Myths

Myth 1: The client device associates with the AP device on each link.

Reality: This is not the case. Rather, a single MLD-level association is established between the higher-layer MLD entities (AP MLD and the non-AP MLD) that encompasses all the links.

Myth 2: The non-AP MLD sends a (Re)Association Request frame on each link that it wants to use.

Reality: This is not the case. Rather, a single (Re)Association Request and (Re)Association Response exchange is performed on one link. The pair of association frames includes the requisite information for all links. The information in the association frames is exchanged between the non-AP MLD and the AP MLD entities.

Myth 3: The link on which the client sends the (Re)Association Request frame is a special, main, or anchor link.

Reality: This is not the case. The client can use any link to send the (Re)Association Request frame to the AP MLD to establish the ML association. The AP MLD sends the (Re)Association Response frame on the same link.

---

10 See P802.11be D5.0, clause 35.3.3.4.

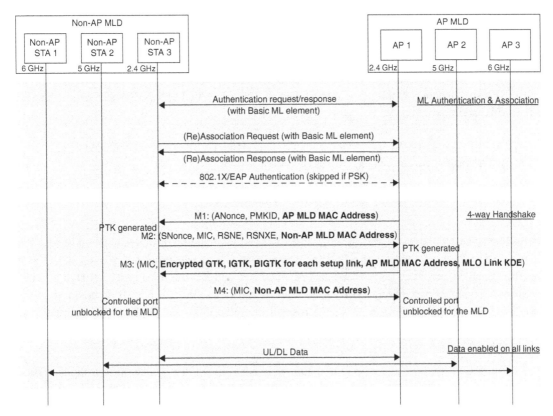

FIGURE 6-9   End-to-End Flow for ML (Re-)Setup Between MLDs

MLO Security

For MLO, the PMK security association (PMKSA) and the PTKSA security context used for the protection of individually addressed frames are established at the MLD level. The group keys (GTK, IGTK, and BIGTK[11]) are established at the link level for each link of the MLD.

The MLD-level authentication establishes PMKSA between peer MLDs. The 4-way handshake procedure establishes PTKSA, where the PTK is tied with the MLD MAC address of the AP MLD and the non-AP MLD. In the PTK generation, the Authenticator Address (AA) is set to the AP MLD MAC address and the Supplicant Address (SPA) is set to the non-AP MLD MAC address. As shown in Figure 6-6, the 4-way handshake messages are enhanced for MLO to include the AP MLD MAC address in the M1 and M3 messages, and the non-AP MLD MAC address in the M2 and M4 messages. The M3 message provides group keys and includes the encrypted GTK, IGTK (if management frame protection is enabled), and BIGTK (if beacon frame protection is enabled) for each of the successfully established setup links. The M3 message also includes an MLO link KDE for each affiliated AP of

11 GTK: group temporal key; IGTK: integrity group temporal key; BIGTK: beacon integrity group temporal key.

the AP MLD providing the RSNE and RSNXE (if present) for the link where the affiliated AP is operating.

For MLO, the PTK generation uses the same KDF as for the non-MLO case, except that the addresses used are the MLD MAC addresses. The PTK generation (for the case when authentication is not FT or FILS) is shown here,[12] highlighting that the PTK generation is tied to the AP MLD and the non-AP MLD MAC addresses:

> PTK = PRF-Length(PMK, "Pairwise key expansion," Min(AA,SPA) ‖ Max(AA,SPA) ‖ Min(ANonce,SNonce) ‖ Max(ANonce,SNonce))

where AA = AP MLD MAC address, SPA = non-AP MLD MAC address, and "‖" indicates concatenation.

The PTKSA is used for cryptographic encapsulation and decapsulation of individually addressed data frames and individually addressed robust management frames (when Protected Management Frame (PMF) is enabled) across all links established between the AP MLD and the non-AP MLD. It provides both confidentiality (encryption) and integrity protection for these frames. For MLO, the same packet number (PN) space is used for a PTKSA across all the setup links. In addition, the construction of additional authentication data (AAD) and nonce (for CCM and GCM) are revised to include the MLD MAC address.

The per-link GTKSA is used for cryptographic encapsulation and decapsulation of group-addressed data frames on that link. When PMF is enabled, the IGTKSA of a link is used to provide integrity protection for group-addressed robust management frames on that link. Similarly, when beacon protection is enabled, the BIGTKSA of a link is used to provide integrity protection for Beacon frames transmitted on that link. For each GTK/IGTK/BIGTK corresponding to a setup link, a single PN/IPN/BIPN space is used by the AP STA and the non-AP STA operating on that link. The GTK/IGTK/BIGTK group keys for an affiliated AP of the AP MLD can be updated over any link using the group key handshake procedure.

MLD Data Plane

Data transmission between MLDs that have established an association encompassing multiple links can occur on one or more of those links subject to TID-To-Link mapping (see the "Link Management" section in this chapter) and multi-link power management. This section covers the enhancements made in the 802.11be amendment for the delivery of individually addressed and group-addressed data frames between peer MLDs.

---

12 See P802.11REVme D5.0, clause 12.7.1.3.

Individually Addressed Data Delivery

An MLD can deliver individually addressed data frames to another associated peer MLD with or without establishing Block Ack (BA) agreements (also known as BA sessions) between the MLDs. Of course, data delivery is more efficient when BA agreements are negotiated between the MLDs—that method allows multiple MPDUs to be acknowledged together by a single Block Ack frame, thereby reducing overhead. The 802.11be amendment defines 512 and 1024 Block Ack bitmap lengths, though support for them is optional on both the AP MLD and non-AP MLD sides. The Wi-Fi Alliance's Wi-Fi 7 certification includes optional support for only the 512 Block Ack bitmap length.

The MLD architecture splits the data plane handling for individually addressed data frames over the MLD Lower MAC and MLD Upper MAC sublayers, as explained in the "Multi-Link Device" section earlier in this chapter. The following enhancements and new functions are added for supporting the delivery of individually addressed data frames between MLDs:

- The assignment of SN and PN parameters and the MPDU encryption/decryption are done at the MLD level by the MLD Upper MAC sublayer, and are applicable across all the setup links.

- Frame reordering, duplicate detection, and replay detection are done at the MLD level by the MLD Upper MAC sublayer.

- The MLD Upper MAC sublayer adds the new functionality of mapping MPDUs to link(s). This is done by considering the TID-To-Link mapping, the per-link power management indications, and likely other implementation-specific factors (e.g., preferentially selecting the 6 GHz link for TIDs carrying low-latency traffic).

- The MLD Upper MAC sublayer adds new functionality for merging the MPDUs of a TID received over multiple links.

- Block Ack agreements are established at the MLD level between the MLD Upper MAC sublayers. The Block Ack scoreboard context is maintained both at the MLD Lower MAC sublayer (link-level BA scoreboard) and the MLD Upper MAC sublayer (unified BA scoreboard).

ML Block Ack Procedure

Between a non-AP MLD and an AP MLD, the BA agreements are established at the MLD level for each TID using the ADDBA Request/Response exchange. Those agreements apply to all the links where data frames for that TID can be exchanged based on the TID-To-Link mapping. A recipient MLD of data frames in a BA agreement maintains a single receive reordering buffer for each <peer MLD, TID> pair across all the links where that TID can be exchanged. The BA scoreboard context is maintained at both the MLD Lower MAC and MLD Upper MAC sublayers, as previously described. Each MLD Lower MAC sublayer maintains an independent, link-level BA scoreboard context per-TID based on the MPDUs received on that link. The MLD Upper MAC sublayer maintains a common unified BA scoreboard per-TID based on MPDUs received across multiple links of that MLD. Some implementations can support sharing of the BA scoreboard context from one link to the other link through the MLD Upper MAC sublayer.

The recipient MLD provides the reception status of MPDUs of a TID transmitted on a link in the Block Ack frame sent over that link. In addition, it might provide the reception status of MPDUs of that TID that are delivered over other links of the MLD.

Figure 6-10 shows the Block Ack procedure between an AP MLD and a non-AP MLD. After a BA agreement has been established for a TID between the MLDs using the ADDBA Request/Response exchange, the Block Ack session applies across all links of the MLD. If the non-AP MLD is capable of exchanging frames simultaneously over multiple links (STR operation; see the "MLO Modes" section), the A-MPDUs for that TID can be transmitted by the AP MLD over multiple links and acknowledged by the non-AP MLD using Block Ack frames. The Block Ack frame on each link, at a minimum, acknowledges the reception of MPDUs successfully sent on that link. An A-MPDU containing MPDUs with sequence number (SN) = 1, 2, and 3 is sent on Link 1 by AP1, and another A-MPDU containing MPDUs with SN = 4, 5, and 6 is sent on Link 2 by AP 2. The Block Ack sent by STA 1 on Link 1 acknowledges MPDUs with SN = 1, 2, and 3, and the Block Ack sent by STA 2 on Link 2 acknowledges MPDUs with SN = 4, 5, and 6.

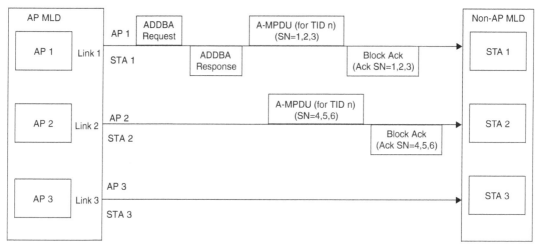

FIGURE 6-10   Block Ack Procedure Between MLDs

MLD Data Path Handling

The Tx and Rx functions for the MLD data plane for individually addressed MPDUs are shown in Figure 6-11.[13] On the Tx/transmission data path, the MLD Upper MAC sublayer receives one or more MSDUs through the MAC SAP from the LLC layer. A-MSDU aggregation may be applied across multiple MSDUs. The SN is assigned to the MPDU, followed by a packet number (PN) for encryption and replay detection. The MPDU is encrypted and integrity protected (by adding a MIC) through the PTK; the same PTK is used for all links. The MPDU is then mapped to one or more links based

13 See P802.11be D5.0, clause 5.1.5.1.

on the mapping of TIDs to links (per TTLM) and sent to one or more MLD Lower MAC sublayers for delivery. In the lower MAC, the MPDU header and CRC are added to the MPDU payload and A-MPDU aggregation is applied. The MPDU or the A-MPDU is then sent to the PHY layer over the PHY SAP for transmission over a link. If an MPDU is sent to multiple MLD Lower MAC sublayers for delivery, when that MPDU is delivered successfully on one of the links, that MPDU is immediately retired at the other links. Additionally, if the transmission of an MPDU (of a specific TID) is unsuccessful on a specific link, the MLD may attempt retransmission of that data frame on any of the links over which that TID is mapped (per TID-To-Link mapping).

On the Rx/receive data path, when the MLD Lower MAC sublayer receives a PSDU from a PHY SAP, the PSDU first goes through the process of A-MPDU de-aggregation and then the MPDU header and CRC fields are validated. The Address 1–based filtering, using the station MAC address of the link, is performed to ensure that the MPDU is relevant to the receiving STA. These functions are the same as the data plane handling for the non-MLO case. The link-level BA scoreboard context for a TID is maintained by the MLD Lower MAC sublayer. The MLD Upper MAC sublayer maintains the common unified BA scoreboard for a TID across the multiple links that the TID is mapped to. The MLD Upper MAC sublayer receives MPDUs for a TID from the lower MACs of multiple affiliated APs; it then merges these MPDUs for duplicate detection and frame reordering based on their SN. Next, the MPDU is decrypted using the PTK, and replay detection is performed based on the PN. Lastly, the upper MAC performs A-MSDU de-aggregation (if MSDU aggregation was applied by the transmitter).

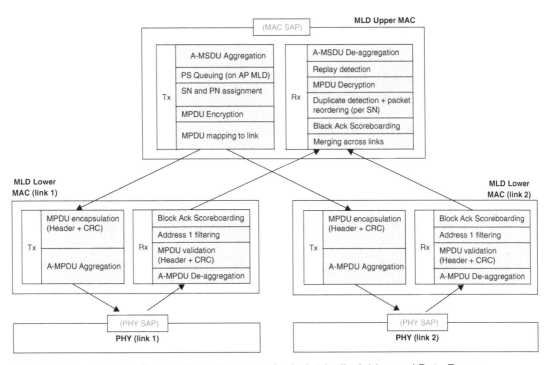

FIGURE 6-11   MLD MAC Data Plane Functions for Individually Addressed Data Frames

Group-Addressed Data Delivery

An AP MLD schedules delivery of buffered group-addressed data frames that are expected to be received by a non-AP MLD on all the enabled setup links of that non-AP MLD. The AP MLD's MLD Upper MAC sublayer receives the group-addressed MSDUs and assigns a SN. The group-addressed MSDUs are then sent to the affiliated APs of the AP MLD and handled by the Non-MLD Upper MAC sublayer (see Figure 6-2), which then performs link-specific encryption for these frames. A link-specific GTK is used to encrypt and apply integrity protection to the group-addressed data frames delivered over MLO links.

An affiliated AP also indicates in the partial virtual bitmap, in the TIM element included in the DTIM Beacon frame, whether each of the other APs affiliated with the same AP MLD has buffered group-addressed frames. This information is useful for the non-AP MLD when some TIDs are mapped to only a subset of links. In those cases, the non-AP MLD would need to retrieve the group-addressed data frames for those TIDs from the link where those group-addressed frames are buffered.

MLO Modes

In MLO, each affiliated STA of an MLD operating on one of the setup links contends for the channel access independently of other affiliated STAs operating on other setup links. The 802.11be amendment defines several modes of operation for MLO (shown in Table 6-1) based on the differing hardware capabilities of devices, and specifically whether they are single-radio– or multi-radio–capable non-AP MLDs. An AP MLD is required to support simultaneous operation across the set of multiple links it supports.

Both the AP MLD and the non-AP MLD signal their MLD-level capabilities and operation parameters in the Basic ML element as follows:

- An AP MLD sets the Maximum Number Of Simultaneous Links field to the number of affiliated APs minus 1.

- A non-AP MLD indicates the maximum number of simultaneous links it supports for transmission or reception using the Maximum Number Of Simultaneous Links field. This field is set to the maximum number of simultaneous radios supported minus 1 by a non-AP MLD. A single-radio non-AP MLD sets this field to 0; a multi-radio non-AP MLD sets this field to a value greater than or equal to 1.

- A multi-radio non-AP MLD indicates whether each pair of links it supports is STR or NSTR by setting the corresponding bit in the NSTR Indication Bitmap field to 0 (indicates STR) or 1 (indicates NSTR). Link pairs where one link is operating on 2.4 GHz and the other link is operating on 5 GHz or 6 GHz must not be indicated as NSTR link pairs.[14]

---

14 See P802.11be D5.0, clause 35.3.16.2.

- Other MLD-level capabilities and parameters are indicated in the MLD Capabilities and Operations field, the Extended MLD Capabilities and Operations field, and the EML Capabilities field.

Five MLO modes are defined in the 802.11be amendment:

- Multi-link single radio (MLSR)

- Enhanced multi-link single radio (EMLSR)

- Simultaneous transmit and receive (STR)

- Non-simultaneous transmit and receive (NSTR)

- Enhanced multi-link multi-radio (EMLMR)

Table 6-1 summarizes the functionality and requirements for these MLO modes.

TABLE 6-1   Summary of MLO Modes

MLO Mode	Number of Radios Required	Functionality	Requirements
Multi-link single radio (MLSR)	Single-radio	Tx or Rx over one link at a time. Link switch happens using PM bit.	Mandatory: for AP MLD and non-AP MLD
Enhanced multi-link single radio (EMLSR)	Single-radio	MLSR operation with capability to listen on multiple links in low-capability mode, one of which is dynamically selected for full-mode Tx and Rx exchange within the TXOP.	Optional: for AP MLD and non-AP MLD
Simultaneous transmit and receive (STR)	Multi-radio (≥2)	Support for simultaneous Tx/Tx, Rx/Rx, and Tx/Rx on a pair of links independent of each other.	Mandatory: for AP MLD for all link pairs Optional: for non-AP MLD
Non-simultaneous transmit and receive (NSTR)	Multi-radio (≥2)	Support for simultaneous Tx/Tx and Rx/Rx on a pair of links with careful PPDU start and end time alignment.	Optional: for AP MLD and non-AP MLD
Enhanced multi-link multi-radio (EMLMR)	Multi-radio (≥2)	MLMR operation with additional capability to dynamically reconfigure spatial multiplexing support over multiple links.	Optional: for AP MLD and non-AP MLD

The client devices that have only a single full-capability radio can support just the first two modes of operation (MLSR and EMLSR). The last three MLO modes (STR, NSTR, and EMLMR) are multi-link multi-radio (MLMR) modes of operation and can be supported only by client devices that have multiple radios capability.

The last column in Table 6-1 specifies the requirements for MLO modes defined in the 802.11be amendment. This amendment mandates AP MLDs to support the MLSR and STR modes of operation and requires that non-AP MLDs, at a minimum, support the MLSR mode. The Wi-Fi Alliance's certification for Wi-Fi 7 mandates further requirements for the support of these MLO modes.

WFA Certification for MLO Modes

The most important MLO modes are MLSR, EMLSR, and STR. In the Wi-Fi Alliance's Wi-Fi 7 certification program, AP MLD devices are required to support these three MLO operation modes. Non-AP MLD devices are required to support MLSR, and either the EMLSR or STR mode of operation as per the next condition stated. If a non-AP MLD supports both two spatial streams and a 160 MHz channel width on at least one of the links, then the non-AP MLD is required to support either the EMLSR or STR operation mode.

Multi-Link Single Radio

Typical entry-level client devices (e.g., laptops, mobile phones, smartwatches) support operation in multiple bands (2.4, 5, and 6 GHz), where the radio hardware is capable of tuning to any channel in any supported band. However, many of these devices have the constraint of supporting only a single radio due to cost considerations. Such devices can associate on multiple links offered by an AP MLD and then use only one link at a time to communicate with the AP MLD. This MLSR mode of operation is the simplest form of ML operation.

In the MLSR mode, a non-AP MLD indicates that it is in power save mode (PM = 1) on all its setup links except one link, using the existing power management signaling (see Chapter 1). The non-AP MLD can contend for and access the channel on the link that is not in power save mode (PM = 0) for UL transmissions, and can receive DL transmissions from the AP on that link. If it wishes to use another link (e.g., when the current link becomes congested) or if it wants to access another preferred link (e.g., a 6 GHz link), then the non-AP MLD indicates that it is in power save mode (PM = 1) on the previous link and not in power save mode (PM = 0) on the new link to the AP MLD, using existing power management signaling. Thus, in the MLSR mode, the active link has the PM = 0 setting and all other links have the PM = 1 setting. The AP MLD transmits to the non-AP MLD only on the one link where the non-AP MLD is not in power save mode.

Figure 6-12 shows MLSR operation between a non-AP MLD and an AP MLD that have set up three links in their ML association (2.4, 5, and 6 GHz links). Initially, the non-AP MLD is in active mode (PM = 0) on the 5 GHz link, and it is in power save mode (PM = 1) on the other two links. All UL and DL transmissions are happening on the 5 GHz link. Then, the non-AP MLD decides to access the 6 GHz link (e.g., for low-latency traffic or because the 5 GHz link becomes congested) and switches its power save mode. It first enters power save mode on the 5 GHz link by indicating PM = 1 to the 5 GHz AP. It then becomes active on the 6 GHz link by indicating PM = 0 to the 6 GHz AP.

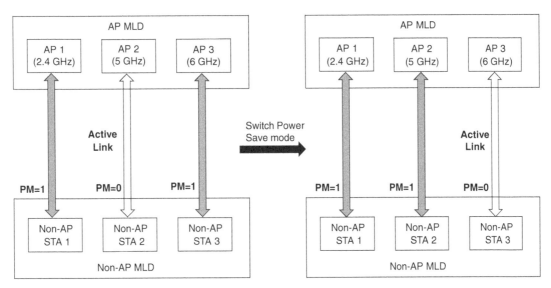

FIGURE 6-12   MLSR Operation

Switching links in the MLSR operation using the power save indication (PM bit) enables flexibility for the client to use and quickly switch to any one of the associated links. The MLSR mode can offer improved latency and reliability by enabling the client to use a less congested link.

Enhanced Multi-Link Single Radio

For single-radio client devices, another nominal mode of operation allows the single high-capability radio to be split into two or more low-capability radios and to listen on two or more links simultaneously. For example, in one possible implementation, a 2×2 high-capability radio could be split and operate as two 1×1 radios in a low-capability mode (i.e., only on 20 MHz and with low MCS rates). Additionally, some client devices with one high-capability radio could have N additional low-capability radios ($N \geq 1$) with minimal cost impact, where each low-capability radio operates on one spatial stream, at 20 MHz and with low MCS rates. Operating in the low-capability mode offers power-saving benefits for the non-AP MLD. Such client devices can operate in the new enhanced MLSR (EMLSR) mode defined in 802.11be to save power and, at the same time, gain some of the benefits of simultaneous multi-link operation from a single-radio device. The EMLSR operation can achieve improved reliability and lower latency, with diversity gains coming from operation over multiple links (e.g., EMLSR on 2.4 + 5 GHz links).

In a nominal implementation of the EMLSR mode, the transceiver modules of a single-radio device are configured to operate as multiple receiving chains listening on multiple links for an initial DL transmission or as the trigger for an UL transmission from the AP MLD. A non-AP MLD can operate in the EMLSR mode on one link, two links, or more than two links based on the number of low-capability receiving chains supported by the non-AP MLD and the power-saving mode selected. The set of links on which a non-AP MLD enables the EMLSR operation are referred to as the EMLSR links. Note that a non-AP MLD may choose to enable EMLSR operation on a single link for further power-saving benefits compared to EMLSR operation on two or more links. However, support for EMLSR operation on a single link is optional for both the AP MLD and the non-AP MLD in Wi-Fi 7.

Support for EMLSR operation is indicated by the non-AP MLD and the AP MLD in the EML Capabilities field in each of their transmitted Basic ML elements. After ML (re-)setup, the non-AP MLD is in power save mode on all links except the current link [where the ML (re-)setup was completed] and EMLSR mode is disabled by default on the setup links. A non-AP MLD sends an EML Operating Mode Notification (EML OMN) frame to explicitly enable the EMLSR mode for the links indicated by the EMLSR Link Bitmap field in the frame. In response, the AP MLD transmits an EML OMN frame to the non-AP MLD when it is ready to serve the non-AP MLD in the EMLSR mode and within the transition timeout period indicated in the EML Capabilities field. The non-AP MLD enters the EMLSR mode after it receives the EML OMN frame response from the AP MLD or after the transition timeout period, whichever happens first.[15]

Once the EMLSR mode is enabled, the non-AP MLD listens in limited Rx capability mode (i.e., NSS of 1, bandwidth of 20 MHz, and support for only low MCS rates of 6, 12, or 24 Mbps) on the indicated set of EMLSR links (which could be just one EMLSR link). The AP MLD sends an initial control frame (ICF) on one of the EMLR links to start the frame exchange sequence with the non-AP MLD. A padding delay is applied to the ICF to allow the non-AP MLD enough time to switch all its radio resources to the link where the ICF is received for full-capability Tx and Rx exchange on that link. The ICF can be either an MU-RTS Trigger frame or a BSRP Trigger frame. In turn, the non-AP MLD sends the corresponding control frame response to the ICF. The link where the ICF frame is sent and a control response is sent back by the non-AP MLD can then perform Tx and Rx with full capability (with NSS ≥ 2), and the other EMLSR links become inactive. After the data exchange is completed on that EMLSR link, the non-AP MLD goes back to the listening mode on all links of its EMLSR link set, after a transition delay.

Figure 6-13 illustrates EMLSR operation between an AP MLD and a non-AP MLD on two setup links (Link 1 and Link 2).

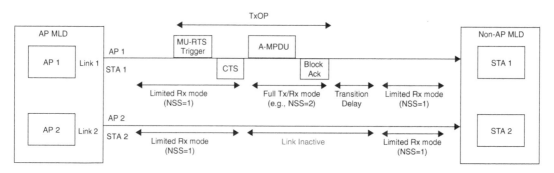

FIGURE 6-13 EMLSR Example Operation

---

15 See P802.11be D5.0, clause 35.3.17.

Simultaneous Transmit and Receive

In the STR mode of operation, an MLD device contains multiple radios, and each radio is capable of operating independently of the other radios. Each radio has its own set of antennas, and the radios are sufficiently isolated from each other inside the device that the activity of one radio does not create unsurmountable EMF interferences on the other co-located radio. The channels used across the multiple links of the MLD have enough separation that these links can operate independently without any self-interference within the device. The AP MLD is required to support STR operation for each pair of links that it supports. However, support for STR operation is not mandated for client devices in the 802.11be amendment, because client devices targeted for different market segments have different hardware and cost constraints. In contrast, the Wi-Fi Alliance's Wi-Fi 7 certification requires that client devices support STR mode (see the earlier sidebar on certification for MLO modes). STR mode enables devices to achieve higher throughput (with link aggregation by simultaneously·sending data over multiple links; e.g., 5 + 6 GHz links) and improved reliability and lower latency (with diversity gains by using any of the available links; e.g., 2.4 + 5 GHz diversity).

An STR link pair is a pair of links on which MLD operates in STR mode and can simultaneously transmit and receive on the two links. This implies that the MLD can be transmitting on one link of the STR link pair and simultaneously receiving on the other link, or that the MLD can be simultaneously transmitting or receiving on the two STR links. As previously described, all pairs of links for an AP MLD (that is not an NSTR mobile AP MLD) are required to be STR link pairs. For an MLD (AP MLD or a non-AP MLD) that supports STR operation on a pair of links, access to the wireless medium is accomplished independently on each of those links.[16] The MLD operates asynchronously and in full-capability mode on each of the links of the STR link pair.

Figure 6-14 illustrates STR operation between an AP MLD and a non-AP MLD with two setup links. The EDCA operation to access the wireless medium and initiate data transfer is happening independently on the two links. Initially, AP 1 is transmitting data to non-AP STA 1 on Link 1; at the same time, AP 2 is receiving data from non-AP STA 2 on Link 2. Later (e.g., during the Ack frame exchange), the transmission directions are reversed.

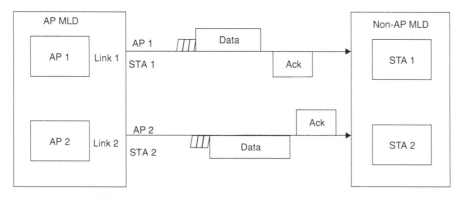

FIGURE 6-14   STR Example Operation

16 See P802.11be D5.0, clause 35.3.16.3.

Non-Simultaneous Transmit and Receive

The STR mode imposes stringent design requirements for radios to be sufficiently isolated from each other. However, in some cases, cross-link self-interference within the device can occur on the links of an MLD. For example, this problem can arise when links are operating on channels that are close to each other and do not have enough frequency separation, or when the physical isolation between the radios inside the device is not sufficient to allow for independent operation. In those circumstances, transmission on one link can jam the reception on the other link due to the in-device interference and result in failure to receive data frames. On such links, simultaneous transmission and reception cannot be supported; thus, the link pairs are called non-simultaneous transmit and receive (NSTR) link pairs. The NSTR link pairs can support simultaneous Tx/Tx and Rx/Rx operations on the two links with careful alignment of the PPDUs' end times on the two links, such that a subsequent Ack frame on one of the links does not overlap with the PPDU transmission or reception on the other link.

The NSTR mode is a much more complicated mode of operation. Support for this mode is optional in the 802.11be amendment, and it is not part of the Wi-Fi Alliance's Wi-Fi 7 certification. As a result, this mode is likely to see little adoption in the market.

For the NSTR mode of operation, the following modified channel access rules are defined:

1. **PPDU end-times alignment[17]:** To avoid in-device cross-link interference between UL and DL transmissions at an NSTR non-AP MLD, the DL PPDUs transmitted on the two NSTR links by the AP MLD to the non-AP MLD need to have their end times aligned. The 802.11be amendment requires the PPDUs' end times to be aligned within ±8 μs of each other. Note that the DL PPDU's start time can be different on the two NSTR links; that is, the main requirement is for the PPDUs' end times to be aligned. For the same reason, the UL PPDUs transmitted on the NSTR links need to have their end times aligned, so that any response frame (e.g., Ack) for one PPDU does not interfere with the other UL PPDU transmission. Figure 6-15 shows the PPDU end-times alignment requirements for NSTR operation in DL.

2. **PPDU start-times sync[18]:** There is no PPDU start-times sync requirement for DL PPDU transmissions. However, when two non-AP STAs operating on an NSTR link pair are performing channel access, if one of the non-AP STAs starts its transmission when its backoff counter reaches zero, the other non-AP STA cannot continue sensing the medium due to cross-interference between the NSTR links. Hence, to support simultaneous UL transmissions on two NSTR links, the start times of PPDUs transmitted by the non-AP STAs on the two links need to be synchronized as well. For example, STA 1 operating on one of the NSTR links would defer transmission when that link's backoff counter reaches zero, and wait for the backoff counter on the other NSTR link to reach zero. When the backoff counter reaches zero on the other NSTR link as well, then the non-AP MLD would initiate simultaneous UL transmissions on the two NSTR links. The PPDU end times of these UL PPDUs also need to be aligned within ±8 μs. Figure 6-16 shows the PPDU start-times sync requirement for NSTR operation.

17 See P802.11be D5.0, clause 35.3.16.5.
18 See P802.11be D5.0, clause 35.3.16.6.

A non-AP MLD identifies a set of NSTR link pairs at the time of association with an AP MLD in the NSTR Indication Bitmap field in the Basic ML element. Support for the NSTR mode of operation is optional for both the AP MLD and the non-AP MLD. An AP or a non-AP STA affiliated with an MLD can choose to not initiate transmission after gaining channel access if that transmission would cause interference to the non-AP STA operating on one of the links of an NSTR link pair.

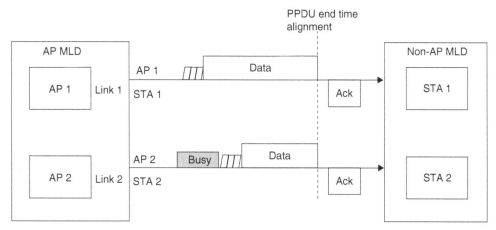

FIGURE 6-15   NSTR Operation: Alignment of PPDU End Times

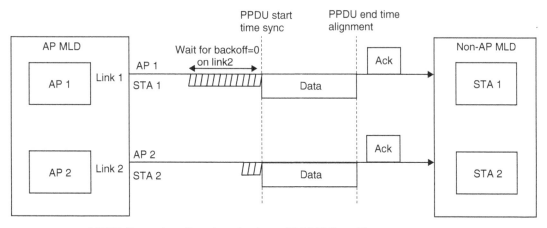

FIGURE 6-16   NSTR Operation: Synchronization of PPDU Start Times

Enhanced Multi-Link Multi-Radio

The EMLMR mode of operation is similar to the STR mode, but with the additional capability of dynamic reconfiguration of spatial multiplexing on multiple links. For example, in the STR mode of operation, a non-AP MLD may have a 2×2 radio on a 5 GHz link and a 2×2 radio on a 6 GHz link. If the 5 GHz link is busy, the 6 GHz link remains operating as a 2×2 radio. However, if the non-AP MLD is operating in the EMLMR mode, and if one link is congested, then it can reconfigure its resources to the other link. Specifically, in the previous example, the 6 GHz link can be reconfigured to operate as a 4×4 radio. EMLMR operation follows a similar procedure as EMLSR operation. In case of congestion or unavailability of certain links, the EMLMR operation mode can achieve higher throughput on other EMLMR links, though these benefits come at the cost of much overhead and implementation complexity.

Support for the EMLMR mode is specified by the non-AP MLD and the AP MLD in the EML Capabilities field in each of their transmitted Basic ML elements by setting the EMLMR Mode field to 1. During the ML (re-)setup, a multi-radio non-AP MLD indicates its support for STR link pairs. The EMLMR mode is disabled by default on the setup links. A non-AP MLD sends an EML Operating Mode Notification (EML OMN) frame to explicitly enable the EMLMR mode. A non-AP MLD specifies the set of MCS, NSS, and channel widths that it supports during EMLMR operation on any of the EMLMR links via the EML OMN frame (using the EMLMR Supported MCS and NSS Set field and the MCS Map Count field). The EMLMR Link Bitmap field in the frame indicates the set of links on which the EMLMR mode is being enabled. In response, the AP MLD transmits an EML OMN frame to the non-AP MLD when it is ready to serve the non-AP MLD in the EMLMR mode and within the transition timeout period indicated in the EML Capabilities field. The non-AP MLD enters EMLMR mode after it receives the EML OMN frame response from the AP MLD or after the transition timeout period, whichever happens first.[19]

After the EMLMR mode is enabled, a non-AP MLD is in listening mode on each of the EMLMR links using its per-link spatial stream capabilities. An affiliated AP operating on one of the EMLMR links transmits a PPDU to the non-AP MLD in the initial frame exchange that contains a padding duration at least as large as that indicated for the EMLMR mode by the non-AP MLD. This padding duration is needed for the non-AP MLD to reconfigure its spatial streams to the value indicated for EMLMR on the link where the initial frame exchange was received. The initial frame exchange for EMLMR between the AP STA and the non-AP STA can take the form of any frame exchange, as long as the EMLMR padding requirement is met in that frame. After the initial frame exchange on an EMLMR link, the non-AP MLD is able to operate with the number of spatial streams indicated for the EMLMR operation for both receiving and transmitting PPDUs on that link within the TXOP. After the frame exchange is completed on one of the EMLMR links, the non-AP MLD switches back to listening operation on its EMLMR links following the EMLMR transition delay period.

EMLMR operation adds further complexity to the STR operation for a non-AP MLD. It may not offer significant benefits over STR operation due to the overhead added by the ICF and response frame. Hence, the EMLMR mode is not being certified by the Wi-Fi Alliance and is unlikely to see wider adoption in the market.

---

19 See P802.11be D5.0, clause 35.3.18.

ML Reconfiguration

In 802.11be, the fundamental way to change the setup links between an AP MLD and a non-AP MLD is through (re)association. Reassociation is an expensive operation, however, because it leads to disconnection. For optimized multi-link operation and better MLO manageability, specifically in enterprise deployments, features are needed that would enable more dynamic reconfigurations of the MLD itself and dynamic setup of multiple links between the peer MLDs. Thus, it is important to address use cases and scenarios that require either the AP MLD or the non-AP MLD to dynamically reconfigure the set of links supported and activated for the MLD. Some of the scenarios that benefit from dynamic ML reconfiguration include the following:

- **Affiliated AP link being removed:** Sometimes, one of the links of an AP MLD goes away for some period—for example, because the AP is being rebooted or the AP software is being upgraded. In other use cases, an AP link may be repurposed for another operation for some extended period—for example, to validate the active state of nearby co-channel APs by acting as a STA that associates to the nearby AP and runs connectivity and feature validation checks. In these scenarios, the AP link removal operation is stateless, and the AP does not maintain the association and other state information for non-AP STAs for that link after the link is added back to the AP MLD. To deal with these use cases, an AP removal mechanism is defined to dynamically advertise the removal of an affiliated AP from the AP MLD.

- **New affiliated AP being added:** In conjunction with the use cases mentioned in the previous bullet, there is the use case of a new affiliated AP being added to an AP MLD—for example, adding back the AP that was removed earlier. For these scenarios, the 802.11be amendment has defined support for dynamically adding one or more affiliated APs to an AP MLD.

- **Dynamic reconfiguration of setup links:** Sometimes, a non-AP MLD needs to add or delete links to its current multi-link setup. The basic way to achieve this goal is by performing reassociation with the added link; however, it is not the preferred approach due to the connection disruption mentioned earlier. The use case for adding a link can arise when an AP MLD removes an affiliated AP (the scenario described in the first bullet) and later adds that AP to the AP MLD. At this point, it is preferable for the non-AP MLD to dynamically add this link to its setup links without reassociation so that it can start using that link again. Another use case is when an AP MLD is recommending links to the non-AP MLD for load-balancing purposes, which can necessitate that a non-AP MLD delete an existing setup link and add another setup link. For such use cases, the 802.11be amendment has defined a link reconfiguration operation for enabling dynamic add/delete links to the setup links of a non-AP MLD.

This section examines the ML reconfiguration features and related procedures. These features have been adopted as part of the Wi-Fi Alliance's Wi-Fi 7 certification program to provide better MLO manageability. The ML reconfiguration operation in the 802.11be amendment collectively refers to the procedures defined for the following features:

- AP addition and AP removal to/from an AP MLD
- Link reconfiguration to dynamically add or delete setup links
- AP recommendation for link reconfiguration

AP Addition and AP Removal

An AP MLD can add a new affiliated AP to the AP MLD at any point in time, triggered by the STA Management Entity (SME) layer. The new affiliated AP gets announced in the Beacon and Probe Response frames transmitted by all other affiliated APs of the AP MLD, as well as in its own Beacon and Probe Response frames. In the Basic ML element, the Maximum Number Of Simultaneous Links field is incremented by 1 when an affiliated AP is added to the AP MLD. The RNR element also includes a new TBTT Information field carrying MLD parameters for the newly added AP.[20]

If a multiple BSSID set is deployed, and if the AP MLD includes a nontransmitted BSSID of the multiple BSSID set, then the addition of the new affiliated AP in the AP MLD is also advertised by the transmitted BSSID of that multiple BSSID. The transmitted BSSID includes the revised Basic ML element (indicating an AP has been added) for the AP MLD in the nontransmitted BSSID profile in the Multiple BSSID element. The transmitted BSSID also includes an entry in the RNR element for the newly added AP.

A non-AP MLD identifies that a new AP has been added to its associated AP MLD from the Basic ML element or RNR element in the Beacon and Probe Response frames. If desired, the non-AP MLD can then initiate the link reconfiguration procedure to add a new link with the new AP in its setup links if that feature is supported on both of the peer MLDs (see the "Link Reconfiguration for Add/Delete Links" section).

Similar to the AP addition case, an AP MLD can trigger removal of an affiliated AP at any point in time, upon trigger by the SME layer. Removal of an AP is advertised in the Beacon and Probe Response frames for some time by including a Reconfiguration Multi-Link element (Reconfig ML element) before the affiliated AP is removed from the AP MLD and its BSS is terminated.[21] This provides for advance notification of AP removal to non-AP MLDs that either have the link corresponding to the AP as one of their setup links or may want to add that link as a setup link. It enables the non-AP MLDs to better plan for the AP removal—for example, by adding another available link to their setup links.

The Reconfig ML element is a new variant of a multi-link element with a type equal to Reconfiguration.[22] The Reconfig ML element provides a common framework for signaling ML reconfiguration

---

20 See P802.11be D5.0, clause 35.3.6.2.
21 See P802.11be D5.0, clause 35.3.6.3.
22 See P802.11be D5.0, clause 9.4.2.312.4.

operations and is used across multiple ML reconfiguration features. For the AP removal operation, the Reconfig ML element includes a Per-STA Profile subelement for the AP being removed; this subelement provides an AP Removal Timer, as shown in Figure 6-17. The AP Removal Timer field specifies the number of target beacon transmission times (TBTTs) of the affiliated AP before that AP is removed, and this value is decremented by 1 in each subsequent Beacon frame. A value of 1 for the AP Removal Timer implies that the AP will be removed from the AP MLD at the next TBTT.

In the somewhat unlikely scenario in which multiple APs of an AP MLD are being removed, multiple Per-STA Profile subelements are included in the Reconfig ML element. Each has its own AP Removal Timer indicating the number of TBTTs before the corresponding affiliated AP will be removed.

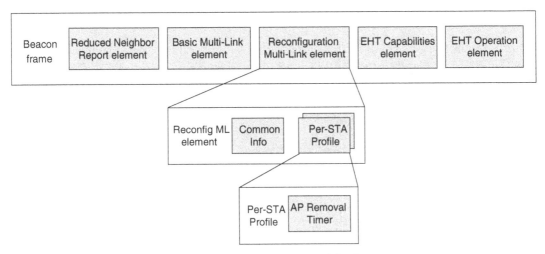

FIGURE 6-17   Beacon Frame Including a Reconfiguration ML Element

In the case of AP removal, the AP also transmits a BSS Transition Management (BTM) Request frame to non-MLD STAs that are associated with the AP being removed. This frame notifies them of the upcoming termination of the BSS of that AP. The BTM request instructs these non-MLD STAs to move to another AP, and informs them that their association on the current AP will be terminated, as indicated by the Disassociation Timer field in the BTM request.

If a non-AP MLD has only one setup link and it was on the link of the AP being removed, then the non-AP MLD is disassociated from the AP MLD when the affiliated AP is removed. If a non-AP MLD has multiple setup links with the AP MLD, then the non-AP MLD remains associated with the AP MLD on the remaining setup links after the AP is removed.

Link Reconfiguration for Add/Delete Links

As described previously in the "ML Reconfiguration" section, a non-AP MLD sometimes needs to add links to or delete links from its setup links. The link reconfiguration operation is defined to enable the dynamic addition and/or deletion of links to/from the setup links without reassociation with the

AP MLD.[23] A non-AP MLD can add and/or delete links by exchanging Link Reconfiguration Request and Response frames with the associated AP MLD. A non-AP MLD can include a request for multiple add link and delete link operations in the same Link Reconfiguration Request frame, which is sent on one of the existing setup links. The AP MLD may accept or reject a requested add link operation based on factors such as load balancing of STAs across links. A non-AP MLD with only one setup link can switch its current link to another link by requesting both a delete link operation (for the current link) and an add link operation (for another link) in the same Link Reconfiguration Request frame. An AP MLD typically accepts a request for a delete link operation, except in some error scenarios—for example, when the non-AP MLD requests that the last setup link be deleted without making another add link request in the same request. The AP MLD sends a Link Reconfiguration Response frame, indicating the status of the add link and/or delete link requests. It also provides group keys (GTK, IGTK, BIGTK) for every successfully added link in the response frame.

Figure 6-18 illustrates the process of link reconfiguration for an add link operation. Initially, Links 1 and 2 are established for the non-AP MLD as part of its setup links. The non-AP MLD then sends a Link Reconfiguration Request frame to the AP MLD requesting to add Link 3 to its setup links. The AP MLD accepts the add link operation and sends a Link Reconfiguration Response frame indicating success. After the successful link reconfiguration, the non-AP MLD now has three links (links 1, 2, and 3) set up with the AP MLD.

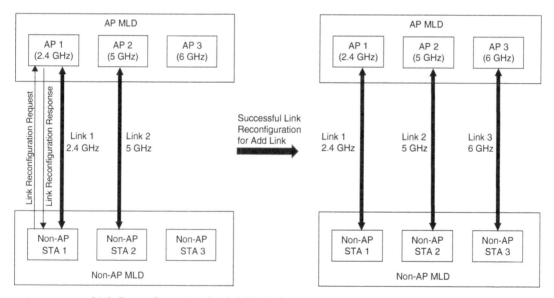

FIGURE 6-18   Link Reconfiguration for Add Link Operation

---

23 See P802.11be D5.0, clause 35.3.6.4.

In other cases, a non-AP MLD may perform both add and delete links operations simultaneously with its associated AP MLD (e.g., when an AP MLD recommends that a non-AP MLD change its set of associated links), or the non-AP MLD may seek to delete one of its links for better link management. Figure 6-19 illustrates the link reconfiguration for simultaneous add link and delete link operations. Initially, Links 1 and 2 are established for the non-AP MLD as part of its setup links. The non-AP MLD then sends a Link Reconfiguration Request frame to the AP MLD requesting to add Link 3 to its setup links and delete Link 2 from its setup links. The AP MLD accepts these operations, and sends a Link Reconfiguration Response frame indicating success for both operations. After successful link reconfiguration for the add link and delete link operations, the non-AP MLD has two links (links 1 and 3) established as part of its setup links with the AP MLD. Note that the protocol allows a non-AP MLD to also send a Link Reconfiguration Request frame requesting to just delete one or more links.

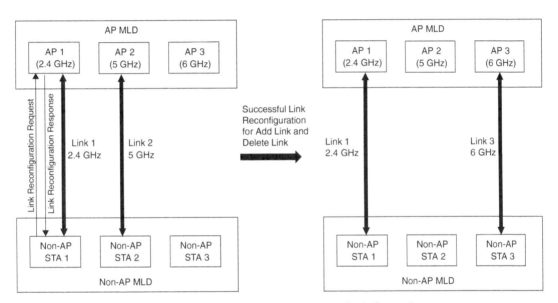

FIGURE 6-19   Link Reconfiguration for Add Link and Delete Link Operations

The Link Reconfiguration Request frame[24] includes a Reconfiguration ML element, which itself includes a separate Per-STA Profile for each requested link reconfiguration operation (Figure 6-20). The add link or delete link operation is indicated using the Reconfiguration Operation Type field in the STA Control field. The Link ID field indicates the link for which the reconfiguration operation is requested. For the add link case, the complete profile for the non-AP STA that will be operating on the added link is provided in the STA Profile. If new NSTR link pairs are created during this operation, then the NSTR Indication Bitmap is included as well.

24 See P802.11be D5.0, clause 9.6.35.13.

> **Note**
>
> When operating channel validation is activated at both the non-AP MLD and the AP MLD, an Operating Channel Information (OCI) element is also exchanged when performing an add link operation as part of the Link Reconfiguration Request/Response exchange. When the OCI element is included, both the AP MLD and the non-AP MLD validate the operating channel information. For simplicity, the OCI element is not shown in Figures 6-20 and 6-21.

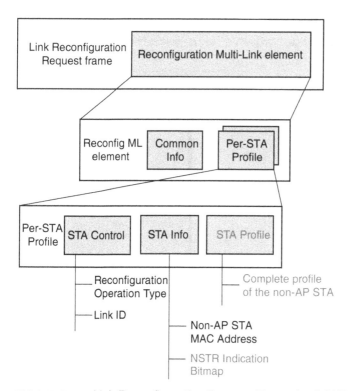

FIGURE 6-20   Link Reconfiguration Request Frame for Add/Delete Links

The Link Reconfiguration Response frame[25] provides the status of the reconfiguration operation for each requested Link ID in the reconfiguration status list, as shown in Figure 6-21. For the case in which one or more links are successfully added, the response frame includes a Basic ML element that provides the complete profile for each of the APs with which a link is added in a Per-STA Profile subelement. The group keys (GTK, IGTK, and BIGTK) for all newly added links are provided in the Group Key Data field in the response frame.

---

25 See P802.11be D5.0, clause 9.6.35.14.

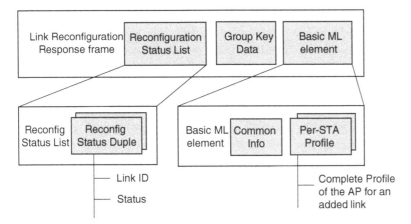

FIGURE 6-21   Link Reconfiguration Response Frame for Add/Delete Links

AP Recommendation for Link Reconfiguration

As described in the "ML Reconfiguration" section, sometimes an AP recommends add link and delete link operations to a non-AP MLD, such as for load-balancing purpose. The 802.11be amendment defines two ways to accomplish AP-based recommendation for link reconfiguration:

- Using the BTM Request frame
- Using the Link Reconfiguration Notify frame

An AP MLD can send a BTM Request frame to a non-AP MLD to recommend a different link set for the multi-link setup to the same AP MLD by including the Basic ML element in the BTM request (see the "BTM for MLDs" section later in this chapter). Relative to the current setup links, the recommended link set includes an extra link (for indicating the add link) and omits an already setup link (for indicating the delete link) in the Basic ML element included in the BTM request. The amendment also defines a new Link Reconfiguration Notify frame[26] to explicitly recommend links to be added and/or deleted from the current setup links. In both approaches, the non-AP MLD follows through by sending a Link Reconfiguration Request frame to update its set of associated links (as explained in the "Link Reconfiguration for Add/Delete Links" section) based on the recommendation provided by the AP MLD.

Link Management

To better optimize use of links across all associated STAs, an AP MLD can spread out (i.e., load balance) all the non-AP MLDs' traffic across two or more available links. For example, it might take this step to avoid one link becoming overly congested or to provide a less congested link for high-priority or

---

26 See P802.11be D5.0, clause 9.6.35.12.

low-latency traffic. An AP MLD has the best knowledge of traffic conditions across multiple links, whereas a non-AP MLD may have only limited understanding of the congestion level for each link, especially if it routinely operates in power save mode. The AP MLD can provide assistance to non-AP MLDs to load-balance traffic across available links so that it can offer the most desirable link conditions to its associated set of non-AP MLDs.

The 802.11be amendment defines a mechanism for dynamic management of how links are used for data frame exchanges of different TIDs via the TID-To-Link mapping (TTLM) feature. The TTLM mechanism allows links of a non-AP MLD to be enabled or disabled by mapping some or no TIDs to those links. The AP MLD may trigger the TTLM mechanism by sending the BTM Request frame to provide basic load balancing without requiring reassociation (see the "BTM for MLDs" section later in this chapter).

Additionally, an affiliated AP link sometimes needs to be deactivated for a short period of time. This scenario can arise when, for example, the AP operates for a short period on another channel for normal radio resource management (RRM) operation. In such cases, the prior state information (association and other state information for non-AP STAs for that link) is maintained when the AP link becomes operational again (i.e., stateful operation in contrast to stateless operation when an AP is removed). The advertised TTLM feature is available to dynamically enable and disable links of an AP MLD to support such cases. When the AP link becomes enabled again, all of the non-AP MLDs that still have this link in their setup links seamlessly resume their connectivity on that link.

TID-To-Link Mapping

The TTLM negotiation feature is defined in the 802.11be amendment as a means of better link management and has also been adopted as part of the Wi-Fi Alliance's Wi-Fi 7 certification program. By default, after the ML (re-)setup is completed between a non-AP MLD and an AP MLD, all TIDs are mapped to all the setup links for both the DL and the UL. This is referred to as the default mapping mode for TTLM. Once this mapping is completed, traffic belonging to any of the TIDs (TIDs 0–7) can be exchanged on any setup links.

An MLD can negotiate a non-default TID-To-Link mapping with its associated peer MLD using the TTLM negotiation exchange, which enables peer MLDs to negotiate how TIDs are mapped to setup links in the DL and UL directions.[27] The TID-To-Link Mapping Negotiation Support field in the MLD Capabilities and Operations field of the Basic ML element indicates whether TTLM negotiation is supported.

Two TTLM negotiation modes are defined in the 802.11be amendment:

- **TTLM mode 1 (TID-To-Link Mapping Negotiation Support = 1):** This mode supports mapping all TIDs to the same subset of links, for both the DL and the UL. It is typically used to support dynamic load balancing across links of a non-AP MLD, by mapping all TIDs to a smaller subset of setup links and later removing that mapping when load balancing is no longer needed.

27 See P802.11be D5.0, clause 35.3.7.2.3.

- **TTLM mode 3 (TID-To-Link Mapping Negotiation Support = 3):** This is the most flexible TTLM mode and allows mapping of any TID to any link set. However, it involves additional implementation complexity and is not likely to see broad market adoption. Support for mode 3 also implies that the MLD supports mode 1.

TTLM mode 2 was also proposed as an intermediate mode between mode 1 and mode 3. The goal with TTLM mode 2 was to allow mapping of only a subset of TIDs (e.g., high-QoS TIDs 4–7) to one of the links and mapping of all TIDs to all other links (for both the DL and the UL), thereby providing a less congested link for TIDs carrying higher-priority and high-QoS traffic. Unfortunately, the proposed mode 2 also entailed higher implementation complexity specifically for single-radio devices; as a result, it did not make it into D5.0 of the 802.11be amendment.

According to the 802.11be amendment, a setup link is considered enabled for a non-AP MLD if at least one TID is mapped to that link in either the DL or UL direction. Conversely, a link is defined as disabled for a non-AP MLD if no TIDs are mapped to that link in either the DL or UL direction. For a non-AP MLD, a TID must always be mapped to at least one setup link for the DL and UL.

The TTLM negotiation is performed between peer MLDs using the TID-To-Link Mapping Request/ Response exchange. Either the AP MLD or the non-AP MLD can initiate this negotiation. The initiator MLD sends a TID-To-Link Mapping Request frame, which includes one or two TID-To-Link Mapping elements specifying the requested TTLM (two elements are included if different TTLMs are suggested for the DL and UL). The responder MLD can accept or reject the proposed TTLM in the request frame; it sends a TID-To-Link Mapping Response frame indicating its decision. If it rejects a request for TTLM negotiation, the responder MLD's response can identify its preferred TTLM by including one or two TID-To-Link Mapping elements. An MLD can also send an unsolicited TID-To-Link Mapping Response frame to suggest a preferred TTLM to its peer MLD. Additionally, the TTLM negotiation can be done as part of the (Re)Association Request/Response exchange between peer MLDs by including a TID-To-Link Mapping element in the (Re)Association Request.

The TID-To-Link Mapping element[28] provides a bitmap indicating how each TID is mapped to the existing setup links via the Link Mapping of TID *n* fields, as shown in Figure 6-22. A Direction field is specified in the element indicating the DL direction, the UL direction, or both the DL and UL directions for which the TTLM is applicable.

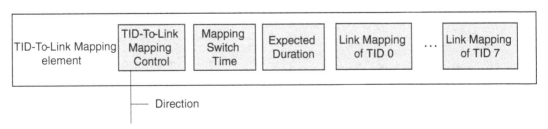

FIGURE 6-22   TID-To-Link Mapping Element

28 See P802.11be D5.0, clause 9.4.2.314.

After a TTLM negotiation is completed between peer MLDs, an enabled link can be used for exchanging individually addressed data frames of the TIDs that are mapped to that link in the direction negotiated in the TTLM. Individually addressed management frames, QoS Null frames, and control frames can be sent on any enabled links, with one exception: A Block Ack Request frame for TIDs not mapped to a link is not sent on that link.

If a link is disabled for a non-AP MLD because no TID is mapped to that link in either DL or UL, then that link is not used for exchanging individually addressed frames. Again, there is an exception: If the link is not globally disabled by the AP MLD, then it can be used in the non-AP MLD initiated procedures for class 1 and class 2 management frames, class 1 control frames, and the TID-To-Link Mapping Request, TID-To-Link Mapping Response, and TID-To-Link Mapping Teardown frames.

After a TTLM is successfully negotiated between two peer MLDs, either of the peer MLDs can tear down the negotiated TTLM. Yet again, there is an exception: When the TTLM was established due to the advertised TTLM, it cannot be torn down by a non-AP MLD (see the "Link Disablement and Enablement" sections for details on advertised TTLM). The TID-To-Link Mapping Teardown frame is used to tear down a TTLM.

Class 1, 2, and 3 Frames

Recall from Chapter 1 that a non-AP STA can be in state 1 (unauthenticated and unassociated), state 2 (authenticated and unassociated), state 3 (authenticated and associated), or state 4 (authenticated, associated, and 4-way handshake completed).

Class 1 frames are permitted to be exchanged from within all four states. They include four types of management frames (Beacon, Probe Request/Response, Authentication, and Deauthentication frames) as well as three types of control frames (RTS/CTS, ACK, and CF-End frames).

Class 2 frames are permitted to be exchanged only in states 2, 3, and 4. They include two types of management frames: (Re)Association Request/Response and Disassociation frames.

Class 3 frames are permitted to be exchanged only in states 3 and 4. They include two types of management frames (Action and Action No Ack frames), five types of control frames (PS-Poll, Poll, Block Ack Request, Block Ack, and Trigger frames), and the data frames.

Note that these classes also include other frame types that are legacy or irrelevant for this chapter—for example, frames used for independent BSS (IBSS) or for personal BSS (PBSS).

Link Disablement and Enablement

The TID-To-Link mapping feature can also be used by an AP MLD to disable a link for a certain amount of time by advertising a TTLM in the Beacon and Probe Response frames; this advertising indicates that no TID is mapped to the link being disabled. This feature is called advertised TTLM[29]

---

29 See P802.11be D5.0, clause 35.3.7.2.4.

in the 802.11be amendment. The advertised TTLM in the Beacon and Probe Response frames is a mandatory TTLM that must be followed by all associated non-AP MLDs. The mandatory TTLM is advertised for a period before the link becomes disabled, to provide advance notification to non-AP MLDs of the impending link disablement.

In the advertised TID-To-Link Mapping element, in the Link Mapping of TID n fields, no TID is mapped to the link being disabled. Additionally, two other fields are used for advertised TTLM: the Mapping Switch Time and the Expected Duration fields (refer back to Figure 6-22). The Mapping Switch Time field indicates the time at which the advertised TTLM becomes established, and the Expected Duration indicates the time duration for which the advertised TTLM is expected to be effective. When the advertised TTLM is still not effective, both the Mapping Switch Time field and the Expected Duration fields are included in the TID-To-Link Mapping element. When the Mapping Switch Time is reached and the advertised TTLM becomes effective, the Mapping Switch Time field is no longer included. In other words, only the Expected Duration field is included, and it indicates the remaining duration for which the advertised (and now established) TTLM is expected to remain effective. The expected duration can be extended if the AP link disablement period needs to be elongated.

After the Expected Duration counts down to 0 and the previously disabled AP link is re-enabled, the corresponding TID-To-Link Mapping element is no longer included in the Beacon and Probe Response frames. In some cases, when an AP MLD disables multiple links at different points in time, it will replace a current advertised TTLM with another advertised TTLM for the future. There are then two advertised TTLM elements in the Beacon and Probe Response frames: One element indicates a currently effective advertised TTLM, and the other element indicates a TTLM that will become effective in the future (per the indicated Mapping Switch Time). Note that when a mandatory TTLM is being advertised by an AP MLD and applies to all the associated non-AP MLDs, a non-AP MLD can still engage in an individual TTLM negotiation with the AP MLD if the individually negotiated TTLM does not conflict with the advertised and mandatory TTLM.

Figure 6-23 shows an example scenario involving advertised TTLM and individual TTLM negotiation between peer MLDs. At the start after the ML (re-)setup, Links 1, 2, and 3 are established between the peer MLDs and the default TTLM mapping is active (i.e., mapping all TIDs to all three links). The AP MLD then advertises TTLM A, which disables Link 3. After the Mapping Switch Time ends, TTLM A is active so only Links 1 and 2 are enabled; for those two links all TIDs are mapped, and no TIDs are mapped to Link 3. The non-AP MLD then initiates a TTLM negotiation for TTLM B, which maps all TIDs to Link 1. Given that TTLM B is a subset of TTLM A, it does not conflict with TTLM A, and the AP MLD accepts TTLM B for the non-AP MLD. After the TTLM negotiation, TTLM B becomes active for the non-AP MLD, so all TIDs are mapped to Link 1 only. Throughout the entire procedure, the non-AP MLD remains associated on Link 3 with the AP MLD. Later, when Link 3 is enabled again, the non-AP MLD can resume operation on that link by performing a TTLM negotiation that maps TIDs to Link 3 (not shown in Figure 6-23).

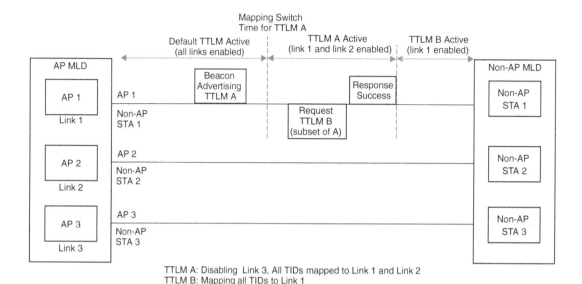

FIGURE 6-23   Advertised TTLM with TTLM Negotiation Between Peer MLDs

Other MLO Enhancements

In this section we provide an overview of some of the other important MAC enhancements related to the MLO feature.

ML Power Management

Each non-AP STA affiliated with a non-AP MLD maintains its own power management mode and power state (awake or doze state). Frames can be exchanged on a setup link between the peer MLDs if the non-AP STA is in the awake state on that link.

When a non-AP MLD is in power save mode on all the links where TID traffic can be delivered, then the AP MLD buffers that TID traffic for the non-AP MLD. The status of buffered individually addressed data for a non-AP MLD is indicated in the partial virtual bitmap of the TIM element by setting the AID bit corresponding to that non-AP MLD to 1. In the typical case when all TIDs are mapped to all links (i.e., the default TTLM mode) or a subset of links (i.e., TTLM mode 1), the bit corresponding to the non-AP MLD will be set in the TIM element transmitted on all enabled links for that non-AP MLD. The non-AP MLD can then detect that it has buffered traffic on one of the links from the TIM element and use the currently defined power-saving mechanism (e.g., PS-Poll or U-APSD) to fetch the buffered data on any of the enabled links.

> **Note**
>
> The 802.11be amendment defines a Multi-Link Traffic Indication (MLTI) element that can be used to set the per-link buffered traffic indication for non-AP MLDs. This consideration applies when all TIDs are not mapped to all enabled links for one or more non-AP MLDs (e.g., when TTLM mode 3 is used). However, due to the complexity of TTLM mode 3 and the MLTI element, both features are unlikely to see widespread adoption.

For MLO, the listen interval operation, which is used for power-saving purposes by clients, is defined at the MLD level. Specifically, the non-AP MLD sets the Listen Interval field in the (Re)Association Request frame at the MLD level. After a successful ML (re-)setup, the AP MLD uses the listen interval that gets agreed during the (Re)Association Request/Response exchange to determine how long to buffer frames for that non-AP MLD. At least one of the affiliated non-AP STAs of the non-AP MLD must transition to the awake state to receive the Beacon frame within the negotiated (MLD level) listen internal.

The AP MLD also applies the Ma Idle Period at the MLD level. If no frames are received from the non-AP MLD on any of the setup links within the MLD-level Max Idle Period, then the AP MLD can consider the non-AP MLD as idle for making disassociation decisions. Similarly, the Wireless Network Management (WNM) sleep mode feature, which enables an extended power save mode for client devices (typically IoT devices), works at the MLD level between the peer MLDs.

BSS Parameters Critical Update

In the 802.11be amendment, the BSS parameters critical update feature is defined as a notification mechanism for critical updates to BSS parameters of affiliated APs of an AP MLD. This mechanism is designed to allow a non-AP MLD to identify when critical updates to BSS parameters of other affiliated APs happen, without the need to periodically scan channels of those APs. Using the BSS parameters critical update procedure, once the non-AP MLD determines that there are critical updates to BSS parameters of any affiliated AP, it can take actions to retrieve the latest set of BSS parameters for that AP.

The BSS parameters critical update is achieved using the BSS Parameters Change Count (BPCC) field, which is maintained separately for each affiliated AP.[30] The BPCC field for an affiliated AP is initialized to 0, and is incremented by 1 whenever a critical update occurs to BSS parameters of that AP. The 802.11 standard had already defined a set of BSS parameter updates as critical updates, but the 802.11be amendment added more BSS parameter updates to that set.[31] The BPCC field for each of the other affiliated APs of the AP MLD is carried in the MLD Parameters of the RNR element corresponding to that AP, identified by the Link ID field (refer back to Figure 6-5). The BPCC field for other

---

30 See P802.1be D5.0, clause 35.3.10.
31 See P802.1be D5.0, clause 11.2.3.14.

affiliated APs of an AP MLD is also included in the (Re)Association Response frame in the Per-STA Profile subelement of the Basic ML element. The BPCC field of the reporting AP is included in the Common Info field of the Basic ML element (refer back to Figure 6-4).

Furthermore, as part of the BSS parameters critical update procedure, a Critical Update Flag (CUF) field is defined in the Capability Information and Status Indication field as a means to identify other MLD-level critical updates. The CUF field is set to 1 in the Beacon and Probe Response frames until and including the next DTIM beacon, when any of the following conditions are satisfied:

- The BPCC field is updated for any AP affiliated with the same AP MLD as the reporting AP.

- A new affiliated AP is added to the AP MLD with which the reporting AP is affiliated, or the removal of an affiliated AP is initiated for the AP MLD with which the reporting AP is affiliated (see the "AP Addition and AP Removal" section).

- An AP affiliated with the same AP MLD as the reporting AP becomes disabled or enabled through advertised TTLM (see the "Link Disablement and Enablement" section).

When a non-AP MLD detects that the CUF field is set to 1, it can use that indication to take actions to retrieve information for critical update events just listed. A non-AP MLD maintains the record of the last received BPCC value for each affiliated AP of the AP MLD it is associated with. When it detects that the latest received BPCC value for an affiliated AP differs from the last recorded BPCC value, the non-AP MLD attempts to receive a Beacon or Probe Response frame from that AP. There is one exception to this rule: When the latest received BPCC value for an affiliated AP is equal to the last recorded BPCC value plus 1 and the All Updates Included field in the MLD Parameters in the RNR element is also set to 1, then the non-AP MLD does not need to take any action because the received frame already includes the updated elements.

BTM for MLDs

For MLO, the BSS transition management procedure applies at the MLD level between the AP MLD and the non-AP MLD. The BSS Transition Management (BTM) Request frame has been enhanced so that it can provide a recommendation for AP MLDs or a subset of their affiliated APs.[32] To achieve MLD-level recommendation, the Neighbor Report element carried in the BTM Request frame includes a Basic ML element. When the AP MLD is providing a recommendation for a reported AP MLD (and not recommending a subset of APs of that AP MLD), it includes a Neighbor Report element for one of the affiliated APs of the reported AP MLD and includes a Basic ML element in the Neighbor Report. The Basic ML element does not include any optional field in the Common Info field; likewise, it does not include any Per-STA Profile subelement in the Basic ML element.

When the AP MLD is providing a recommendation for a subset of affiliated APs of a reported AP MLD, then it follows the same logic as just described and includes the Basic ML element in the

---

32 See P802.1be D5.0, clause 35.3.23.

Neighbor Report. In this case, the Common Info field in the Basic ML element includes the Link ID Info field, which is set to the Link ID of the AP reported in the Neighbor Report element. If the AP MLD is recommending other APs of the AP MLD, it includes in the Basic ML element a Per-STA Profile subelement for each of the recommended APs affiliated with the AP MLD.

Note that an AP MLD can send a BTM Request frame recommending a subset of links to the same AP MLD (i.e., to itself) to achieve load balancing of associated non-AP MLDs across its links. When the non-AP MLD receives such a BTM Request frame, and if it has more links in its setup links than the set of recommended links, it can initiate a TTLM negotiation with the AP MLD to reduce its set of enabled links to the recommended set of links. This feature is included in the Wi-Fi Alliance's Wi-Fi 7 certification as the basic load balancing feature.

Summary

Multi-link operation is one of the most important MAC layer enhancements introduced in the 802.11be amendment. MLO is intended to offer seamless operation over multiple links for client devices. MLO devices can achieve much higher throughput by using link aggregation, and can achieve reduced latency and improved reliability by using any of the available links for frame exchanges. In this chapter, we reviewed the architectural enhancements, new functionalities, operation modes, and procedures defined for the MLO feature. We specifically examined the new multi-link device (MLD) architecture and the details of MLD discovery and multi-link association procedures. Multiple MLO operation modes are defined in the 802.11be amendment to accommodate different radio capabilities of client devices (single-radio versus multi-radio), with the three most important MLO modes being MLSR, EMLSR, and STR.

For optimized MLO and improved MLO manageability, it is important to enable dynamic ML reconfiguration operations. Thus, in this chapter we reviewed the procedures for adding and removing affiliated APs to/from an AP MLD, and for dynamically adding and deleting links to/from the setup links of a non-AP MLD. We also reviewed dynamic link management operations using TID-To-Link mapping and the advertised TTLM feature, which enables an AP to achieve load balancing and dynamic enablement/disablement of AP links. Finally, we reviewed the ML power management aspects, the BSS parameters critical update mechanism defined for the MLD, and enhancements made to the BTM Request frame for MLDs.

This chapter has provided a thorough and comprehensive review of the MLO feature, which is one of the most important MAC layer enhancements added in the 802.11be amendment. In the next chapter, we detail other key MAC layer enhancements and features included in this amendment.

Chapter 7

EHT MAC Operation and Key Features

extremely high throughput (EHT) basic service set (BSS): [EHT BSS] A BSS in which the transmitted Beacon frame includes an EHT Operation element.

P802.11be, Draft 5.0, Clause 3

This chapter describes MAC layer operation and other MAC feature enhancements for 802.11be, also called Extremely High Throughput (EHT). These enhancements complement those described in Chapter 6, "EHT MAC Enhancements for Multi-Link Operations," which dealt with multi-link operation (MLO). First, we will review basic operation of EHT APs and EHT non-AP STAs within an EHT BSS. Next, we will explore some of the key MAC features and related procedures defined in 802.11be, including enhanced stream classification service (SCS) with QoS characteristics, restricted target wake time (R-TWT), triggered TXOP sharing, and Emergency Preparedness Communications Service (EPCS) priority access. The last section of this chapter describes the Wi-Fi security aspects unique to Wi-Fi 7.

EHT BSS Operation

A STA (an AP STA or a non-AP STA) that supports (mandatory and possibly optional) features defined in the 802.11be amendment is an EHT STA. Each affiliated AP of an AP MLD is an EHT AP, and each non-AP STA affiliated with a non-AP MLD is an EHT non-AP STA. For BSS operation, the AP first needs to start the BSS and start advertising the BSS's operation parameters in beacon and probe responses. For EHT, the Station Management Entity (SME) manages starting each of the affiliated (EHT) APs of the AP MLD, through MLME primitives defined to start an AP (MLME-START. request). Calling the MLME primitive for an AP initiates the process of creating an EHT BSS for that AP. The MLME primitive provides all the operation parameters needed to start the BSS. Once the EHT BSS is created, the MAC layer notifies the SME about the result by sending an MLME-START. confirm primitive to the SME. The SME then initiates the MLME-START.request primitive for each affiliated AP of the AP MLD, which tells it to start the EHT BSS for that AP. Once started, an EHT

BSS can provide connectivity for EHT non-AP STAs; it also provides connectivity for HE, VHT, and HT non-AP STAs based on the operating radio band of that EHT BSS. After the affiliated APs BSS are started, the non-AP MLD can then associate with the AP MLD using the ML setup after scanning and discovery.

Each generation of Wi-Fi defines a set of operation parameters specific to that generation in an operation element defined for that generation. An EHT AP indicates EHT-specific operation parameters for its EHT BSS in an EHT Operation element. The EHT Operation element of an AP is advertised in the Beacon and Probe Response frames and provided in the (Re)Association Response frames. In addition to the EHT Operation element, the operation of an EHT BSS is controlled by a combination of HT, VHT, and/or HE Operation elements, depending on the operating radio of the EHT AP:

- If operating in 2.4 GHz, the EHT BSS operation is controlled by the EHT Operation element, the HT Operation element, and the HE Operation element.

- If operating in 5 GHz, the EHT BSS operation is controlled by the HT Operation element, the VHT Operation element, the HE Operation element, and the EHT Operation element.

- If operating in 6 GHz, the EHT BSS operation is controlled by the HE Operation element and the EHT Operation element.

Note

An EHT STA operating in the 2.4 GHz band is also an HE STA and an HT STA. An EHT STA operating in the 5 GHz band is also a VHT STA. An EHT STA operating in the 6 GHz band is also an HE STA.

Basic BSS Operation

An EHT AP advertises, in the EHT Operation element, the set of basic operational parameters that are required to be supported by any EHT non-AP STA for operation in that EHT BSS. Figure 7-1 shows the main set of BSS operation parameters included in the EHT Operation element.[1] First, the AP advertises the Basic EHT-MCS and NSS Set field, which specifies the basic <EHT-MCS, NSS> tuples that must be supported by all EHT STAs in that BSS for both transmission and reception. Note that the 802.11be amendment defines two new EHT MCSs: MCS 12 and MCS 13 for 4K-QAM (as explained in Chapter 5, "EHT Physical Layer Enhancements").

The maximum number of spatial streams (NSS) that must be supported for reception and transmission are signaled for EHT-MCS 0–7, EHT-MCS 8–9, EHT-MCS 10–11, and EHT-MCS 12–13 as shown in Figure 7-1. For example, the Rx Max NSS That Supports EHT-MCS 12–13 field signals the maximum number of spatial streams supported for MCS 12 and MCS 13 for reception, and the Tx Max NSS

---

1 See P802.11be D5.0, clause 9.4.2.311.

That Supports EHT-MCS 12–13 field signals the maximum number of spatial streams supported for MCS 12 and MCS 13 for transmission. The basic <EHT-MCS, NSS> sets must be supported for all of the mandatory bandwidths for EHT; those bandwidths include 20, 40, and 80 MHz for an EHT-STA that is not a 20 MHz-only STA. An EHT STA must not join an EHT BSS if it cannot support all of the basic <EHT-MCS, NSS> sets advertised in the EHT Operation element. So, a non-AP MLD should attempt to associate only on affiliated APs/links of an AP MLD for which it can support the basic <EHT-MCS, NSS> sets.

The EHT AP announces its BSS operating channel width in the Channel Bandwidth field in the EHT operation information if it is announcing a different bandwidth for the EHT STAs than the non-EHT STAs. If no separate EHT channel bandwidth is announced, then the operating bandwidth of EHT BSS is determined based on the HE, VHT, and/or HT operating bandwidth announced in the respective operation element, per the radio band of the EHT BSS. If an EHT channel bandwidth is announced, then the operating channel center frequency (CCF) is indicated through the CCFS0 and CCFS1 parameters.

The Disabled Subchannel Bitmap field is included when there are punctured channels. It provides a list of the subchannels (of 20 MHz width) that are punctured in the BSS operating bandwidth (see the "Preamble Puncturing" section later in this chapter). MCS 15 is a longer-range MCS mode that was introduced in 802.11be, but is not certified in Wi-Fi 7. The MCS 15 Disable field indicates whether the AP has disabled or enabled the reception of the EHT PPDU with EHT MCS 15 in both the Data field and the EHT-SIG field.

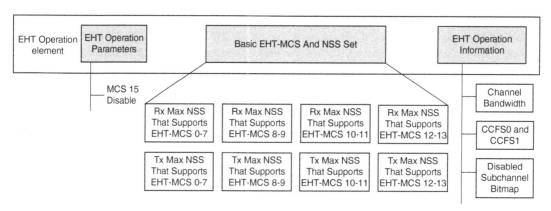

FIGURE 7-1   EHT Operation Element Content

A closely related element is the EHT Capabilities element.[2] This element defines the capabilities of an EHT STA rather than those of the EHT BSS. The EHT Capabilities element is transmitted by both the AP and the non-AP EHT STA. A non-AP STA declares that it is an EHT STA by transmitting the EHT Capabilities element in the (Re)Association Request frame. An AP includes the EHT Capabilities element in the Beacon, Probe Response, and (Re)Association Response frames. The EHT Capabilities

2 See P802.11be D5.0, clause 9.4.2.313.

element advertises the set of EHT capabilities supported by the EHT STA, including the EHT MAC capabilities, the EHT PHY capabilities (covered in Chapter 5), the Supported EHT MCS and NSS Set, and optionally the EHT PPE thresholds.

The set of supported EHT MAC features is identified in the EHT MAC Capabilities Information field by both the AP and the non-AP STA, as shown in Figure 7-2. Capabilities related to new EHT MAC features include SCS traffic description support, restricted TWT support, triggered TXOP sharing feature-related capabilities, and capabilities related to EPCS priority access. These MAC features are described later in this chapter. The maximum MPDU length that can be supported for 2.4 GHz is indicated by the Maximum MPDU Length field. For the 5 GHz and 6 GHz bands, the maximum MPDU length supported is the same as that indicated in the VHT Capabilities and HE Capabilities elements, respectively. Some other capabilities are signaled in fields shown under the miscellaneous category in Figure 7-2; the names of these fields are self-explanatory (OM = operating mode, TRS = triggered response scheduling, BQR = bandwidth query report).

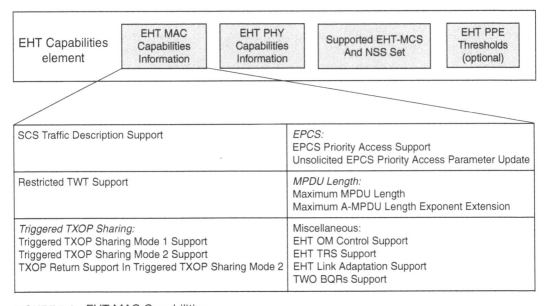

FIGURE 7-2 EHT MAC Capabilities

Besides the basic set of <EHT-MCS, NSS> tuples that must be supported by all the non-AP STAs connecting to an EHT BSS, an EHT STA can support other combinations of <EHT-MCS, NSS> at different bandwidths. This capability is indicated by the Supported EHT-MCS and NSS Set field in the EHT Capabilities element. As shown in Figure 7-3, the Supported EHT-MCS and NSS Set field indicates the supported combination of <EHT-MCS, NSS> at different PPDU bandwidths: 20 MHz only for the 20 MHz-only STA, and ≤ 80 MHz, 160 MHz, and 320 MHz for reception (Rx) and transmission (Tx). For each channel width, the EHT STA indicates support for a maximum NSS for different EHT MCSs for Rx and Tx in the corresponding EHT MCS Map field for that channel width. For example, for 80 MHz, the STA indicates support for its maximum NSS for MCS 12 and MCS 13 for reception in the Rx Max NSS That Supports EHT-MCS 12–13 field. Similarly, it indicates support for its maximum NSS for MCS 12 and MCS 13 for transmission in the Tx Max NSS That Supports EHT-MCS 12–13 field. Both are included in the EHT-MCS Map (BW <= 80 MHz) field. A 20 MHz-only STA supports the basic <EHT-MCS, NSS> set. The EHT PPE Thresholds field provides the nominal packet padding for an EHT PPDU (see Chapter 5 for details).

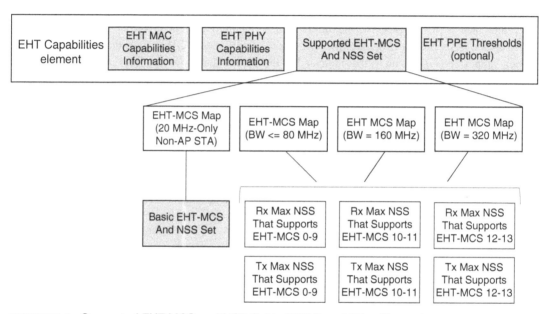

FIGURE 7-3 Supported EHT-MCS and NSS Set in EHT Capabilities Element

In addition to signaling the operation parameters and capabilities, an EHT STA supports operating mode updates to dynamically update its operating mode parameters. To do so, it uses the Operating Mode Notification (OMN) or Operating Mode Indication (OMI) mechanism; these mechanisms are described in the next section, "Operating Mode Updates."

An EHT STA determines the maximum receive NSS for an EHT-MCS that can be supported by a peer EHT STA (corresponding to different PPDU bandwidths) as the smaller of the values indicated in the Supported EHT-MCS and NSS Set and the maximum receive NSS indicated through the operating mode update procedure[3] by that STA. An EHT STA uses the maximum Rx NSS determined for a peer STA and its own maximum Tx NSS to determine the maximum NSS to be used for transmitting EHT PPDUs to that STA. Similarly, an EHT STA determines the maximum transmit NSS for an EHT-MCS that can be supported by a peer STA as the smaller of the values indicated in the Supported EHT-MCS and NSS Set and the maximum transmit NSS (Tx NSTS) indicated through the OMI update (see the "Operating Mode Updates" section).

---

3 See P802.11be D5.0, clause 9.4.2.313.4.

Operating Mode Updates

Sometimes, an AP or a non-AP STA may change its receive and/or transmit NSS and operating bandwidth as a means of saving power or improving performance. The AP may also update its BSS operating bandwidth for better performance. Such use cases are addressed by enabling dynamic updates to operating mode parameters.

The 802.11ac/VHT amendment introduced the operating mode notification (OMN) mechanism, whereby a STA (either an AP or non-AP STA) can notify another STA of changes in its operating mode (maximum NSS and channel width) by transmitting an Operating Mode Notification frame. The Operating Mode field included in the frame indicates the maximum number of spatial streams that the STA can receive (Rx NSS) and its supported channel widths (20, 40, 80, 160, or 80 + 80 MHz). An AP can notify another STA of changes in its maximum Rx NSS by sending either a broadcast or an individually addressed Operating Mode Notification frame, or alternatively by including an Operating Mode Notification element (which includes the Operating Mode field) in the Beacon frames for a period. When changing its operating bandwidth, an AP indicates the new operating channel width in the operation elements included in the Beacon and Probe Response frames. The OMN feature adds the flexibility to dynamically update the operating mode parameters for both the AP and the non-AP STA, and also applies to HE STAs and EHT STAs. However, this mechanism has a notable limitation: The scope of operating mode changes is limited to receive mode only.

The 802.11ax/HE added another, faster mechanism for indicating operating mode updates called the operating mode indication (OMI). OMI supports updates to both receive and transmit operating mode parameters. This mechanism enables an OMI initiator (either an AP or a non-AP STA) to signal a change in its receive operating mode and/or its transmit operating mode, in-band within a data or management frame. Specifically, an OM Control field is included in the A-Control field in the HT Control field in the MAC header of that frame. The OM Control field indicates the channel width and maximum number of NSS that the OMI initiator supports for receiving and transmitting PPDUs. The changes to operating mode parameters for Rx and/or Tx apply only after the TXOP in which the acknowledgment from the OMI responder is received for the frame carrying the OM Control subfield. A non-AP STA can also use the OM Control field to dynamically update its transmit operation to single-user (SU) versus multiuser (MU) UL OFDMA operation. A STA can set the UL MU Disable field and/or UL MU Data Disable field in the OM Control to signal that it has suspended trigger-based UL MU transmissions and will not respond to a trigger frame, and can set these fields to 0 to indicate that UL MU transmissions are enabled by the STA.

In EHT, the OMI mechanism is extended to add support for signaling the 320 MHz channel width as well as support for dynamic updates for the operating mode. EHT adds a new EHT OM Control field that works in conjunction with the HE OM Control field to indicate an EHT STA's channel width plus its receive and transmit NSS.

Preamble Puncturing

As you learned in Chapter 3, "Building on the Wi-Fi 6 Revolution," and Chapter 5, "EHT Physical Layer Enhancements," the preamble puncturing feature added in 802.11ax enables more efficient utilization of available channel bandwidth when an AP is operating on a wider bandwidth and interference (e.g., incumbents) is present in one or more 20 MHz channel blocks within the AP's operating bandwidth. This scenario is particularly applicable to the 6 GHz spectrum, where there is good chance that incumbents will be present. Such incumbents may include providers of primary services being offered on that channel and typically have narrowband signals (e.g., 10 MHz wide). As described in Chapter 3, the automated frequency control (AFC) system identifies channel widths where incumbent systems are in use. The AP can use this information to determine which channels should be punctured in a wider channel bandwidth. Other sources of constant interference could include unmanaged Wi-Fi networks or unmanaged 5G (NR-U) systems, and their presence could feed into the determination of which channels will be punctured from a BSS with a wide operating channel bandwidth.

EHT STAs puncture channels by skipping preamble transmission on those channels in DL and UL transmissions (hence the name *preamble puncturing*). Moreover, no data RUs are transmitted on the punctured channels. The 802.11ax amendment and Wi-Fi 6 enabled preamble puncturing only for OFDMA transmissions. In EHT, preamble puncturing is supported for both OFDMA and non-OFDMA transmissions, and support for puncturing is mandatory for non-OFDMA SU transmissions. This approach enables improved channel efficiency for both SU and MU transmissions when wider channel widths are deployed in an EHT BSS. In addition, puncturing is supported for the EHT sounding procedure.

In EHT, puncturing is supported for 80 MHz and higher-bandwidth PPDUs (20 and 40 MHz PPDUs cannot be punctured). For an 80 MHz PPDU, only 20 MHz is allowed to be punctured. For higher bandwidths PPDUs, larger channel bandwidths can be punctured (40, 80, 40 + 80 MHz). Table 7-1 summarizes the allowed puncturing bandwidths for different operating bandwidths of an EHT BSS. Figure 7-4 provides examples of preamble puncturing for each of the bandwidths (80, 160, and 320 MHz). Refer to Chapter 5 for details on all possible preamble puncturing patterns and large MRU allocations for different patterns.

TABLE 7-1 Allowed Preamble Puncturing Bandwidths

BSS Bandwidth	Puncturing Bandwidth Allowed
20 and 40 MHz	Puncturing not allowed
80 MHz	20 MHz
160 MHz	20 or 40 MHz
320 MHz	40, 80, or 40 + 80 MHz

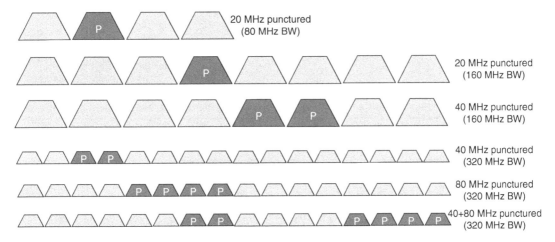

FIGURE 7-4   Preamble Puncturing Examples for Different BSS Bandwidths

An EHT AP signals the punctured channels of the BSS in the EHT Operation element. The AP indicates the set of 20 MHz subchannels that are punctured in the Disabled Subchannel Bitmap field in the EHT Operation element, which the AP transmits in Beacon and Probe Responses. The EHT STAs in the BSS do not use the set of punctured 20 MHz channels advertised by the AP for any PPDU transmissions within the BSS. In some scenarios, an EHT STA may also puncture other subchannels, besides the subchannels indicated in the Disabled Subchannel Bitmap, in an EHT MU PPDU or a non-HT duplicate PPDU as per the rules defined in the 802.11be amendment.[4] An EHT AP updates the advertised puncturing pattern in the EHT Operation element if the BSS's set of punctured subchannels changes.

EHT Sounding

As described in the "Sounding Procedure" section in Chapter 3, transmit beamforming and DL MU-MIMO operations require the AP to have knowledge of the channel state, so that it can organize its downstream transmissions and optimize the signal for each receiver. To that end, sounding allows a beamformer (typically the AP) to send a sounding PPDU—a null data packet (NDP)—to one or more beamformees (typically a STA or a group of STAs). The receiving STAs respond with compressed steering matrices that indicate the state of the channel for a subset of the tones. (The "Sounding and Beamforming" section in Chapter 5 provides PHY-related details for the sounding exchange and the returned steering matrix.)

The goals of sounding are the same in 802.11be as in 802.11ax, with the addition of two use cases:

- Sounding for 320 MHz bandwidths (in 6 GHz).

- Sounding for partial bandwidth. This provision is necessary to allow for sounding within a puncturing context, where one or more 20 MHz subchannels have to be bypassed.

4 See P802.11be D5.0, clause 35.15.2.

Figure 7-5 depicts the format of the EHT sounding PPDU.

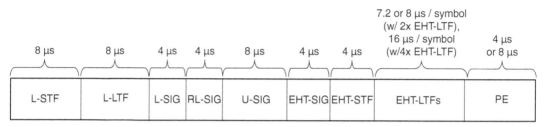

FIGURE 7-5   Format of the EHT Sounding NDP

The general structure of the EHT sounding NDP is similar to that of the previous generation's (802.11ax) sounding NDP, including the L-STF, L-LTF, L-SIG, and RL-SIG fields; the PE field; and a variable number of protocol-specific LTFs. However, whereas 802.11ax included the HE-SIG-A, HE-STF, and HE-LTFs fields, 802.11be includes the U-SIG, EHT-SIG, EHT-STF, and EHT-LTFs fields:

- The U-SIG field indicates that the NDP is EHT (802.11be) and identifies the NDP bandwidth. 802.11ax allowed up to 160 MHz transmissions, and 802.11be adds the 320 MHz option for the 6 GHz band. The U-SIG field also carries a Punctured channel Indication field, which indicates if some of the RUs are punctured. The field includes a code (a number between 0 and 24, depending on the NDP bandwidth[5]) that represents which 20 MHz, 40 MHz, and/or 80 MHz RU(s) or MRU(s) is/are punctured (when applicable). There is no puncturing for 20 MHz or 40 MHz transmissions. The 80 MHz bandwidth allows for one 20 MHz puncture; a code from 1 to 4 indicates which 20 MHz segment is punctured. The 160 MHz bandwidth allows for 20 MHz or 40 MHz punctures, so in this case the codes range from 1 to 12 (one of 8 possible 20 MHz segments, or one of 4 possible 40 MHz segments). The 320 MHz bandwidth allows for 40 MHz, 80 MHz, or concurrent 80 MHz and 40 MHz puncturing (24 possibilities).[6]

- The EHT-SIG field contains only a Common Field, which indicates if spatial reuse modes are allowed for this transmission, and identifies the number of spatial streams in effect (if spatial reuse is allowed). The Common field also indicates if the NDP is beamformed, and specifies the number of EHT-LTF symbols included in the frame.

- The EHT-STF has the same role as the STF in previous generations, albeit this time in an 802.11be transmission context: It is intended to improve the automatic gain control estimation in a MIMO transmission.

5 See P802.11be D5.0, Table 36-30.
6 See Chapter 5, Table 5-3 and Table 5-4, for the list of codes and their puncturing representation.

Just like 802.11ax, 802.11be allows for trigger-based and non-trigger-based sounding, for a group of STAs or a single STA, respectively. These two modes in 802.11be follow the same choreography and message exchange as in 802.11ax (illustrated in Figure 3-5 in Chapter 3). Just as in 802.11ax, the STAs' response can be single-user (SU) feedback, multi-user (MU) feedback, or channel quality indication (CQI) feedback.

The 802.11be amendment includes other possibilities for EHT sounding. In particular, EMLSR mode allows a client device to implement a single full-capability radio, and one or more low-capability radios (see the "Enhanced Multi-Link Single Radio" section in Chapter 6, "EHT MAC Enhancements for Multi-Link Operations," for more details). When a STA receives an initial control frame on one of its links, the STA in EMLSR mode switches its full radio to that link and operates there using all antennas. Therefore, before starting the sounding procedure with STAs operating in EMLSR mode, the AP sends a MU-RTS frame to ensure that the STA is operating in its full capability mode. The AP then sends the EHT NPDA and EHT sounding NDP.

The upper part of Figure 7-6 illustrates this scenario, for non-triggered sounding, where the compressed beamforming/CQI is sent directly after the EHT sounding NDP. Naturally, there is an alternative where multiple non-EMLSR STAs can participate in the conversation. In this case, the MU-RTS is first sent to the STAs in EMLSR mode. Each of the STAs then switches its main radio to that channel and replies with a CTS. Next, the AP sends the subsequent EHT NDPA frame, then the EHT sounding NDP, to the group. At that point, the STAs are triggered and reply with their individual EHT compressed beamforming/CQI information.

Many other sounding scenarios are possible with EMLSR operations. The main requirement is the presence of an initial control frame to ensure that the EMLSR STAs are operating in full-capability mode before they receive the EHT NDPA frame. At the bottom of Figure 7-6, you can see an example where n STAs are in the EMLSR mode, and other STAs ($n + 1$ to x) are not in the EMLSR mode. The initial control frame is a BSRP trigger frame; this frame brings clients 1 to n to operate with full-capability radio on the channel. It is followed by the corresponding PPDUs transmission by those STAs. This transmission is then followed by the EHT NDPA. The AP can also group the STAs (e.g., STAs in EMLSR mode in one group and non-EMLSR STAs in another) when it triggers the group with the BFRP Trigger frame to retrieve the STAs sounding matrices, as shown in Figure 7-6.

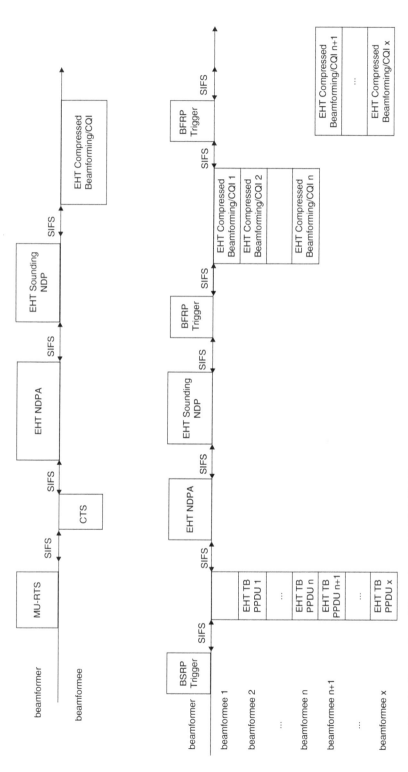

FIGURE 7-6   Examples of Sounding with EMLSR Operation

Enhanced SCS

Around the time frame when the 802.11be amendment was being developed, adoption of some existing applications accelerated and some new applications and use cases began emerging across different vertical segments (enterprise, home, industrial) that demanded more predictable and lower latency and improved reliability. Use of video collaboration applications across enterprises exploded during the COVID pandemic and post-COVID era, adding requirements for more consistent latency and jitter so that users could achieve better productivity during these sessions. Industrial IoT applications of robotics saw increasing growth at warehouses and factories—for example, with automated guided vehicles (AGVs) and autonomous mobile robot (AMR) devices, which required more determinism. The emergence of extended reality (XR) devices and experiences—including virtual reality (VR), augmented reality (AR), and mixed reality (MR) capabilities—required end-to-end low latency of approximately 20 ms to meet the target motion-to-photon latency for VR experiences and avoid VR-induced sickness. In many environments, requirements for these applications could not be met with the existing QoS mechanism adopted from 802.11e with four access categories (AC_VO, AC_VI, AC_BE, and AC_BK). See Chapter 4, "The Main Ideas in 802.11be and Wi-Fi 7," for more details on the challenges related to meeting QoS for these existing and emerging applications.

One of the goals of the EHT project in IEEE 802.11 was to define at least one mode of operation that offered improved latency and jitter, so as to address the QoS requirements for applications that demanded improved QoS. The 802.11be designers started by evaluating the existing protocols and mechanisms that could be leveraged to achieve improved QoS. Previously, the 802.11e amendment had defined a TSPEC (traffic specification) element[7] that provided a way for a STA to signal traffic requirements for its flows to the AP using the ADDTS request/response exchange. The flow classification was done using one or more TCLAS elements in the ADDTS exchange that specified the admission control for these flows. Although this mechanism was defined in 802.11e in 2005, the use of TSPEC to signal the traffic flow requirements was never widely adopted by Wi-Fi STAs because of the complicated set of parameters that had to be provided in the TSPEC element. All the parameters shown in Figure 7-7 were required to be provided in a TSPEC element, and it was extremely hard (if not impossible) for an application to characterize all these parameters for its flow. Due to the complexity involved, the TSPEC and ADDTS request/response mechanism was not much used outside single-function devices.

Well aware of these limitations and challenges faced by the TSPEC and ADDTS mechanisms in the past, the 802.11be designers sought to develop a new scheme for QoS improvements that would be much simpler to implement and, therefore, would have a higher chance of being adopted for Wi-Fi devices. After the 802.11e amendment, the IEEE 802.11 group had proposed some further enhancements to improve multimedia streaming performance as part of the 802.11aa amendment. This included defining intra-access category prioritization and the stream classification service (SCS) to enable prioritization of DL flows at the AP. SCS turned out to be a good fit for adding QoS-related enhancements in 802.11be, as you will learn next.

---

7 See P802.11-REVme D5.0, clause 9.4.2.28.

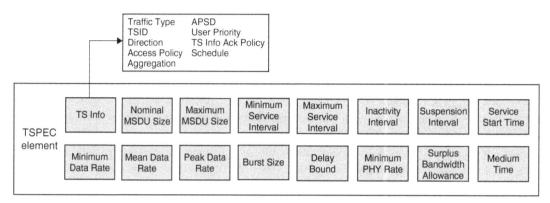

FIGURE 7-7   TSPEC Element Definition

SCS to Prioritize DL Flows (Pre-802.11be)

The 802.11aa amendment defined two new features for QoS-based prioritization (see Chapter 2, "Reaching the Limits of Traditional Wi-Fi," for the background context). The first feature was intra-access category prioritization (IACP), which provided a way to differentiate between audio/video streams within the same AC category for AC_VO and AC_VI. One of the main use cases envisioned for this enhancement was differentiating between a video-conferencing flow and a streaming flow, where both could be using the AC_VI category. The IACP feature enabled differentiation of traffic streams within the same AC for voice and video by splitting the transmit queue for each of these ACs into two queues—a primary queue and an alternate queue—leading to a total of six transmit queues (Figure 7-8). The alternate queues were tagged as A_VO or A_VI. The six queues were still mapped to four EDCA functions (EDCAF), just as before, for the four ACs. An implementation-specific scheduler was used to select packets from the primary or alternate queue to pass to the EDCA function. This scheduler was configured to select frames from the primary queue with higher probability than the frames from the alternate queue.

The 802.11 standard defines mapping of user priority (UP) received from the upper layer to ACs and the two transmit queues (primary and alternate), as shown in Table 7-2. For VI, UP 4 is mapped to the alternate queue (A_VI) and UP 5 is mapped to the primary queue (VI). Frames with UP 5 are served with higher probability than frames with UP 4, which made sense in terms of low to high priority order for UPs. However, for VO, the mapping is reversed; that is, the lower UP 6 is mapped to the primary queue and the higher UP 7 (used for network control traffic per the 802.1D designation) is mapped to the alternate queue. Hence, for VO, frames with UP 6 are served with higher probability than frames with UP 7. This seems counterintuitive, but there was a good reason for mapping them this way. Given that UP 7 in 802.1D is used for network control traffic (e.g., switch-to-switch traffic) and such traffic is rarely sent over Wi-Fi, UP 7 is hardly used but was kept in 802.11. The most frequently occurring voice traffic is mapped to UP 6, so it made sense to use the primary voice queue to prioritize the most prevalent voice traffic. In fact, no differentiation can be provided for voice traffic with the alternate A_VO queue because the common voice traffic is not mapped to UP 7. This was a limitation of the alternate queuing scheme (see Chapter 2 for more details).

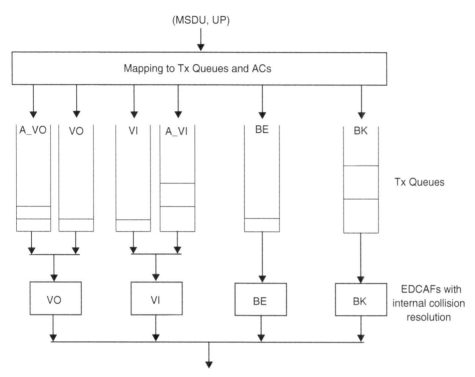

FIGURE 7-8   Intra-Access Category Priority

Supporting two alternate queues required hardware changes in most Wi-Fi devices, adding to their costs. The QoS differentiation benefits provided by having two alternate queues did not prove to be strong enough to ensure wide market adoption of this feature, mainly due to the cost impact and additional complexity added.

TABLE 7-2   UP-to-AC-to-Tx Queue Mapping

Priority	UP	AC	Tx Queue	Tx Queue with IACP	Designation
Lowest	1	AC_BK	BK	BK	Background
	2	AC_BK	BK	BK	Background
	0	AC_BE	BE	BE	Best effort
	3	AC_BE	BE	BE	Best effort
	4	AC_VI	VI	A_VI	Video (alternate)
	5	AC_VI	VI	VI	Video (primary)
	6	AC_VO	VO	VO	Voice (primary)
Highest	7	AC_VO	VO	A_VO	Voice (alternate)

The second QoS feature defined in the 802.11aa amendment was the stream classification service (SCS). The main functionality provided by this feature was to enable STAs to request specific QoS treatment from the AP for downlink flows. This allows a STA to indicate to the AP that certain application flows desire specific QoS mapping in the downlink to meet their QoS requirements. For an application, it is desirable to retain the QoS markings (e.g., DSCP in the IP header) as the packets traverse the Internet, so that prioritization can happen on each network node along the way. To achieve equivalent prioritization within Wi-Fi, the AP needs to have the correct QoS marking when packets arrive from the DS. However, one of the challenges for achieving end-to-end QoS was that the DSCP QoS marking was typically reset by routers along the way (due to trust reasons), resulting in packets being marked as "best effort." Hence, when the packets reach the Wi-Fi AP, their DSCP markings are typically not intact (as intended by the application), so the packets are delivered using the AC_BE category over Wi-Fi. SCS signaling from the STA addressed this issue. For example, a STA could request marking of a video-conferencing flow in DL to the UP 5 (AC_VI) category, independent of how the packets coming from the DS for that flow are tagged.

The SCS feature defines the SCS request/response exchange for negotiating the DL QoS classification and matching the QoS treatment to application flows. A STA can request a specific UP, drop eligibility, and/or transmit queue (primary or alternate) selection for flows (or streams) that match the indicated classification. The stream/flow classification is indicated by using one or more TCLAS elements, which can specify the IP-level classification for packets (e.g., a 5-tuple frame classifier based on the IP header). If multiple TCLAS elements are included, then a TCLAS Processing element is also included to indicate how multiple TCLAS elements need to be processed for classifying the packets (e.g., ANDed or ORed). An Inter-Access Category Priority element was defined to indicate the access policy for the desired QoS treatment for the matching flow.

As shown in Figure 7-9, the SCS Request frame includes one or more SCS Descriptor elements,[8] each requesting specific DL QoS treatment for a traffic flow. The key information provided in the SCS Descriptor element includes the SCS ID (assigned by the STA; identifies the SCS stream), the Request Type (for Adding, Changing, or Removing an SCS stream), the Intra-Access Category Priority element, one or more TCLAS elements, and optionally a TCLAS Processing element. The Intra-Access Category Priority element[9] specifies the user priority (UP) to be used for mapping the DL frames for the specific SCS stream. The Alternate Queue field indicates whether the primary or alternate queue should be used for the SCS stream. The Drop Eligibility field, when set to 1, indicates that, in case of insufficient resources, the packets for this stream can be dropped in preference to other SCS streams that have their Drop Eligibility field set to 0. An AP can accept or reject a request for each SCS stream based on its policy, resources, and/or identification of the SCS stream. For example, an enterprise AP might have a network policy to accept only SCS requests for DL QoS mapping for flows that are considered high priority/critical in that given deployment. The AP then indicates the accept or reject status for each request, identified by its SCS ID, in the SCS Response frame, as shown in Figure 7-9.

8 See P802.11-REVme D5.0, clause 9.4.2.120.
9 See P802.11-REVme D5.0, clause 9.4.2.119.

If an SCS stream is accepted by the AP, the AP processes the matching MSDUs (per-flow classification of that stream based on TCLAS) and assigns them to the signaled UP in the Intra-Access Category Priority element. It uses the Alternate Queue and Drop Eligibility values (if indicated) to process the matching MSDUs accordingly.

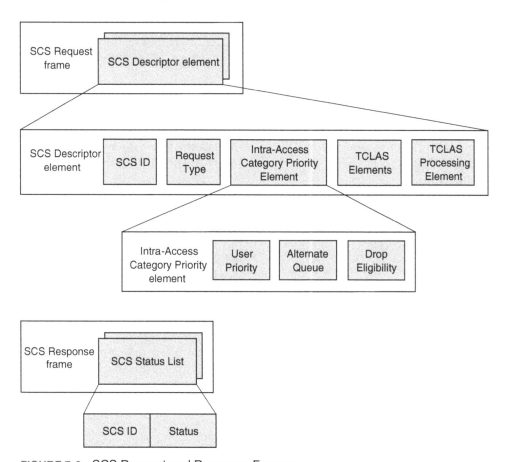

FIGURE 7-9   SCS Request and Response Frames

Figure 7-10 illustrates an SCS request/response exchange between a STA and an AP. In this example, the STA sends an SCS Request frame to request classification and QoS treatment for a DL application flow. The AP, per its policy, accepts this request from the STA and sends an SCS Response frame to signal Success status to the STA. After the successful SCS Request and SCS Response frames exchange, the AP processes downlink MSDUs that match the flow classification and applies the access policy specified in the Intra-Access Category Priority element.

The SCS protocol also provides flexibility by allowing either endpoint (the AP or the STA) to terminate an already established SCS stream. For example, a STA should remove an SCS stream if the application flow is no longer active, and an AP may terminate an SCS stream if its policy changes so that it no longer prioritizes that particular traffic flow. To remove an SCS stream, either the AP or the STA can send an SCS Request frame with a "Remove" request type for that SCS ID. This terminates the flow classification and special QoS treatment for the DL flow.

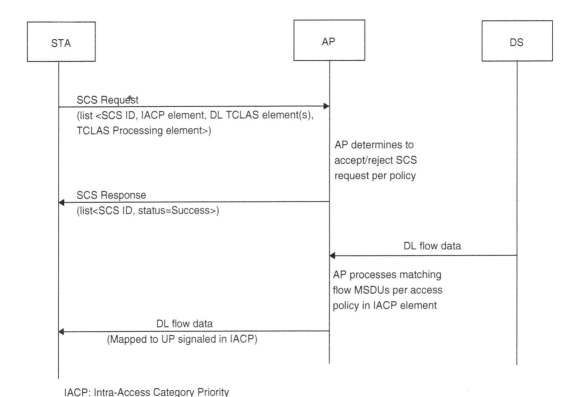

IACP: Intra-Access Category Priority

FIGURE 7-10   SCS Request/Response Exchange

As you might expect, the SCS feature for DL traffic classification and prioritization is an optional feature in the 802.11 standard for both the AP and the STA. The Wi-Fi Alliance's QoS Management (Release 2) certification program also includes this feature as an optional feature, which enhances the market visibility and adoption opportunities for this feature, particularly in the enterprise AP market segments.

Although the SCS feature from 802.11aa was a promising direction for providing prioritized treatment for DL traffic, some challenges remained unaddressed. Specifically, the SCS did not address the issue of differentiating between flows that fall into the same UP/AC but require different treatments based on their unique traffic characteristics (e.g., latency, data rate, burst size). Also, Wi-Fi 6 introduced trigger-based scheduling for the UL. The AP had to perform BSR polling (BSRP) to learn BSR information from STAs to determine how to schedule STAs for the UL using UL OFDMA. Constant BSRP/BSR exchanges added overhead to the network. It became clear that the SCS mechanism defined in 802.11aa needed to be enhanced to address these issues.

SCS with QoS Characteristics (802.11be)

While working on QoS-specific enhancements, the 802.11be designers naturally sought to leverage and extend the existing mechanisms. As explained in the previous section, the SCS mechanism was a great step toward solving QoS challenges, and it made sense to build on top of that success. The key issue that needed to be addressed was how a STA can request resources by providing a simple set of traffic characteristics at finer granularity, which can then be used by the AP for better QoS-based scheduling. Any solution should enable better scheduling for QoS flows not only in the downlink, but also for the uplink, to ensure QoS-based scheduling prioritization can be enforced in both directions for the desired end-to-end experience. The solution should also accommodate the reality that the QoS requirements of an application can be different in DL and UL. For example, for an XR application with distributed rendering, the DL flow to an XR device or head-mounted display (HMD) delivers rendered video. This rendered video has different characteristics for delay, throughput, and periodicity as compared to the UL flow, which delivers pose and/or IMU (inertial measurement unit) data that has a much shorter delay and periodicity needs and requires generally lower throughput.

Recall that the 802.11e-defined TSPEC proved to be extremely difficult to implement, because it mandated use of a large set of traffic parameters (see Figure 7-7) and applications could not supply that information. Given that challenge, the goal in 802.11be was to define a minimal set of QoS parameters to be provided that could adequately specify the application's QoS needs for better scheduling in the DL and UL. This led to the definition of a new QoS Characteristics (QC) element[10] in 802.11be (Figure 7-11). The QC element requires only four parameters related to traffic characteristics to define the scheduling needs for the flow—a major simplification from the TSPEC, which required 15 parameters for the traffic characteristics. In the QC element, the STA indicates the periodicity of its traffic flow using the Minimum Service Interval and the Maximum Service Interval parameters. The two other required traffic characteristics are the Minimum Data Rate and Delay Bound requirements for the flow. The QC element is delivered as part of SCS request/response exchange, as described later in this section.

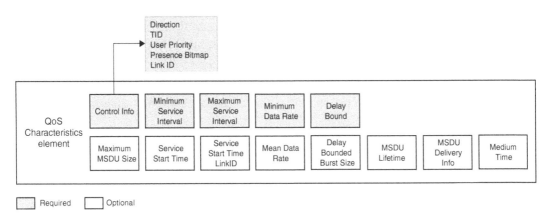

Required Optional

FIGURE 7-11   QoS Characteristics Element

10 See P802.11be D5.0, clause 9.4.2.316.

The Control Info field includes key control information. This field specifies the direction for the flow and indicates whether the flow is uplink, downlink, or direct link (for peer-to-peer [P2P] flows). The TID and User Priority fields are set to indicate the TID and UP of the data frames for the flow, respectively. (The TID field is set to the same value as the User Priority field in the 802.11be amendment.) The presence of optional traffic parameters in the QC element is signaled using the Presence Bitmap field. Also, when the QC element is specified for a direct link, the link ID indicates the link of the AP MLD upon which the direct link transmission will happen.

> **Note**
>
> An often-asked question is why both the TID and User Priority fields are needed in the QC element, given that both fields are set to the same value. The answer is to accommodate any possible future expansion of the TID space in use, beyond the current eight TIDs (0–7). Having two separate fields would allow for carrying a TID value independent of the UP value of the flow.

The four required traffic parameters of the QC element are as follows:

- **Minimum Service Interval:** Specifies the minimal interval, in microseconds, between the start of two consecutive service periods allocated for the uplink, downlink, or direct-link frame exchanges for the traffic flow. For a downlink, this parameter may be unspecified by setting the value to 0.

- **Maximum Service Interval[11]:** Specifies the maximum interval, in microseconds, between the start of two consecutive service periods allocated for the uplink, downlink, or direct-link frame exchange for the traffic flow. For a downlink, this parameter may be unspecified by setting the value to 0.

- **Minimum Data Rate:** Specifies the lowest data rate at the MAC SAP, in kilobits per second (kbps), for the traffic flow. For a direct link, this parameter may be unspecified by setting the value to 0.

- **Delay Bound:** Specifies the maximum amount of time, in microseconds, targeted to transport an MSDU/A-MSDU for the traffic flow. For an uplink or direct link, this parameter may be unspecified by setting the value to 0.

---

11 For periodic traffic, the Minimum Service Interval and the Maximum Service Interval can be set to the same value.

The QC element also includes eight optional traffic parameters. A STA may provide none, some, or all of these parameters for an application flow, based on its knowledge of the application. These parameters are as follows:

- **Maximum MSDU Size:** Specifies the maximum size, in octets, of an MSDU belonging to the traffic flow.

- **Service Start Time:** Specifies the anticipated start time, in microseconds, when the traffic is expected to start for the traffic flow. It is expressed as the four lower octets of the TSF timer of the link indicated by the Service Start Time Link ID.

- **Service Start Time Link ID:** Indicates the link identifier of the link for which the TSF timer is used to indicate the Service Start Time.

- **Mean Data Rate:** Specifies the average data rate at the MAC SAP, in kbps, for the traffic flow.

- **Delay Bounded Burst Size:** Specifies the maximum burst, in octets, of the MSDUs/A-MSDUs belonging to the traffic flow that arrives at the MAC SAP within any time duration equal to the value indicated in the Delay Bound field.

- **MSDU Lifetime:** Specifies the maximum amount of time, in milliseconds, since the arrival of the MSDU at the MAC SAP beyond which the MSDU is not useful, and hence can be discarded. The value of this field is larger than the Delay Bound.

- **MSDU Delivery Info:** Specifies the MSDU delivery information in two subfields: MSDU Delivery Ratio and MSDU Count Exponent. The MSDU Delivery Ratio indicates the percentage of MSDUs that are expected to be delivered successfully (ranging from 95% to 99.9999%) within the Delay Bound, computed over the set of MSDUs determined based on the MSDU Count Exponent field. The number of MSDUs used for computing the MSDU delivery ratio is equal to $10^{\text{MSDU Count Exponent}}$.

- **Medium Time:** Specifies the average medium time, in units of 256 μs, needed by the STA every second for direct-link frame exchanges for the traffic flow. This field is used only for direct links.

The 802.11be task group chose to enhance the existing SCS request/response exchange by including the QoS Characteristics element, which provides traffic characteristics for applications flows. This enabled a STA to request resources for DL and UL traffic flows from the AP; that, in turn, enabled better QoS scheduling at the AP. The SCS Descriptor element was enhanced to include a QoS Characteristics element,[12] as shown in Figure 7-12. Both the AP and the STA indicate support for transmitting and receiving an SCS Descriptor element that contains a QoS Characteristics element through the SCS Traffic Description Support field in the EHT Capabilities element. The QoS Characteristics element

12 See P802.11be D5.0, clause 9.4.2.120.

can specify the traffic characteristics of a DL flow, a UL flow, or a direct-link/P2P flow, using the Direction field as described earlier in this section. A STA can also send a single SCS Request frame indicating traffic characteristics for both the DL and UL flows of an application; this frame includes two SCS Descriptor elements, each one containing a QC element for the traffic flow in a particular direction (UL or DL). Each SCS stream created (identified by its SCS ID) for which traffic characteristics were provided using the QC element, now has the corresponding QC element maintained with the SCS stream.

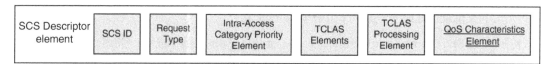

FIGURE 7-12   SCS Descriptor with QoS Characteristics Element

For a DL SCS stream (with a QC element for downlink), the DL traffic flow continues to be identified using the TCLAS element(s) (and optionally TCLAS Processing element) as was defined in 802.11aa. However, the use of SCS for indicating the requirements for an UL traffic flow, by signaling a QC element for the uplink, was introduced in 802.11be for the first time. For an UL SCS stream, the 802.11be amendment mandates that no TCLAS element(s) are provided in the SCS Request frame. This means that the AP cannot identify the traffic flows (and corresponding application) for which STA is requesting UL resources in the SCS Request frame. This limitation made it difficult to apply the network policy at the AP to the UL SCS streams. This was recognized as a gap in the 802.11be TG, but the group did not reach consensus on supporting TCLAS element(s) for UL SCS streams.

The AP's scheduling behavior differs for DL versus UL traffic flows. For a DL traffic flow, an SCS exchange based on the QoS Characteristics element is illustrated in Figure 7-13. The SCS Request frame requests an SCS stream for one or more flows, each with a DL TCLAS element (or DL TCLAS elements + TCLAS Processing element) and a QC element with its Direction field set to downlink. At a minimum, the following set of traffic characteristics must be provided: the minimum and maximum service interval, minimum data rate, and delay bound. Note that the minimum and maximum service interval parameters are not required for DL scheduling and can be set to unspecified. The AP decides whether to accept or reject the SCS request based on its policy and resource availability.[13] In the example, AP accepts the request and sends an SCS Response frame with Success status. The AP then processes the matching DL MSDUs for that traffic flow, taking into account the DL traffic characteristics indicated in the QC element. For example, the AP's DL scheduling logic attempts to meet the delay bound and minimum data rate requirements specified for the DL flow.

---

13 How the AP determines resource availability is implementation-specific and is outside the 802.11be specification.

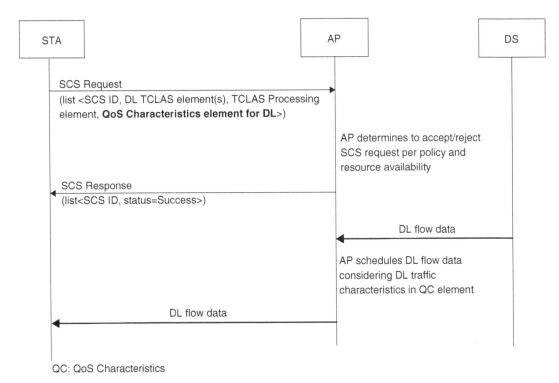

QC: QoS Characteristics

FIGURE 7-13   SCS Exchange with QoS Characteristics for DL Traffic Flow

For an UL traffic flow, an SCS exchange based on the QoS Characteristics element is illustrated in Figure 7-14. The SCS Request frame includes a QC element with its Direction field set to uplink and includes, at a minimum, the following set of traffic characteristics: the minimum and maximum service intervals, minimum data rate, and delay bound. The delay bound may be unspecified, as it is not required for UL scheduling at the AP. Note that the TCLAS elements, TCLAS Processing element, and Intra-Access Category Priority element are not included. The AP decides to accept or reject the SCS request based on its policy and resource availability. In this UL flow example, the AP accepts the request and sends an SCS Response frame with Success status. The AP then periodically schedules UL trigger frames while taking into account the UL traffic characteristics indicated in the QC element. For example, the AP's UL scheduling logic attempts to schedule trigger frames at an interval that falls within the minimum and maximum service intervals and attempts to meet the minimum data rate for the UL flow.

A STA can send a single SCS Request frame with two SCS Descriptor elements to request resources for both the downlink and uplink flows. Thus, the SCS exchanges depicted in Figure 7-13 and Figure 7-14 can also be performed via a single SCS request/response exchange.

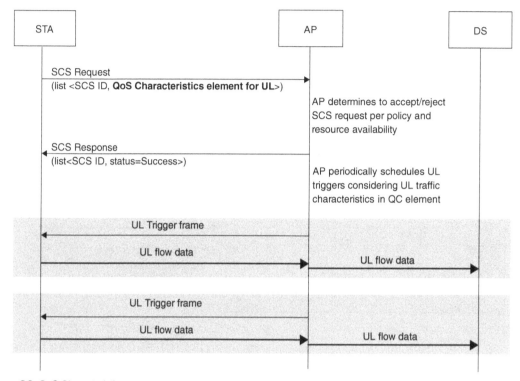

QC: QoS Characteristics

FIGURE 7-14   SCS Exchange with QoS Characteristics for UL Traffic Flow

Another important aspect of the SCS feature in 802.11be is that it is defined at the multi-link device (MLD) level, rather than at the individual link level. As a result, a traffic flow corresponding to a DL SCS stream can be scheduled on any enabled link where the TID corresponding to the flow is mapped in the downlink (see the "TID-to-Link Mapping" section in Chapter 6), so as to meet the flow traffic characteristics specified in the QC element. Similarly, an AP can send triggers for an UL SCS stream on any link where the corresponding TID is mapped in the uplink and STA is not in power save mode. This provides scheduling flexibility to the AP for UL and DL SCS streams, improving the chances that the AP will be able to meet the QoS requirements for the flows.

Overall, the SCS protocol, as enhanced by the QoS Characteristics element in 802.11be, solves the key issue of providing traffic characteristics for flows at finer granularity to the AP, thereby facilitating better QoS-based scheduling in DL and UL. This feature shows promise in addressing the QoS requirements of existing and emerging applications such as video-conferencing collaboration, XR apps on HMD devices with remote rendering, and robotics applications in the industrial IoT arena. The SCS with QoS Characteristics element feature is an optional feature in 802.11be for both the AP and the STA. The Wi-Fi Alliance's QoS Management (Release 3) certification program also included this feature as an optional feature, opening up opportunities for this feature to be adopted more widely by AP and STA devices across a variety of market segments (e.g., enterprise, residential, and industrial).

Restricted TWT

Recall from the "Target Wake Time" section in Chapter 3 that 802.11ax/Wi-Fi 6 adopted the TWT feature from the 802.11ah amendment mainly for its power-saving benefits. The 802.11be amendment built on top of that feature by defining a new feature called restricted TWT to provide more predictable latency performance for latency-sensitive traffic flows. This section first provides an overview of relevant TWT aspects from 802.11ax (see the "Target Wake Time" section in Chapter 3 for more details), and then covers the motivation behind and design aspects for the new restricted TWT feature.

One of the main goals of the TWT feature from 802.11ax was to provide power-saving benefits to the STA when the STA's traffic patterns can be predictable. TWT enabled a STA and the AP to agree on time durations called TWT service periods (SPs), during which the STA will be awake for DL and UL traffic exchanges; the STA will likely be dozing outside those SPs to save power. TWT also enables better management of a STA's activity within the BSS and minimizes contention by scheduling STAs to operate at different times through different TWT SPs. The AP can manage and influence when STAs are awake as part of the negotiation of TWT SPs. The negotiation to establish TWT SPs (also called a TWT agreement) is done by exchanging a TWT element between the STA and the AP using TWT Setup frames.[14] Negotiation of a TWT agreement can be initiated by either the STA or the AP. The STA or AP sends a TWT Setup frame with the TWT Request field set to 1, indicating a TWT request. The 802.11ax amendment also allows for a mode in which an AP can send an unsolicited TWT Setup frame to a STA to establish a TWT agreement by implementing the Accept TWT command in the setup frame.

The 802.11ax amendment defined two modes of TWT operation—individual TWT and broadcast TWT (see Chapter 3 for more details). For an individual TWT, a TWT agreement is established between a STA and the AP, and either peer can initiate negotiation by sending a TWT Setup frame. An individual TWT agreement is always set up as an implicit TWT (implying a periodic TWT) and can be either trigger enabled or non-trigger enabled. The broadcast TWT feature enables an AP to establish a given TWT schedule with multiple STAs. The AP can also recommend that a STA join an existing broadcast TWT schedule (by setting the Broadcast TWT ID field to 1) in a TWT Setup frame sent as a response to a TWT Setup frame requesting the establishment of an individual TWT agreement.

> **Note**
>
> One terminology clarification deserves a mention here. The standard uses the term "TWT agreement" for individually negotiated TWT SPs between a STA and the AP. Multiple individual TWT agreements can overlap in time, enabling an AP to trigger multiple STAs for UL transmission. For broadcast TWT SPs, the standard uses the term "broadcast TWT schedule"; such a schedule can be established between the AP and a group of one or more STAs.

---

14 The 802.11ax amendment also defines a way for a STA to initiate a TWT agreement as part of a (Re)Association Request frame. However, the AP is not mandated to respond with a TWT element in the (Re)Association Response frame.

The restricted TWT feature is built on top of the broadcast TWT mode of TWT operation. It is important to first understand the details of the signaling and procedures for the broadcast TWT mode of operation, if we are to better appreciate the enhancements added for the restricted TWT feature. For a broadcast TWT, an AP can create a sequence of TWT SPs (the TWT schedule) and advertise that TWT schedule in Beacon frames using a TWT element. This broadcast TWT schedule can be shared by a group of STAs associated with the AP. A STA can send a TWT Setup frame (with the Negotiation Type field set to 3) to request to join the advertised broadcast TWT schedule. If the STA receives a TWT Setup frame from the AP with success status, then the STA becomes a member of the broadcast TWT schedule. A STA can also send a TWT Setup frame to request the creation of a new broadcast TWT schedule with specified TWT parameters, and the AP can accept or reject this request. The AP can also allocate a broadcast TWT schedule to a STA by sending an unsolicited TWT response to the STA (using the Accept TWT setup command), which makes that STA a member of the broadcast TWT schedule.

Each broadcast TWT (schedule) is uniquely identified by the <Broadcast TWT ID, AP MAC address> tuple. An AP can create a broadcast TWT schedule that applies to all STAs; such a schedule has a Broadcast TWT ID value of 0. Other broadcast TWT schedules that are intended for groups of member STAs have Broadcast TWT ID values greater than 0. The AP includes a broadcast TWT element in its Beacon frames to advertise one or more broadcast TWT schedule(s). Figure 7-15 shows the format of a broadcast TWT element.[15] The Negotiation Type field determines whether the TWT element information is for an individual TWT (value = 0), for a wake TBTT interval (value = 1), or for broadcast TWT (value = 2 or 3). A broadcast TWT element advertising broadcast TWT schedules in Beacon frames uses a Negotiation Type value of 2, whereas a broadcast TWT element used in TWT Setup frames for negotiating or managing membership of the broadcast TWT schedules with an individual STA uses a Negotiation Type value of 3.

> **Note**
>
> A STA can use TWT to negotiate the wake TBTT of the first Beacon frame and the wake interval between subsequent Beacon frames it intends to receive. This negotiation is performed using a TWT request/response exchange between the STA and the AP, where the Negotiation Type field is set to a value of 1. After successful negotiation, the STA can go to doze state until the next negotiated wake TBTT.[16]

---

15 See P802.11-REVme D5.0, clause 9.4.2.198.
16 See P802.11-REVme D5.0, clause 26.8.6.

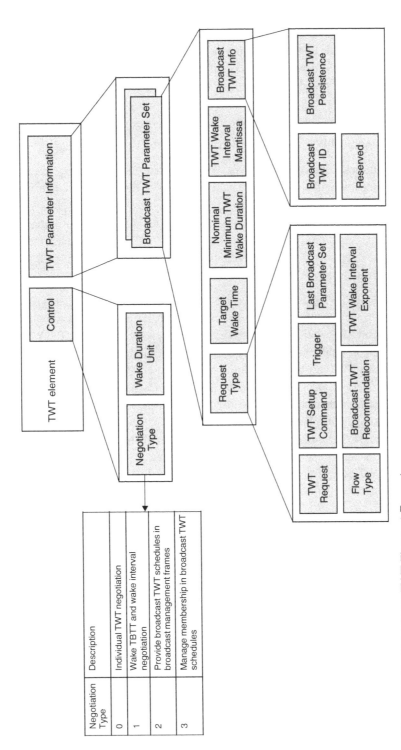

FIGURE 7-15   Broadcast TWT Element Format

As shown in Figure 7-15, a broadcast TWT element includes one or more Broadcast TWT Parameter Set fields, and the Last Broadcast Parameter Set field is set to 1 for the last Broadcast TWT Parameter Set field included. The Broadcast TWT Recommendation field provides recommendations for the types of frames that are transmitted between STAs and the AP during the broadcast TWT SPs. As you will learn soon, this field is used to indicate a restricted TWT. Four fields are used to indicate TWT SPs:

- **Target Wake Time**: Indicates start of the TWT schedule.

- **Nominal Minimum TWT Wake Duration:** Indicates the TWT SP duration.

- **TWT Wake Interval Mantissa** and **TWT Wake Interval Exponent:** Together these fields indicate the time between the start of consecutive TWT SPs (the TWT wake interval).

The Broadcast TWT Info field holds the Broadcast TWT ID, an identifier for the broadcast TWT. The Broadcast TWT Persistence field identifies the number of TBTTs during which the corresponding broadcast TWT SPs are present.[17] The reserved bits in the Broadcast TWT Info field are extended for the restricted TWT feature, as you will learn soon.

The restricted TWT (R-TWT) feature is designed to provide predictable latency performance for low-latency flows. It achieves this by defining enhanced channel access protection and a resource reservation mechanism for R-TWT SPs. An EHT STA declares its support for R-TWT by setting the Restricted TWT Support field in its transmitted EHT Capabilities element to 1. The EHT STA must also set the Broadcast TWT Support field in its transmitted HE Capabilities element to 1 because R-TWT makes use of the broadcast TWT feature. An R-TWT schedule is negotiated between an AP and a STA using the same mechanism as a broadcast TWT schedule. A STA can request creation of an R-TWT schedule using a TWT Setup frame exchange (the same as for the broadcast TWT schedule) with the AP, and the AP can either accept or reject this request. Additionally, an AP can create and advertise one or more R-TWT schedules, and other STAs can request to become members of these schedules using a TWT Setup frame exchange. Like broadcast TWT schedules, R-TWT schedules are link specific. All active R-TWT schedules that are set up on a link are advertised by the AP operating on that link in its Beacon frames. The 802.11be amendment recommends that an R-TWT schedule be set up as a trigger-enabled TWT schedule, as that approach supports efficient MU operation across members of an R-TWT schedule.

The R-TWT feature provides the following key enhancements on top of the broadcast TWT mechanism[18]:

- **Identification of TIDs for latency-sensitive traffic:** With R-TWT, one of the goals is to prioritize latency-sensitive traffic during the R-TWT SPs. When setting up an R-TWT schedule, a STA can include a set of R-TWT UL and/or DL TIDs to indicate the TIDs that are carrying latency-sensitive traffic and that need to be prioritized during the corresponding R-TWT SPs.

17 A value of 255 in the Broadcast TWT Persistence field indicates that the corresponding broadcast TWT SPs are present until explicitly terminated.

18 See P802.11be D5.0, clause 35.8.

- **SP Start Time Protection rule:** An R-TWT–supporting STA is required to end its TXOP before the start of any active R-TWT SPs advertised by its associated AP on that link. This ensures that the medium is free for the AP and the R-TWT member STAs to acquire at the start of each R-TWT SP, which greatly reduces the number of collisions and, in turn, ensures more predictable latency performance for R-TWT member STAs.

- **Overlapping quiet interval:** An R-TWT scheduling AP can optionally schedule a quiet interval of 1 TU (announced in Beacon and Probe Response frames) that overlaps with the R-TWT SP and starts at the same time as the R-TWT SP, with the aim of quieting the legacy (non-EHT) STAs. However, many Quiet elements are needed for typical service intervals, and the EHT STAs are not required to honor these overlapping quiet intervals. Given that this behavior is optional, in the Wi-Fi Alliance's Wi-Fi 7 certification, this aspect of the R-TWT feature is not tested.

- **Traffic prioritization during R-TWT SPs:** The 802.11be amendment defines rules such that the delivery of QoS traffic for the identified R-TWT TIDs is prioritized during the R-TWT SPs. This provides the resource reservation functionality for traffic carried over R-TWT TIDs during R-TWT SPs. An AP or a member STA that exchanges data frames during an R-TWT SP must ensure that the QoS data frames of the R-TWT TIDs are delivered first during the R-TWT SP. In a trigger-enabled R-TWT SP, the AP is required to trigger R-TWT member STAs first to facilitate delivery of their UL data frames of the R-TWT UL TIDs. The R-TWT member STAs must deliver QoS data frames of the R-TWT UL TIDs in response to the trigger frame.

As shown in Figure 7-16, the broadcast TWT element has been enhanced for R-TWT.[19] First, the Broadcast TWT Parameter Set has been enhanced to indicate that the parameter set corresponds to an R-TWT. This is achieved by using a new value for the Broadcast TWT Recommendation field (value = 4) to indicate an R-TWT schedule, and such a Broadcast TWT Parameter Set is referred to as a Restricted TWT Parameter Set. Next, a new Restricted TWT Traffic Info field has been added to indicate the UL and DL TIDs (referred to as R-TWT UL/DL TIDs) of the traffic flows that must be prioritized during the R-TWT SPs. The R-TWT UL/DL TIDs are specified using the Restricted TWT DL TID Bitmap and Restricted TWT UL TID Bitmap fields. The Restricted TWT Traffic Info field is included only in the individually addressed TWT Setup frames for establishing an R-TWT schedule to negotiate the set of R-TWT UL and/or DL TIDs. This information is not included in the R-TWT schedules that are advertised in the TWT element in its Beacon frames. The presence of the new Restricted TWT Traffic Info field is indicated by the Restricted TWT Traffic Info Present field in the Broadcast TWT Info.

Finally, the Restricted TWT Schedule Info field provides additional information for the corresponding R-TWT schedule included in the Restricted TWT Parameter Set, as shown in Table 7-3. This field indicates whether an R-TWT schedule is idle, active, or full, and whether the schedule corresponds to a nontransmitted BSSID or a co-hosted BSSID. Based on this information advertised in the Beacon frame, a STA can determine whether it is permitted to set up membership for an advertised R-TWT schedule. For example, a STA should not request to set up membership for an R-TWT schedule that is indicated as full.

19 See 802.11be D5.0, clause 9.4.2.198.

TABLE 7-3   Restricted TWT Schedule Info Field Definition

Restricted TWT Schedule Info Field Value	Description
0	The corresponding R-TWT schedule is idle: Either it does not have any member STAs or the schedule is suspended for all member STAs.
1	The corresponding R-TWT schedule is active: It has at least one member STA.
2	The corresponding R-TWT schedule is full: The AP is not likely to accept a request from a STA to become a member of that R-TWT schedule.
3	The corresponding R-TWT schedule is active and is for a nontransmitted BSSID or a co-hosted BSSID.*

* The Beacon frame of an AP corresponding to the transmitted BSSID of a multiple BSSID set also advertises any R-TWT schedules for the nontransmitted BSSIDs of the same multiple BSSID set, because the nontransmitted BSSIDs do not transmit any Beacon frames. Based on the same logic, the advertised R-TWT schedules in the Beacon frames of an AP also include any R-TWT schedule corresponding to any co-hosted BSSIDs of that AP.

> **Note**
>
> A broadcast TWT element can contain a mix of the Broadcast TWT Parameter Set fields corresponding to the broadcast TWT schedules and the Restricted TWT Parameter Set fields corresponding to the R-TWT schedules. If a broadcast TWT element contains only Restricted TWT Parameter Set fields, then it is referred to as a restricted TWT element.

An AP must advertise R-TWT schedules that have any membership set up in the Beacon frames by including the Restricted TWT Parameter Set in a broadcast TWT element. Based on the R-TWT schedules advertised in these Beacon frames, an R-TWT–supporting STA can determine the starting time of the R-TWT SPs and use that to determine when it needs to end its TXOPs to comply with the SP Start Time Protection rule for R-TWT.

Figure 7-17 shows an example operation for R-TWT. Three STAs are associated with AP 1 (STA 1, STA 2, and STA 3), and all are R-TWT–supporting STAs. STA 1 establishes an R-TWT schedule with AP 1 by sending a TWT request (a TWT Setup frame with the TWT Request field set to 1). AP 1 responds with a TWT response (a TWT Setup frame with the TWT Request field set to 0) indicating Accept TWT; this response establishes the R-TWT schedule between AP 1 and STA 1. This R-TWT schedule is advertised in the Beacon frame in the TWT element. STA 2 learns about the R-TWT schedule from the Beacon frame and sends a TWT request to AP 1 to become a member of that R-TWT schedule. The AP accepts the TWT request from STA 2 by sending a TWT response. The R-TWT schedule now has both STA 1 and STA 2 as members. Note that the membership information of an R-TWT schedule is not advertised in the Beacon frame—which is the same behavior as for broadcast TWTs. STA 3 determines the start time of each R-TWT SP from the advertised R-TWT schedule and ends its current TXOP based on the start time of the R-TWT SP to comply with the SP Start Time Protection rule of R-TWT. Within the R-TWT SP, the AP sends a Basic Trigger frame to trigger both STA 1 and STA 2 to send their UL data. AP 1 then initiates MU DL data transmissions to both STA 1 and STA 2. During the R-TWT SP, the data exchanges for R-TWT UL and DL TIDs are prioritized by AP 1, STA 1, and STA 2.

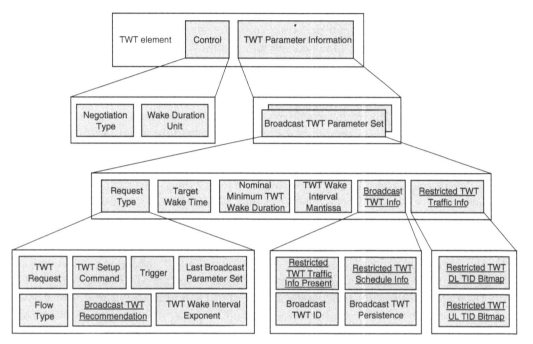

FIGURE 7-16   Restricted TWT-Related Enhancements (Underlined) to a Broadcast TWT Element

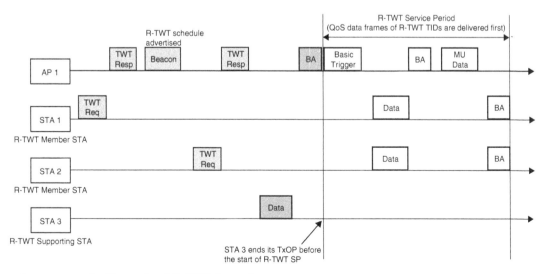

FIGURE 7-17   An Example of R-TWT Operation

Triggered TXOP Sharing

Some of the emerging use cases requiring or benefiting from peer-to-peer (or STA-to-STA) communications were considered when designing the scope of 802.11be, including VR and AR applications. The channel over which a STA chooses to perform peer-to-peer frame exchanges is largely determined by each individual STA and is subject to interference from other STAs that select the same channel. Such peer-to-peer exchanges can also interfere with the communication between AP and STAs. The 802.11be designers sought a better approach for enabling peer-to-peer frame exchanges between an associated STA and another STA—specifically, a method that would be more coordinated by the AP, and as a result would lead to less interference from such peer-to-peer exchanges. In addition to investigating a more efficient way for the peer-to-peer exchanges, the 802.11be group considered the situation in which the AP has a portion of its TXOP unused and is aware of buffered traffic at another STA (e.g., based on BSR). In such a case, the STA can benefit if the AP shares part of its TXOP time with the STA and thereby enables the STA to transmit its data. If the AP triggers the STA, then the STA can send data only to the AP. The key benefit of the AP sharing a portion of its TXOP with the STA is that the STA can send data either to the AP or to another peer-to-peer STA, thereby providing more flexibility to the STA.

The 802.11be amendment defines a new feature called triggered TXOP sharing (TXS) that enables an AP to share a portion of its TXOP with an associated STA, so that the STA can transmit non-TB PPDUs.[20] The TXS feature enables a STA to use the shared TXOP for frame exchanges with another STA and/or with the AP based on the mode under which the TXOP is shared. Two operation modes are defined for the Triggered TXOP sharing feature :

- **Triggered TXOP sharing mode 1:** Allows an AP to allocate a portion of its TXOP duration to an associated STA to transmit non-TB PPDUs to the AP.

- **Triggered TXOP sharing mode 2:** Allows an AP to allocate a portion of its TXOP duration to an associated STA to transmit non-TB PPDUs to other STAs or to the AP.

One of the key differences between the TXS feature and the triggered UL access mechanism (which was first introduced in 802.11ax) is that, in TXS, the STA exchanges non-TB PPDUs. In contrast, in triggered UL access, a STA transmits TB-PPDUs in response to the Trigger frame from the AP. Triggered TXOP sharing is initiated by the MU-RTS Trigger frame, whereas Triggered UL access is initiated by the Basic Trigger frame.

The 802.11ax defined the MU-RTS Trigger frame to allow the AP to protect the medium for a DL or UL multiuser (MU) exchange with associated STAs (see the "Trigger Frames" section in Chapter 3 for details on the MU-RTS Trigger frame). The duration field in the MU-RTS Trigger frame is set to the duration of the frame exchanges to be protected, and the frame is sent as a non-HT duplicate PPDU on each 20 MHz subchannel of the expected transmission. The MU-RTS Trigger frame includes one or more User Info fields. Each of these fields contains the AID value for a STA that is expected to

20 See P802.11be D5.0, clause 35.2.1.2.

respond with a CTS (if the channel is clear at the STA) in the allocated RU, which is identified in the RU Allocation field of the User Info field. All STAs that have an AID specified in the User Info field and whose channels are clear will respond with a CTS in parallel, at SIFS duration after receiving the PPDU containing the MU-RTS Trigger frame.

For the TXS feature, the AP uses a variant of the MU-RTS Trigger frame called the MU-RTS TXS Trigger frame to allocate a portion of its TXOP to an associated STA. Figure 7-18 shows the format for the MU-RTS TXS Trigger frame, which is largely the same as that for the MU-RTS Trigger frame.[21]

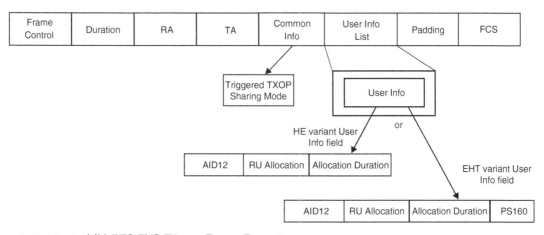

FIGURE 7-18   MU-RTS TXS Trigger Frame Format

The Common Info field in the MU-RTS TXS Trigger frame includes a Triggered TXOP Sharing Mode field. The values for this field (shown in Table 7-4) are used to specify the mode for triggered TXOP sharing. An MU-RTS Trigger frame that has the Triggered TXOP Sharing Mode field set to a non-zero value is referred to as an MU-RTS TXS Trigger frame.

TABLE 7-4   Triggered TXOP Sharing Mode Field Definition

Triggered TXOP Sharing Mode Field Value	Description
0	MU-RTS Trigger frame that does not initiate the TXS procedure.
1	MU-RTS Trigger frame that initiates the TXS procedure: The STA that is allocated the TXOP can transmit only MPDU(s) to its associated AP.
2	MU-RTS Trigger frame that initiates the TXS procedure: The STA that is allocated a portion of the TXOP can transmit MPDU(s) to its associated AP or to another STA.
3	Reserved.

21 See P802.11be D5.0, clause 9.3.1.22.9.

The User Info List in the frame contains a single User Info field that indicates the AID for the associated STA to which the AP is allocating the portion of the TXOP. The 802.11be amendment defines two variants of the User Info field for the MU-RTS TXT Trigger frame: an HE-variant User Info field and an EHT-variant User Info field (Figure 7-18). The EHT-variant User Info field must be used if the channel bandwidth of the MU-RTS TXS Trigger frame is 320 MHz or the channel bandwidth is punctured. Otherwise, the AP can choose to use either variant of the User Info field. The STA to which the portion of the TXOP is allocated is identified by the AID12 field within the User Info field. The RU Allocation field indicates the RU allocated for the transmission of the CTS frame in response to the MU-RTS TXS frame. The Allocation Duration field specifies the TXOP time allocated to the associated STA in units of 16 ms. The PS160 field is set to 1 to indicate a 320 MHz channel bandwidth.

TXS is an optional feature in 802.11be. The AP and the STA declare support for this feature in the EHT MAC Capabilities Information field of the EHT Capabilities element by setting the three capability fields as shown in Table 7-5. For example, an AP that supports transmission of an MU-RTS TXS Trigger frame that allocates time to an associated STA for transmitting non-TB PPDUs to the AP or to another STA, sets the Triggered TXOP Sharing Mode 2 Support field in the EHT Capabilities element to 1. If it does not support the transmission of such an MU-RTS TXS Trigger frame, it sets the value of that field to 0. A STA indicates its support in the same way. In regard to the TXOP sharing, a TXOP return support capability is defined (the TXOP Return Support In Triggered TXOP Sharing Mode 2 field) for the AP that supports operating in Triggered TXOP sharing mode 2. The TXOP return support capability indicates whether the AP can receive/accept a frame signaling the return of the TXOP from the STA (see Table 7-5). In its last data frame transmission during the allocated time, the associated STA to which a TXOP is allocated in mode 2 may set the RDG/More PPDU subfield to 0 in the CAS Control field in the A-Control field; this setting indicates that the STA has finished its transmission and is returning any remaining TXOP time to the AP.

TABLE 7-5   Triggered TXOP Sharing Capability Fields in EHT Capabilities Element

Field Name	Description
Triggered TXOP Sharing Mode 1 Support	Indicates support for transmitting (for the AP) or responding to (for the STA) an MU-RTS TXS Trigger frame with the Triggered TXOP Sharing Mode field equal to 1 (see Table 7-4).
Triggered TXOP Sharing Mode 2 Support	Indicates support for transmitting (for the AP) or responding to (for the STA) an MU-RTS TXS Trigger frame with the Triggered TXOP Sharing Mode field equal to 2 (see Table 7-4).
TXOP Return Support in Triggered TXOP Sharing Mode 2	Indicates support by the AP for receiving a frame signaling TXOP return from the STA in the Triggered TXOP Sharing Mode 2. The TXOP return is signaled by setting the RDG/More PPDU subfield in the CAS Control field of an A-Control in a QoS Data or a QoS Null frame.

> **Note**
>
> One terminology aspect of the triggered TXOP sharing feature could be somewhat confusing. The EHT Capabilities element uses two separate capabilities fields—called Triggered TXOP Sharing Mode 1 Support and Triggered TXOP Sharing Mode 2 Support—to signal capability support for the two modes of TXS operation. However, the MU-RTS TXS Trigger frame uses a single field, called Triggered TXOP Sharing Mode, which can be set to a value of either 1 or 2 to indicate whether TXS sharing mode 1 or mode 2, respectively, is initiated by the MU-RTS TXS Trigger frame.

A STA that is allocated a portion of a TXOP by the AP using the TXS mechanism must ensure that its transmissions and any expected responses fit entirely within the allocated time. Additionally, if the STA successfully transmits QoS Data frames of one or more ACs in a non-TB PPDU to its associated AP during the allocated time, it must update its EDCA parameters to the MU-EDCA parameters advertised by the AP (see the "MU EDCA Channel Access" section in Chapter 3). The STA with allocated TXOP time also ignores the intra-BSS NAV for the allocated time period or until the allocated time is returned to the AP. Meanwhile, the AP must comply with the following rules during the allocated TXOP time for mode 1 and mode 2:

- **Triggered TXOP sharing mode 1:** If an MU RTS TXS Trigger frame for mode 1 is successfully sent to a STA, the AP must not transmit any PPDU during the allocated time, unless (1) the PPDU is an immediate response solicited by the STA or (2) the AP determines, via the carrier sense (CS) mechanism, that the medium is idle PIFS time after the immediate response sent to the STA or after a frame transmission from the STA that does not require an immediate response.

- **Triggered TXOP sharing mode 2:** If an MU RTS TXS Trigger frame for mode 2 is successfully sent to a STA, the AP must not transmit any PPDU in the allocated time, unless (1) the PPDU is an immediate response solicited by the STA or (2) the AP receives an indication of TXOP return in one of the frames transmitted by the STA (using the fields identified in Table 7-5), in which case the AP can transmit a PPDU SIFS after the frame signaling the TXOP return from the STA.

When the allocated time ends, if the AP has a non-zero amount of TXOP time remaining, then it can transmit a PPDU to another STA during that time. When doing so, it follows these rules:

- If the AP detects that the medium is idle (based on the CS mechanism) at the end of the allocated time, then the AP can transmit PIFS after the end of the allocated time.

- If the last PPDU transmitted by the AP (as an immediate response to the STA) ended less than a PIFS duration before the end of the allocated time, then AP can transmit SIFS after the last PPDU transmission.

■ If the last PPDU transmitted by the STA to the AP did not solicit an immediate response from the AP and ended less than a PIFS duration before the end of the allocated time, then the AP can transmit SIFS after the end of that last PPDU.

When a STA receives an MU-RTS TXS Trigger frame from its associated AP with the Triggered TXOP Sharing Mode field value equal to 1 (mode 1), the STA can use the allocated time for transmitting one or more non-TB PPDUs addressed to the AP. Figure 7-19 shows an example of TXS operation for mode 1. Before starting the TXOP sharing, the AP has the option of sending a CTS-to-self transmission to protect the duration of its acquired TXOP. The AP then sends an MU-RTS TXS Trigger frame with the Triggered TXOP Sharing Mode field equal to 1. The AID of STA 1 is indicated in the User Info field of the Trigger frame. Accordingly, STA 1 responds with a CTS frame. STA 1 then performs data transmission via a non-TB PPDU to the AP. As shown in the figure, STA 1 sends a second non-TB PPDU to the AP. At that point, STA 1 has finished with its transmissions. The AP detects that the medium is idle PIFS after transmitting the Block Ack to STA 1's second non-TB PPDU. The AP then transmits data to another associated STA during the remaining portion of its TXOP.

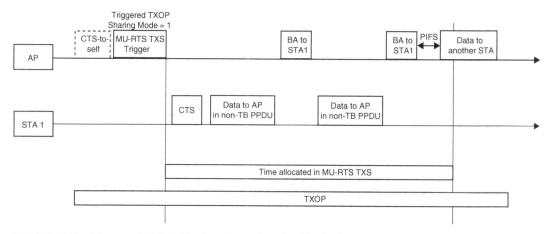

FIGURE 7-19 Triggered TXOP Sharing Operation for Mode 1

When a STA receives an MU-RTS TXS Trigger frame from its associated AP with the Triggered TXOP Sharing Mode field value equal to 2 (mode 2), it can use the allocated time for transmitting one or more non-TB PPDUs addressed to the AP or to another STA. Figure 7-20 shows an example of TXS operation for mode 2. Similar to the mode 1 example, before starting the TXOP sharing, the AP may send an optional CTS-to-self transmission to protect the duration of its acquired TXOP. The AP then sends an MU-RTS TXS Trigger frame with the Triggered TXOP Sharing Mode field equal to 2. The AID of STA 1 is indicated in the User Info field of the trigger frame. In response, STA 1 sends a CTS frame. As shown in the figure, STA 1 next performs data transmission in a non-TB PPDU to the AP and, in return, receives a BA from the AP. At that point, STA 1 sends data to STA 2 in a non-TB PPDU and, in return, receives a BA from STA 2. The AP determines that the medium is idle at the end of the allocated time, and it transmits a PPDU to another associated STA at a PIFS duration after the end of the allocated time, in the remaining portion of the TXOP.

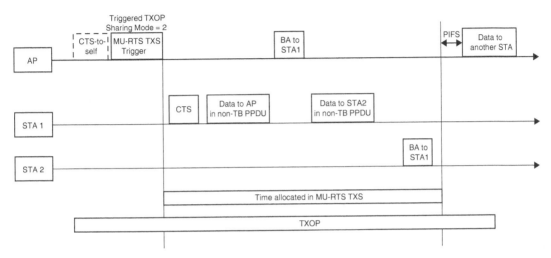

FIGURE 7-20 Triggered TXOP Sharing Operation for Mode 2

In summary, the TXS mechanism enables an AP to share a portion of its TXOP with an associated STA to facilitate transmission of non-TB PPDUs to the AP or another STA for peer-to-peer exchanges. However, the TXS operation adds significant amount of complexity to the AP and STA. The AP still needs to determine when and with which STA it can share part of its TXOP, and the 802.11be amendment does not define specific rules or guidance for that. For TXS mode 1, it is not clear whether there are additional benefits on top of the Triggered UL access feature from 802.11ax. TXS mode 2 enables peer-to-peer exchanges, but the AP does not have a clear way to determine on its own when to allocate TXOP using mode 2 unless an SCS stream was established by the STA for direct links. Given these limitations and the implementation complexity associated with this feature, TXS is not likely to see widespread adoption.

EPCS Priority Access

With many years of continuous and rapid growth, Wi-Fi acquired the reputation of being ubiquitous indoors. The ubiquity argument may be an exaggeration, but it is true that Wi-Fi is present in many buildings, including in locations where cellular coverage is limited or unavailable. In that context, there has been a long history of pushing the 802.11 standard to integrate provisions to facilitate emergency services communications. The 802.11be amendment allows an automatic connection for EPCS 802.11-capable devices using the EPCS priority access feature.

Emergency Services Access Before 802.11be

The idea of using 802.11 for emergency services largely predates 802.11be. The developers of 802. 11u-2011, when designing a protocol for interworking with external networks, focused on the use case of a hotspot. By its very nature, this type of network is expected to operate in public areas. Many people in range of the hotspot network are likely to be visitors, who may or may not be familiar with the local WLAN.

One goal of 802.11u was to allow for automatic and secure connection to the hotspot network without the need to have hotspot-specific local credentials. The 802.11 device would be configured with one or more general credentials (e.g., with the smartphone's cellular operator). The hotspot owner would create connection agreements with various identity providers. Then, as a visitor's 802.11 device discovered the hotspot network, 802.11u would allow the device and the AP to perform exchanges through a general mechanism called the generic advertisement service (GAS). GAS frames can carry Access Network Query Protocol (ANQP) messages, through which the network can announce the identity providers it supports. The visitor's device could then recognize a provider for which it has credentials, and authenticate through the hotspot network with the identity provider. In turn, the identity provider would send to the hotspot a PMK for the visitor's device. This is just like the standard 802.1X/ EAP process,[22] except that the RADIUS server is not local or managed by the local network provider. Figure 7-21 illustrates this process.

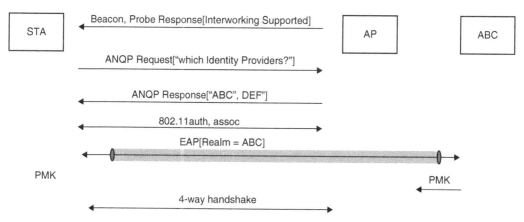

FIGURE 7-21   Automatic Secure Connection with Passpoint

The 802.11u amendment allows the AP to provide, to the visiting device, information that could be useful. That is, the device asks questions, and the network provides answers when available (e.g., "Does this network connect to the Internet?"[23], "Does this network support IPv6?"). This pre-association exchange is where 802.11u integrated three functions related to emergency services:

- A smartphone may have subscribed (automatically or not) to emergency alert systems (EAS). These services allow local authorities to send text alerts to all (subscribed) phones in an area

---

22 See Chapter 2, "Reaching the Limits of Traditional Wi-Fi."
23 Remember, this was 2011—the answer was not always "yes."

where an emergency event occurs (e.g., earthquake or other natural catastrophic event; child abduction or other public safety event). Visitors can use GAS and ANQP to automatically receive EAS alerts through the AP. Each alert includes a specific hash (which makes it unique), allowing the visitor's smartphone to determine which alerts match the categories that were subscribed to, and whether there is a new alert. The alert message is then downloaded and displayed on the smartphone screen (usually with a specific ringtone or vibration pattern).

■ Foreign visitors may not know the local emergency call number. The STA can query the AP for its support of interworking services. Upon receiving a positive response, it can query the AP for the Emergency Call Number ANQP element to receive the local emergency number.

■ Foreign visitors may not have global credentials for the local network. For example, if a visitor's identity provider is the visitor's cellular network service provider with national coverage (e.g., "ABC"), and the visitor wants to connect to a new hotspot on the other side of the world, the local hotspot may have no knowledge of, and no agreement with, a foreign service provider called ABC. In that case, the visitor's device may not have any credentials to connect. If the visitor's device cannot connect, they cannot place emergency calls over Wi-Fi. The 802.11u amendment allows the STA to query the AP for the emergency network access identifier (NAI). The emergency NAI is another ANQP element, which a STA can learn about (before association) and retrieve through the 802.11u GAS exchange carrying the ANQP protocol. The emergency NAI realm element carries temporary credentials, which a STA needing to place an emergency call can use to automatically and securely connect to the local hotspot network. In the ANQP protocol, the AP has a policy (not defined by the 802.11 standard) that limits the activity of devices connecting with the NAI emergency realm credentials, ensuring that their credentials are used to place emergency calls and for nothing else.

Figure 7-22 illustrates the first two mechanisms.

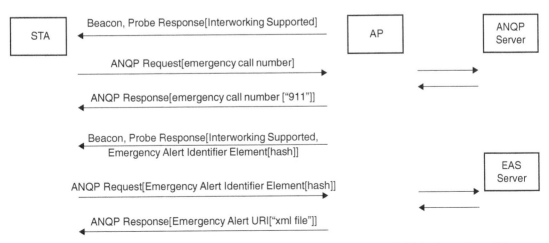

FIGURE 7-22 ANQP Exchanges for Emergency Services in 802.11u-2011 for Locations Where 911 Is the Emergency Call Number

EPCS Priority Access in 802.11be

At the time when 802.11be was developed, the 802.11 standard included strong provisions to allow users to receive EAS messages, learn the local emergency telephone number, and automatically connect to a local hotspot network to place emergency calls, even when the device did not have preconfigured credentials. However, one aspect of the emergency exchange was ignored: the case of emergency responders rushing to a location to provide assistance. Just like individuals calling about an emergency, responders may need to provide assistance in places where the cellular network provides insufficient service or is not available, or where cellular network operation has been disrupted. To solve this part of the equation, 802.11be introduced a prioritized access mechanism for the devices of emergency service personnel, part of a group of devices requiring what 802.11 calls the Emergency Preparedness Communication Service (EPCS). This is an umbrella term that, for instance, maps to the National Security and Emergency Preparedness (NS/EP) service in the United States.

The goal of EPCS priority access[24] in 802.11be is to provide prioritized access to emergency services devices. Imagine an emergency responder rushing to assess a fire alert in a public building. The responder's smartphone automatically connects to the local Wi-Fi network, and the responder needs to coordinate the emergency action with voice–over–Wi-Fi (VoFi) calls or video calling. Unfortunately, because of the emergency, many people are also connected over the same Wi-Fi network, and either the responder's calls fail or their quality is too low to be useful. With EPCS, the AP can provide privileged EDCA parameters to EPCS devices, providing them with prioritized access on the Wi-Fi connection relative to "non-emergency" devices.

An AP MLD and a non-AP MLD declare their support for EPCS priority access in the EHT Capabilities element. An AP MLD advertises its EPCS priority access support in its Beacon and Probe Response frames. Naturally, any device that is requesting EPCS privileged access needs to be authorized for such access, so as to avoid abuse of this mechanism. During the (re)association process for a non-AP MLD that indicates support for EPCS priority access, the AP MLD acquires the information to verify the authorization of that non-AP MLD to use this feature. The AP MLD may acquire the information from a government entity or have other methods to acquire the information to validate the authorization of EPCS priority access for devices—for example, it may obtain the necessary information from a service provider supporting EPCS. The 802.11be amendment (and the Wi-Fi 7 program) clarifies that this validation for EPCS device identity and authorization for EPCS priority access for non-AP MLDs belongs to the upper layers and is the responsibility of other standardization efforts.

> **Note**
>
> The validation of the EPCS device's identity and authorization for EPCS priority access is crucial and was the subject of intense debates in and around the 802.11be task group. As soon as a device is given privileged conditions in the BSS, the risk of abuse needs to be considered. If 20 emergency responders' devices connect to the BSS, and all of them use EPCS privileges, is

---

24 See P802.11be D5.0, clause 35.16.

it acceptable if other (non-EPCS) devices in the BSS see their performance severely degraded to the point of being unusable? A spontaneous answer may be "yes." However, a closer look at the details reveals that, once privileged access is granted, the AP does not verify the type of traffic sent by the EPCS device (all traffic is given privileged access). One can imagine a scenario in which EPCS devices are given privileged access and then send noncritical traffic, even as a non-EPCS device is denied access to make an emergency call. Therefore, 802.11be allows EPCS devices to have EPCS privileged access only during specific periods when EPCS priority access is explicitly enabled, and not beyond that time span.

Once a non-AP MLD is associated with an AP MLD, the non-AP MLD can initiate setup of the EPCS priority access when triggered by a higher layer. The non-AP MLD sends, through any of its affiliated STAs, an EPCS Priority Access Enable Request frame (an action frame) requesting that EPCS priority access be enabled. The receiving AP MLD validates the authorization for the non-AP MLD to use the EPCS priority based on locally stored information (previously obtained during the association) or based on information obtained from a service provider or government entity. At that point, the AP MLD returns an EPCS Priority Access Enable Response frame, which provides a success or failure status for the EPCS priority access setup.

EPCS priority access can also be enabled by an AP MLD, which can send an EPCS Priority Access Enable Request frame to a non-AP MLD in an unsolicited way. The AP MLD first verifies that the non-AP MLD is authorized to use EPCS priority access before initiating the unsolicited setup. For example, this situation might happen because the non-AP MLD was identified as an EPCS device and authorized for EPCS priority access based on higher-layer triggers received at the AP MLD.

During the EPCS priority access setup, the AP MLD provides an EPCS Priority Access Multi-Link element to the non-AP MLD in the EPCS Priority Access Enable Response frame (if the setup was initiated by the non-AP MLD) or in the EPCS Priority Access Enable Request frame (if the setup was initiated by the AP MLD). The EPCS Priority Access Multi-Link element specifies the EDCA (and optionally MU EDCA) parameters that the non-AP MLD can use on each of the setup links (AIFSN, CWMin, CWMax, TXOP Limit, or MU EDCA Timer for each of the four ACs) during the EPCS priority access. The values of these EDCA and MU EDCA parameters are left to AP implementations. In most cases, the values will provide prioritized access to the EPCS STA for some or all of the ACs. Here again, the local network would know which type of traffic EPCS STAs are likely to require and may prioritize the matching ACs (and not the other ACs) accordingly.

Once the EPCS priority access request/response exchange is successfully completed between an AP MLD and a non-AP MLD, the EPCS priority access is in the enabled state for that non-AP MLD on all the setup links. During the period while EPCS priority access remains enabled, the non-AP MLD uses the EDCA and the MU-EDCA parameters provided for EPCS priority access by the AP MLD. Alternatively, the AP MLD could send an EPCS Priority Access Enable Response frame in an unsolicited fashion to the non-AP MLD to update the EDCA and/or MU EDCA parameters to be used while EPCS priority access remains enabled. Such a parameter update can be triggered by a higher-layer function.

The non-AP MLD updates its EDCA and MU EDCA parameters based on the received parameters for its subsequent EPCS priority access.

As soon as the emergency activity by the EPCS devices is complete, the higher layer can trigger a teardown of the EPCS priority access. Either side (the AP MLD or the non-AP MLD) can send an EPCS Priority Access Teardown frame to terminate EPCS priority access. This frame simply indicates that EPCS privileged access is no longer granted, so the non-AP MLD must go back to using the default EDCA and MU EDCA parameters. As you might expect, when a non-AP MLD disassociates, that also causes a teardown of the EPCS priority access state for that non-AP MLD.

The EPCS general procedure is illustrated in Figure 7-23. The non-AP MLD and the AP MLD enable EPCS priority access by exchanging EPCS Priority Access Enable Request/Response frames. The AP MLD determines whether the non-AP MLD is authorized for EPCS priority access during the association procedure. The EPCS traffic exchange happens using the EDCA and/or MU EDCA parameters for EPCS priority access specified by the AP MLD. The AP MLD then updates the EPCS EDCA or MU EDCA parameters by sending an unsolicited EPCS Priority Access Enable Response frame. The subsequent EPCS traffic exchange uses the updated EDCA and/or MU EDCA parameters. At the end, the non-AP MLD tears down the EPCS priority access by sending an EPCS Priority Access Teardown frame, which changes the EPCS state to "torn down" at both the non-AP MLD and the AP MLD.

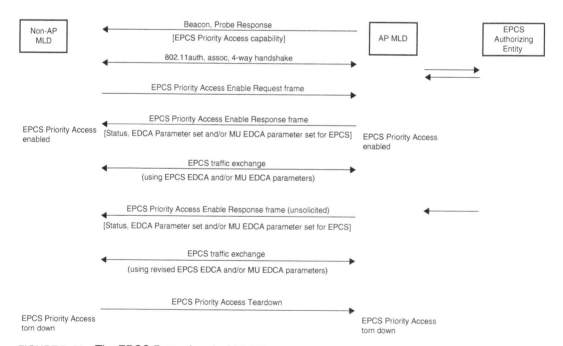

FIGURE 7-23   The EPCS Procedure in 802.11be

As noted earlier, the EPCS procedure was defined in 802.11be. It is also included in the Wi-Fi 7 certification as an optional feature for the AP MLD and the non-AP MLD.

Wi-Fi 7 Security

The 802.11be amendment was not intended to produce a completely new security scheme. It continues to use the previously defined WPA3 (Wi-Fi Protected Access 3) and adds some new security requirements, which are discussed in this section. The security aspects of the multi-link connection of MLO were covered in the "MLO Security" section in Chapter 6. For Wi-Fi 7, complexities related to the default security modes in different bands create some specific considerations.

WPA3 Versus WPA2

WPA3 was released in 2018 and offered several improved Wi-Fi security schemes relative to WPA2. Notably, compared to its predecessor, it included stronger encryption, more robust password-based authentication, protection of management frames, and security for open public networks. As noted in Chapter 2, Wi-Fi security relies on an authentication component (802.1X, or a pre-shared key) intended both to verify that the STA and AP have valid credentials, and to generate keying material that allows them to encrypt their communications. In some scenarios, the STA and AP need to confirm that they are in possession of a specific key value (e.g., the PMK identified by a PMKID) or need to send a value that should not be tampered with between sender and receiver. The Hash-Based Message Authentication Code (HMAC) allows the sender to produce a hash of the value that is then presented to the other side for verification.

The 802.11 standard supports several different types of algorithms for the encryption and the hashing tasks. The AP signals the combination(s) it supports in its beacon and probe responses in the Robust Security Network element (RSNE), expressed in the form of authentication and key management (AKM) suites and cipher suites for group data and pairwise encryption.[25] WPA2 authorized two modes: enterprise (with 802.1X authentication) and personal (with a pre-shared key). WPA2-Enterprise mode mandates the use of 802.1X for the authentication process, AES-CCMP (128 bits) for the encryption algorithm, and SHA-1 for the HMAC; this combination forms AKM:1. WPA2-Personal leverages the same encryption and HMAC, but uses a pre-shared key; this combination forms AKM:2.

WPA3 similarly supports an enterprise mode and a personal mode. In addition to AES-CCMP, WPA3 allows the use of a stronger (and faster) encryption algorithm, AES-GCMP. SHA-256 is used instead of SHA-1 for the HMAC part. The WPA3-Enterprise mode for these parameters is defined as AKM:5 and, in its Fast Transition (FT) form, AKM:3. AES-GCMP can be used with 128-bit keys, but also with 256-bit keys, in what is called WPA3-Enterprise 192-bit mode. In that case, it is used in combination with SHA-384; this version is known as AKM:12. For the personal mode, simultaneous authentication of equals (SAE) replaces the pre-shared key scheme. The WPA3-Personal mode with SAE authentication and SHA-256 HMAC is identified as AKM:8 and, in its FT form, as AKM:9.

25 An AKM suite is formally identified using the nomenclature "OUI 00-0F-AC" followed by a number—for example, 00-0F-AC:1. More commonly, though, the suite is referred to by its number directly—for example, AKM:1. See 802.11-2020, Table 12-10, for a list of AKM schemes. A cipher suite uses the same format—that is, OUI followed by a number indicating the encryption algorithm. For example, 00-0F-AC:4 indicates CCMP128.

WPA3-Personal mode also uses AKM:24 when SAE is accompanied with a variable hash algorithm chosen from SHA-256, SHA-384, or SHA-512. The FT version of AKM:24 is AKM:25.

The hardware (and software) needed to encrypt or decrypt transmissions at the line rate with GCMP (or process SHA-256) implies that older devices are unlikely to become WPA3-compatible with a software upgrade. To facilitate the transition to WPA3, a WPA2/WPA3 transition mode is defined that allows devices supporting WPA2 and devices supporting WPA3 to coexist in the same BSS. The AP uses WPA3 with the devices that support WPA3 (AKM:5 in enterprise mode; AKM:8 in personal mode), WPA2 with the WPA2-only devices (AKM:1 in enterprise mode; AKM:2 in personal mode), and WPA2 for the common messages (protected broadcast). The Wi-Fi Alliance is presently exploring an alternative way to signal a hybrid WPA2 + WPA3 BSS that can address deployment complications when certain legacy WPA2-only clients are present.

Wi-Fi 7 Security Procedure

The Wi-Fi 7 program mandates support for WPA3, including AKM:24 (in addition to AKM:8 in personal mode). The AP and STA must also support the Wi-Fi Alliance's Enhanced Open scheme based on opportunistic wireless encryption (OWE) when operating in open networks. Enhanced Open is defined in RFC 8110 (and labeled as AKM:18) and was released at the same time as the WPA3 program. The Enhanced Open mode is designed to provide encryption and privacy for non-password-protected open networks in public places. Wi-Fi 7 (and the 802.11be amendment) also mandate support for the GCMP-256 cipher for the encryption of both pairwise individually addressed frames (as the pairwise cipher suite) and group data frames (as the group data cipher suite), as a means to promote stronger encryption. In addition, Wi-Fi 7 devices are required to support GMAC-256 for the group management cipher.

The Wi-Fi 7 program mandates support for protected management frames (PMF), as defined in 802.11w-2009. This protocol allows for the encryption of some management frames (e.g., unicast management frames to associated STAs), supports the integrity and authenticity validation of other management frames (e.g., multicast or broadcast management frames sent to associated STAs), and limits the use of management frames that cannot be protected (e.g., broadcast management frames that can be replayed). Support for this feature is not new. PMF is a requirement for WPA3 and, therefore, is mandated for any program that mandates WPA3.

Additionally, Wi-Fi 7 requires beacon protection for both the AP and the STA. The Beacon frame includes a management MIC element (MME), which holds a message integrity code (MIC) generated over the Beacon frame using the BIGTK key. The STA is required to validate the MIC in the Beacon frame. Beacon protection ensures that the Beacon frames cannot be forged by an attacker trying to impersonate a legitimate AP.

With MLO, security needs to be established across all the links of a multi-link association. As noted in the "MLO Security" section in Chapter 6, for MLO the security context (PMKSA and PTKSA) used for the protection of individually addressed frames is established at the MLD level using the MLD MAC address, and the same context applies for all the setup links established between peer MLDs. The group keys (GTK, IGTK, and BIGTK) are still established at the link level for each link of the MLD.

Multiband Challenges

WPA3 provides increased security, and MLO is easily integrated into the WPA3 scheme. However, the integration of MLO with existing Wi-Fi 6 and 6E deployments, along with the requirement for backward compatibility (or at least coexistence), create new challenges. The mandate for Wi-Fi 7 APs and STAs to support WPA3 does not mean that they cannot support WPA2. In fact, all Wi-Fi 7 devices will support WPA2, as support for Wi-Fi 6, 802.11ac (Wi-Fi 5), and 802.11n (Wi-Fi 4) is mandated for Wi-Fi 7. It might be logical to suppose that a Wi-Fi 7 AP will be configured exclusively for WPA3 (which offers better protection than WPA2), as WPA3 is required. However, the reality of operations may be more complicated. On the one hand, some non-Wi-Fi 7 and non-Wi-Fi 6 clients may need to associate to the AP. These older clients may not support WPA3. In such a hybrid client scenario, it is tempting for the network owner to configure the clients for WPA2 or WPA3/WPA2 transition mode. Some networks may also include pre-Wi-Fi 6 APs, which do not support WPA3. In such hybrid deployments, a Wi-Fi 7 client cannot perform fast roaming (FT) between an MLO AP with WPA3 and another AP (pre-Wi-Fi 6 or Wi-Fi 6 AP) that offers WPA2 only. The same is true when a Wi-Fi 7 client is moving in the other direction, from an AP that supports only WPA2 to a Wi-Fi 7 AP that supports WPA3. In both of these scenarios, the client would need to perform full authentication and establish a new association when moving across the APs.

An additional difficulty is that Wi-Fi 6E mandated the use of WPA3 in the 6 GHz band. In consequence, the WPA2/WPA3 transition mode can become "WPA2/WPA3" on the 2.4 GHz and 5 GHz radios, but can become only "WPA3" on a 6 GHz radio. In some deployments that incorporate legacy devices, 2.4/5 GHz radios may deploy only WPA2 to avoid interoperability issues. In that case, if the client is moving from a Wi-Fi 7 AP with an MLO association using WPA3 to a Wi-Fi 6E AP, which is required to support WPA3, then the client can potentially use fast roaming when moving between the APs, if those APs are configured as part of the same FT mobility domain. However, such fast roaming becomes impossible when the client is moving from an MLO AP (with WPA3) to a 2.4 or 5 GHz Wi-Fi 6 AP offering WPA2 only. These differences create roaming challenges and deployment complications in a brownfield deployment. For Wi-Fi 7, just as for previous Wi-Fi generations, several years may be needed before network owners can deploy the technology without needing to carefully consider the complications presented by backward-compatibility demands.

Summary

This chapter reviewed the basic operation of EHT APs and EHT non-AP STAs within an EHT BSS. Topics covered included EHT operation parameters and EHT capability signaling, the selection of EHT operation parameters, EHT operating mode updates, preamble puncturing, and the EHT sounding procedure.

The key features introduced in 802.11be (besides MLO) were described as well. For example, enhancements to SCS based on the QoS Characteristics element enable applications to provide their traffic characteristics and QoS requirements to the AP for improved resource allocation and scheduling to meet applications' QoS needs, including low latency, a minimum data rate, and traffic periodicity. The restricted TWT feature is designed to provide predictable latency performance for low-latency flows, by defining enhanced channel access protection and a resource reservation mechanism for R-TWT service periods. The triggered TXOP sharing (TXS) feature allows an AP to share a portion of its TXOP with an associated STA to enable that STA to transmit non-TB PPDUs to the AP or to another STA for a peer-to-peer exchange, facilitating better performance for peer-to-peer use cases over Wi-Fi. In addition, 802.11 supports Emergency Preparedness Communication Services (EPCS) through the new EPCS priority access feature, which is designed to provide prioritized access to authorized emergency services devices.

Wi-Fi 7 security aspects were explored, too. These include the requirement to support WPA3 (same as for Wi-Fi 6E at 6 GHz), along with the mandatory support for GCMP-256 and for Beacon protection. Despite these updates, network operators continue to face deployment and roaming challenges in Wi-Fi 7 brownfield deployments.

In summary, Wi-Fi 7 supports much higher throughput (in keeping with its name—Extremely High Throughput) through the MLO aggregation, 320 MHz, and 4096-QAM features. It also brings the benefits of lower latency and improved reliability through MLO diversity, SCS with QoS character-istics, and restricted TWT. Collectively, these features can be harnessed to promote better and more deterministic QoS for existing and emerging applications (e.g., video-conferencing, AR/VR/XR, and robotics).

Chapter 8

Wi-Fi 7 Network Planning

link margin: ratio of the received signal power to the minimum required by the STA.
The STA might incorporate rate information and channel conditions, including interferences, into its computation of link margin. The specific algorithm for computing the link margin is implementation dependent.

802.11-2020, Clause 3

802.11be and Wi-Fi 7 bring along many new features that promise to improve the efficiency of 802.11 deployments. The 6 GHz band adds new constraints and opens up new possibilities that apply to Wi-Fi 6E, but that most designers have waited for Wi-Fi 7 to address. This chapter is dedicated to the professionals tasked with making Wi-Fi real, whether it be at home, at work, or at play. Designing and deploying Wi-Fi networks can be a daunting task, filled with unexpected discoveries. However, with the aid of proven practices and a little math, radio planning is more in the field of science than a black art. This chapter covers the general principles of 802.11 WLAN design, and includes the key elements you need to successfully incorporate Wi-Fi 7 requirements and new features into your design.

WLAN Design Principles

At its heart, WLAN design is about positioning and configuring APs to provide a continuous coverage experience for all the intended STAs. This goal means that each BSS should be designed to accommodate all its target STAs and their traffic. Each STA should be able to move between BSSs while maintaining a satisfactory quality of experience. Your task, then, is to design a first coverage area (which this chapter calls a *cell*), and subsequently to duplicate this design across the entire floor.

> **Note**
>
> The term *BSA* has been used in other chapters of this book, and the term *cell* will be used in this chapter as if they were virtually interchangeable. The *cell* terminology dates back to the early days of outdoor cellular network design. In that era, the coverage of a radio base-station (the cellular equivalent of an access-point device) was logically constrained (or geo-fenced) to a hexagonal area generally considered to approximate a naturally circular area (because RF generally propagates omnidirectionally in the horizontal azimuth plane). The term *cell* has since become commonly used when referring to the design of the BSS (primarily focusing on the AP configuration), its associated BSA (the physical area resulting from the configuration choices, such as the data rates enabled/disabled and the power of the AP radio), and the logical combination of multiple BSSs next to each other.

User Device Requirements

Designing a cell is not a deterministic task. The channel width, AP power, and allowed MCSs and data rates depend on the requirements driven by the expected client mix and traffic. In an ideal scenario, the client population is known, and you have a detailed list of each client device specs and the list of applications they are likely to run. In the real world, such a list usually does not exist. The intended applications may be known, but the list and each application's characteristics are likely to evolve over time. The client's devices are likely to suffer from the same limitations. Some enterprises have a strict list of allowed devices, but most environments accept the connection of uncontrolled devices (possibly to guest SSIDs, on the same AP radios as the corporate SSIDs), making the strict characterization of client devices impossible.

In such a case, your only option may be to define one or a few predominant typical devices, and design the WLAN around the performance targets of these clients. Creating such design requires understanding the following elements:

- Which protocol family does the user device support? Is it still supporting only Wi-Fi 5 (802.11ac), or has it been upgraded to Wi-Fi 6 (802.11ax)? Or are you planning for Wi-Fi 7 (802.11be) devices?

- Which MIMO capability does the user device support? Is it an IoT device supporting only one spatial stream (1SS)? A phone supporting 2SS? Or a laptop supporting 3SS?

- Which channel width can the equipment support (the STAs and the planned APs)? Which channel width would your deployment support? In other words, what level of co-channel-interference (CCI) is acceptable?

- Which area of the BSS needs to support which data rate?

A simple way of capturing these basic requirements is in a table similar to Table 8-1. This information can usually be extracted from the specification pages of your target client devices' models. You should also document their expected count in the network (and each BSS).

TABLE 8-1   Example Device Requirements

Device Type	Protocol	Ratio	Expected Rate
1SS (20 MHz, 2.4 GHz)	11an	20%	10 Mbps
1SS (80 MHz, 5 GHz)	11ac	20%	100 Mbps
2SS (160 MHz, 6 GHz)	11ax	55%	250 Mbps
3SS (320 MHz, 5/6 GHz)	11be	5%	1 Gbps

Table 8-1 shows a commonly used Wi-Fi device distribution for an enterprise or even a home environment. You should also document which devices will connect to the network and not move, as they may impact the design. For example, many IoT devices (e.g., printers, building control) are limited to 2.4 GHz/20 MHz operation because their intended operation is low speed (e.g., remote control, bulk file download). Their manufacturers want these devices to operate from any position in a home, even if there is only a single AP—hence the choice of 2.4 GHz for their longer usable range. Many networks still include many legacy 5 GHz 802.11ac/Wi-Fi 5 devices, such as smartphones and tablets, that will be upgraded only as consumer demand and mobile operator plans allow. In 2024, the bulk of new devices were likely to be 802.11ax and 6 GHz-capable, as the Wi-Fi 6E certification had been available for two years. As the availability of Wi-Fi 7 devices ramps up, you should expect to encounter them at low, but slowly increasing concentrations (except if you plan a massive device refresh).

Once you have an idea of the device mix and expected count, you can look at the peak or desired data rate for each of these clients. A reasonable approach is to evaluate the required throughput of the most demanding application that each device is likely to run, and apply a ratio to that throughput. For example, you might determine that the most demanding application is AR/VR for some devices (with a nominal bandwidth of 25 Mbps for a 4K display), and video-conferencing for other devices (with a nominal bandwidth of 12 Mbps). Each device might run other processes in the background, so that overall it consumes 1.5 times the nominal bandwidth. At the same time, you might estimate that only 20% of the devices expected in the cell will be using the peak application at any point in time, with most other devices being idle or running lower-bandwidth applications (e.g., web browsing, with an average bandwidth of 800 kbps). Those numbers are, of course, projections, and each design will lead to project-specific numbers. Even so, these numbers can help you estimate the desirable throughput for each expected device.

The achievable data rate and throughput decrease as the distance to the AP increases. This decrease occurs because the minimum received signal strength (RSS) and the signal to noise ratio (SNR) requirements are higher for complex (high MCS, high throughput) modulations than for simpler modulations.

In the days of 802.11n, it was perfectly reasonable to design a cell so that the maximum 802.11n rate of 150 Mbps (for a single-antenna device on a 40 MHz channel) would be achieved at the edge of the cell, and thus the maximum data rate was achievable anywhere in the cell. The most complex modulation in 802.11n was 64-QAM.[1]

Later, 802.11ac introduced a more complex modulation (256-QAM, in MCS 8 and MCS 9), but with requirements for such high RSS and SNR that the rate could be achieved over only a few tens of feet (and about 10 meters) away from the AP, representing a small coverage area (e.g., 12,500–2500 sq ft). Designing cells of such small size was unrealistic, as only a part of the BSA was considered in the design. The center area of the BSA would be considered as the "useful area." In some parts of the BSA, the highest MCS could be achieved, but only lower data rates were realizable in other parts of the BSA. Beyond a certain distance from the AP, STAs could still connect and function normally. However, the data rates achievable at these distances were no longer considered useful. Thus, other APs were installed nearby to allow the STA to roam to a BSS where the desirable data rates would be achievable again.

Since 802.11ac, chipset advances have allowed 802.11ac MCSs 8 and 9 to reach farther distances. But then 802.11ax introduced 1024-QAM, repeating the same scenario again. As a result of this race to higher data rates (at the cost of shorter coverage), most WLANs are designed based on the concept of useful area in the BSA, the zone at which an optimal data rate can be achieved, as shown in Figure 8-1. The designer selects that data rate based on their requirements.

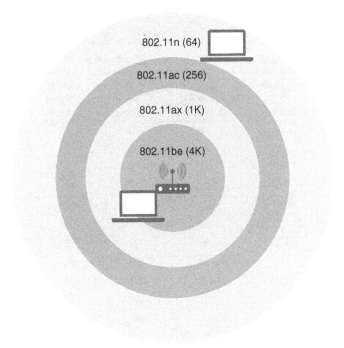

FIGURE 8-1   Cell Area Versus Maximum QAM Support

1 See Chapter 1, "Wi-Fi Fundamentals," for details on modulations.

As a designer, once you have listed the expected devices and their characteristics, the next step is to evaluate the minimum target data rate that should be achieved by all devices. You can then design the "useful part of the cell" to stop at that point. Beyond that point, devices can still connect and communicate with the AP, but another AP in the neighboring cell offers this minimum data rate, allowing each client to roam if needed to maintain a consistent throughput.

Link Budget and Cell-Edge Data Rate

Determining the minimum data rate achievable at the edge of the cell presupposes knowledge of each client's physical characteristics, and of the protocols the client supports. Fortunately, the 802.11 standard mandates minimum sensitivity levels for each MCS. Each new-generation 802.11 amendment with new PHY capabilities incorporates the requirements from previous versions of the standard, adding the requirements of the added PHY. Therefore, the 802.11be draft[2] is a good source in which to find the minimum sensitivity for all amendments since 802.11n.

Specifically, the standard mandates that receivers should be able to successfully receive 90% of frames—that is, the packet error ratio (PER) should be less than 10%—when the measured receiver input sensitivity is at the levels listed in Table 8-2. The test supposes 4096-byte-long PSDUs. The receiver minimum input level sensitivity in the table is the receive signal strength indicator (RSSI) measured by a STA or AP adapter in dBm (decibel milliwatts).

TABLE 8-2   Receiver Minimum Input Level Sensitivity

MCS	Modulation	Rate	Minimum Sensitivity (20 MHz PPDU) (dBm)	Minimum Sensitivity (40 MHz PPDU) (dBm)	Minimum Sensitivity (80 MHz PPDU) (dBm)	Minimum Sensitivity (160 MHz PPDU) (dBm)	Minimum Sensitivity (320 MHz PPDU) (dBm)
0	BPSK	½	−82	−79	−76	−73	−70
1	QPSK	½	−79	−76	−73	−70	−67
2	QPSK	¾	−77	−74	−71	−68	−65
3	16-QAM	½	−74	−71	−68	−65	−62
4	16-QAM	¾	−70	−67	−64	−61	−58
5	64-QAM	⅔	−66	−63	−60	−57	−54
6	64-QAM	¾	−65	−62	−59	−56	−53
7	64-QAM	⅚	−64	−61	−58	−55	−52
8	256-QAM	¾	−59	−56	−53	−50	−47
9	256-QAM	⅚	−57	−54	−51	−48	−45
10	1024-QAM	¾	−54	−51	−48	−45	−42
11	1024-QAM	⅚	−52	−49	−46	−43	−40
12	4096-QAM	¾	−49	−46	−43	−40	−37
13	4096-QAM	⅚	−46	−43	−40	−37	−34

2 See P802.11be D5.0, clause 36.3.21.2.

Most clients, and almost all APs, can achieve better performance than these target minimums.[3] If you know the characteristics of the APs and the clients expected in your network, use the devices' spec sheets instead of this table, basing your design on the weakest element (the device with the lowest performance). If you expect unknown devices in your WLANs, the values in Table 8-2 are a prudent choice, as you know that all devices will at least achieve these sensitivity numbers.

The figures in Table 8-2 refer to the received signal. The RF engineering practice uses a link budget to determine whether the signal quality or the signal to interference plus noise ratio (SINR) is high enough at either the AP (uplink) or the STA (downlink) to achieve the desired data rate. Table 8-2 supposes a low level of background noise (–94 dBm). In general, the noise component is a result of background RF activity, rather than being specifically related to the WLAN itself.

For reasons that will become clear later in this chapter, designers generally ignore the other Wi-Fi cells' co-channel interference component of the link budget. The focus is on the simpler signal to noise ratio (SNR) case that considers only the noise in the general environment, and not the noise from other APs in the ESS. Factors that influence SNR include the following:

- The transmit power (TXP) of AP or STA (in dBm). Higher TXP increases the received signal at any given point, and thus increases the SNR, if the environmental noise value is constant.

- Environmental noise (N) (in dBm or dBm/Hz). Higher environmental noise decreases the SNR at the receiver, for a given RSSI value.

- Antenna gain (in dBi) for the receiver (G_{RX}) and/or the transmitter (G_{TX}). Both increase the signal level and thus the SNR, if the environmental noise is constant.

- Pathloss (PL) between the AP and the STA (in dB). The PL increases with the distance, and the presence of obstacles, between the transmitter and the receiver.

- Shadow fade (F) loss margin (in dB). The SNR decreases as the fade loss margin increases.

- Noise figure (NF), the noise inherent to the receiver circuit (in dB). The SNR decreases as the noise figure decreases.

Formally, the relationship between these elements can be expressed as follows:

$$SNR_{dB} = TXP + G_{TX} - PL - F - N + G_{RX} - NF$$

SNR^A and SNR^D

The link budget SNR is measured from the point of view of the device's digital modem (i.e., post RF processing based on the MCS), because it includes the receiver's noise figure (NF). The version of the link budget SNR is often referred to as SNR^D (signal to noise and distortion ratio).

3 For example, it is common for enterprise-grade APs to be able to receive MCS 0 over 20 MHz at –98 or –99 dBm, more than 16 dB below the 802.11 standard minimum.

This is distinct from the RSSI or SNR as measured or estimated by an analog RF planning tool, which is often referred as SNR^A (analog SNR).

So, if a virtual site survey predicted the RSSI and SNR at a certain distance or range from the AP, this prediction would be the SNR^A, but would not tell you the associated supported data rate. You would need to deduct the NF from SNR^A to arrive at SNR^D and estimate the actual data rate supported by the device. This is one reason that "SNR" and "RSSI" requirements for various minimum cell edge data rates can vary significantly from design tool to design tool, based on which definition is used. Unless stated otherwise, this chapter refers to the SNR^D when mentioning "SNR."

Another difficult element that contributes to different readings between devices (and site survey applications) is the fade (F) loss margin of the environment. This value is a log-normal variable that expresses the variability of the RSSI around the mean. With multipath and the limitations in RSSI measurement accuracy, each frame measured at the receiver may have a slightly different RSSI.[4] With tens of frames per second, it would be very difficult to display the RSSI (and changes in it) in real time. Site survey and other Wi-Fi utilities tend to display a stable number, which is the mean of the measured RSSI over some interval. However, that mean value can change if you measure at the same location at another time. Therefore, designers commonly account for an additional fade loss margin.

Typically, this margin is chosen to be the 90th percentile that, if subtracted from the link budget, would ensure that 90% of the received frames at a given location are received at the required RSSI or better. This chapter assumes F = 6 dB, which is a commonly accepted figure in 802.11 design. Therefore, if your site survey utility displays a raw mean of –65 dBm RSSI at a given location, you would record –59 dBm (–65 + 6). This margin supposes that the site survey displays the raw RSSI, and does not already include a fade margin. In the end, the margin decision is yours. Feel free to account for this fade, use a lower figure, or ignore it entirely, depending on the reliability requirement of your design.

In most environments, the environmental noise (N) can be estimated from the thermal noise formula:

$$N_{dBm} = 10log_{10} (1000K_BTB)$$

where K_B is Boltzmann's constant and equal to 1.380649×10^{-23}, T is the temperature in kelvins, and B is the bandwidth in hertz. Table 8-3 shows the thermal noise (in dBm) for all the relevant 802.11be channel widths and an ambient temperature of 68°F (20°C). For example, the noise power for a 20 MHz-wide channel at 68°F (20°C) can be calculated as follows:

$$N = 10 \; log_{10}\big((1000)(1.380649 \times 10^{-23})(20 + 273.15)(20 \times 10^6)\big)$$
$$= -100.91797$$

---

4 The receiver evaluates the power of each frame preamble, and produces an RSSI result that evaluates this power over all radio chains that received the signal. Variations in the multipath can easily change the power received at each radio chain, and/or cause some of the radio chains to receive one preamble but not another one.

Although temperature affects the noise figure, the effect is not very strong in the range of normal indoor temperatures (say, 62–77°F, or 17–25°C). However, you can calculate that at 86°F (30° C), the noise power for 20 MHz reaches –100.77 dBm.

TABLE 8-3 Thermal Noise Constants

Bandwidth (MHz)	Noise Power (dBm) @ 20°C
20	–100.92
40	–97.91
80	–94.90
160	–91.89
320	–88.88

The environmental noise does not represent the background noise measured in the receiver. To calculate the background noise value, you need to add the noise figure (NF), which is the noise added by the receiving circuit. The total can vary depending on the receiver circuit's quality and temperature. In most cases, an 802.11 design supposes a background noise (N + NF) of approximately –94 dBm.

A key component to determine is the path loss, which measures how the received signal decreases as the distance increases. In a theoretical open space, distance is the only factor. In an indoor setting, walls and other obstacles increase the loss. If you are interested in estimating pathloss (PL) by hand for any environment, note that it can be defined in terms of dB as follows:

$$PL_{dB} = \gamma * 10 * log_{10}(d) + 20 * log_{10}(f) - 147.55$$

where d is distance (m) and f is the center frequency (Hz) of the signal (the channel). The pathloss exponent (γ) depends on the environment: It is 2 in free-space or pure line-of-sight (LOS) environments, and between 3.3 and 3.8 for indoor environments.

These figures, and Table 8-2, suppose a single spatial stream (1SS). It is common to consider that separating 2SS requires 3 dB. Therefore, MCS 0 with 2SS may double the data rate (to 13 Mbps instead of 6.5 Mbps), but at the cost of a higher minimum sensitivity level (e.g., –79 dBm instead of –82 dBm). The same logic applies as you add more spatial streams. In contrast, if MIMO and the closely related beamforming and maximum ratio combining (MRC) techniques can be used to increase the effective link speed, the SNR can be lowered to achieve the same rate. When two or more spatial streams are combined, the SNR required to reach a given MCS is lower than the SNR needed in a single spatial stream scenario. Table 8-4 provides an example of the SNR needed to achieve 1Gbps (MCS 6), based on the number of spatial streams that the STA and the AP enable, in the case of an uplink transmission.

TABLE 8-4   Example Uplink SNR Gains with MIMO

STA MIMO	AP MIMO	SNR for MCS 6 (160 MHz)
1	2	36 dB
2	2	30 dB
1	4	27 dB
2	4	22 dB
3	4	21 dB

As you can see, the number of MIMO/MRC antennas supported by the AP or the STA makes a significant difference in the SNR needed to achieve the exact same data rate. Incorporating the spatial streams into your design may mean increasing the useful size of the BSA, allowing your APs to operate at lower power, and/or increasing the MCS achievable at a given distance from the AP. However, keep in mind that the beamforming techniques (MIMO, TxBF, and MRC) do not offer guaranteed results, because they depend quite heavily on the RF environment. If the RF environment is unstable,[5] their benefits may not be achievable. Make sure you thoroughly understand the RF environment before deciding to incorporate these features as a key part of your design. A prudent approach may be to design for 1SS, using the values in Table 8-2, the specification documents for your APs and STAs, or a site survey utility,[6] and to consider the gains obtained by the beamforming techniques as an occasional operational bonus.

Cell Capacity

The design goal of the first cell is not to provide a minimum data rate for a single target device, but rather to provide that minimum required data rate to all intended clients. This goal is commonly called designing the *cell capacity*. The cell capacity design answers a practical question: Given a mix of devices and their capabilities (e.g., 802.11ax versus 802.11ac), and given backward compatibility (an 802.11ax is capable of communicating at 802.11ac MCS if 802.11ax data rates are not achievable), what percentage of clients will achieve what type of data rate in which area of the cell? To answer this question, you need to consider the following objectives and parameters:

- A minimum data rate for any device (either in each band, or in a desired target band, such as the 6 GHz band)

- The maximum channel bandwidth deployable in each band (or in the desired target band)

- The MIMO capabilities of the AP and the client devices (usually, the limiting side is the client device, and two spatial streams can be assumed)

From these objectives, you can determine the BSA size (and hence the AP density) from a combination of equipment provider performance specifications and site characteristics (from a virtual or physical

---

5 A rotating metallic fan in the RF environment vicinity may be sufficient to render MIMO and similar techniques unusable.
6 In this last case, incorporating the considerations of this section on noise and raw RSSI measurements limitations.

site survey). Different types of antennas will result in different coverage areas. If you increase the AP power, the coverage area increases in tandem. But before focusing on the AP power, you should start from the mix of clients and their data rate requirements, as described in the "User Device Requirements" section earlier in this chapter. Next, you should validate the type of cell capacity that each channel width can offer. You can then compare this cell capacity projection to your needs, and choose the channel width design that will fulfill your clients' throughput requirements.

The cell capacity is the average of the data rates available to the majority of users in the BSS. For example, if only one STA is connected at 1 Gbps (2SS, MCS 6 in 160 MHz), then the capacity of the cell is 1 Gbps. However, if a second STA is added to the same BSS and is able to achieve only 500 Mbps (2SS, MCS 4 in 160 MHz) while the achievable data rate of the first STA stays unchanged, then the cell capacity is 750 Mbps (($[1 \times 1000] + [1 \times 500]$)/2).

The achievable data rate depends on the distance of each STA to the AP, as well as the number of clients in the cell. In most cases, it is difficult to guess the future number of STAs in the cell, and where they will be positioned, with any accuracy. Even so, several methods are frequently used to estimate the area capacity, including integral methods and Monte Carlo simulation. Both have the advantage of assuming that a theoretically infinite number of potential users are evenly distributed through the cell area (represented as a circle). The data rate of each STA, based on the RSSI and SNR at each position, is estimated to compute the average capacity. Although your BSSs will not have "an infinity" of STAs, the conclusions from these mathematical methods can be used to estimate the average capacity of the cell—for example, to compute the data rate that will be available to 50% or more of the users in the BSS.

Compared to the integral methods, the Monte Carlo simulation is easier to visualize and can be parameterized with a mix of devices (e.g., 2SS phones + 3SS laptops) using commercial planning tools. Both methods effectively ask the question: What percentage of the cell area can support each target data rate, given some cell-edge requirement? For example, if you select MCS 0 (the lowest data rate) as the cell-edge requirement, then you know that 100% of the STAs in the BSS can operate at this minimum MCS—because all your STAs support MCS 0, and the STAs closer to the AP will likely operate at a higher MCS. A smaller percentage of the STAs (closer to the AP) in the BSA will operate at the higher data rate of MCS 1, and an even smaller percentage will operate at the 802.11be maximum of MCS 2 to MCS 13. If the STAs are deployed uniformly in the BSA, the percentage of STAs that can perform at each MCS can be calculated easily.

Table 8-5 shows the percentage of STAs that operate at each MCS, based on an ideal circular cell in an indoor environment and a cell-edge requirement of MCS 0. You can use this table as quick reference. Note that the ratios are the same regardless of the transmit power and the cell size. If you increase the AP power, the cell area also increases. However, if the edge of the cell stays at MCS 0, then the *percentage* of clients achieving each MCS (given that some are at the edge of the cell, whereas others are closer to the AP) is the same—all that matters is the choice of the cell-edge data rate (in this case, MCS 0). The limit of this model is naturally when the number of clients exceed the AP capacity, which will be addressed later in this chapter.

TABLE 8-5   MCS by Coverage Area (Large Cell)

MCS	Percentage of Cell
0	100%
1	67%
2	52%
3	35%
4	21%
5	12%
6	11%
7	9%
8	5%
9	4%
10	3%
11	2%
12	1%
13	1%

From Table 8-5, you can see that the rate available to 50% or more of the users in the BSS is MCS 2. Therefore, you can conclude that the cell mean capacity is MCS 2.

One important observation is that the highest MCSs (MCS 12 and MCS13, with 802.11be 4K QAM) are viable in only 1% of the cell. Very few users will likely benefit from these MCSs in the cell. Instead, the majority of STAs will operate at MCS 2 or less. If you want to improve the overall cell capacity, you should choose a higher cell-edge MCS. A direct consequence of that choice is a higher average data rate throughout the cell, but also a smaller cell (BSA). For example, if the cell-edge requirement is a 1 Gbps minimum data rate (MCS 6, 160 MHz, 2SS), supposing at least a 29 dB SNR at the cell, or roughly –65 dBm RSSI, then the call capacity and the data rate distribution are approximately shown in Table 8-6.

TABLE 8-6   MCS by Area (Small Cell)

MCS	Percentage of Cell
6	100%
7	88%
8	45%
9	35%
10	24%
11	18%
12	12%
13	8%

In this indoor small cell case, 1 Gbps (MCS6) is available throughout the cell and the peak data rate, 2.882 Gbps (MCS 13), is now available throughout 10% of the cell area, for an average data rate (and thus a cell capacity) of approximately 1.3 Gbps (MCS 8). The cell size (–65 dBm) corresponds roughly to a 2500 sq ft cell with a 14 dBm AP transmit power. Beyond the useful area of the BSA, the AP beacons may still be detected, and STAs may still be able to associate at lower MCSs (unless you disable lower MCSs, which is generally not a good idea[7]). However, the STAs should also be able to find neighboring APs at an equivalent or better signal level to associate to.

The values in Tables 8-5 and 8-6 were derived by computing the SNR delta between the edge MCS and each higher MCS, and then using a pathloss equation with a pathloss exponent (y) to estimate the range of each concentric circle that represents the coverage area of that MCS. In indoor environments, $y = 3.5$ is a good choice. For example, if your starting point is an edge MCS 0, you should first determine from Table 8-2 that this MCS requires an SNR of 12 dB for a 20 MHz channel (–82 dBm RSSI, supposing a –94 dBm noise floor; see the "Link Budget and Cell-Edge Data Rate" section). As this MCS is available from the edge of the cell and all the way to the AP, 100% of STAs in the BSA can achieve that MCS (provided they have the capability). The next possible MCS is 1, which has an SNR target of 15 dB (3 dB more than the SNR required for MCS 0). This delta (d) results in a relative cell coverage area of

$$\left(10^{\left(\frac{-d}{(10.y)}\right)}\right)^2$$

The same logic is applied to each of the other cell MCS targets (2–13), comparing each target MCS SNR to the SNR required for the cell-edge MCS (MCS 0, in this example). Table 8-7 shows the cell capacities matching different cell-edge MCSs.

In this table, the cell-edge MCS target is indicated in the top row, and the percentage of area coverage by each MCS, up to and including that target MCS, is shown in the matching column. For example, if the cell-edge target is MCS 8, then column with the MCS 8 header contains the coverage area for each MCS from 8 up to the last MCS of 13. In that example, you can see that MCS 13 is achievable in 18% of the cell area, and the mean cell capacity is MCS 10.

These tables assume there is a uniform type of STAs—that is, that all STAs have the same support for MCS and spatial streams. In a practical design, you will likely design for different types of STAs. For each of them, you would then perform the same evaluation (cell-edge minimum data rate, then percentage of the cell available to each MCS). You can then apply the device mix distribution (e.g., x% of device type A, y% of devices type B) to calculate the average cell capacity. For example, if you

7 Legacy data rates (6, 12, 18, 24, 36, 48, and 54 Mbps) could be configured to be Basic (mandatory support), Supported, or Disabled on the AP, and the AP would communicate these requirements to associating STAs. In the early days of Wi-Fi, this system allowed vendors to manage client support for various modulations (BPSK, QPSK, QAM). This method was not carried forward with the introduction of MCSs in 802.11n, and then 802.11ac/ax/be. Practically, this means that you can disable an MCS on the AP, but the STA can still use that MCS to communicate with the AP. The AP just cannot answer at that MCS. This setting tends to create more cell asymmetry than cell area control and is in most cases not recommended.

TABLE 8-7   MCS Coverage Area

MCS	0	1	2	3	4	5	6	7	8	9	10	11	12
0	100%												
1	67%	100%											
2	52%	77%	100%										
3	35%	52%	88%	100%									
4	21%	31%	52%	59%	100%								
5	12%	18%	35%	35%	59%	100%							
6	11%	16%	24%	31%	52%	88%	100%						
7	9%	14%	18%	27%	45%	77%	88%	100%					
8	5%	7%	12%	14%	24%	40%	45%	52%	100%				
9	4%	6%	8%	11%	18%	31%	35%	40%	77%	100%			
10	3%	4%	6%	7%	12%	21%	24%	27%	52%	67%	100%		
11	2%	3%	4%	6%	9%	16%	18%	21%	40%	52%	77%	100%	
12	1%	2%	3%	4%	6%	11%	12%	14%	27%	35%	52%	67%	100%
13	1%	1%	2%	3%	6%	7%	8%	9%	18%	24%	35%	45%	67%

estimate that 40% of the STAs will support 576.5 Mbps on average (i.e., MCS 5 is available in 50% or more of the BSA for these devices that support 2SS) and 60% of the STAs will support 864.7 Mbps (MCS 2 is available in 50% or more of the BSA for these devices that support 8SS), then the cell capacity is 749.42 Mbps [(40 × 576.5 + 60 × 864.7)/100]. You can apply the same mix-logic calculation to each other area and percentages of the BSA.

Cell Size and Channel Reuse

At this point, you have written down the requirements for your different clients and key target applications, established the minimum data rate required at the edge of the cell for these clients and applications, and computed the cell capacity for the entire population of clients in your cell.

Cell Size

You also know that the physical size of the cell depends on the power at which the STAs and the AP transmit (larger power = larger cell). It may be tempting to set the AP to maximum power, in the hope of maximizing the useful part of the BSA. However, such maximum power presents two major downsides:

- The maximum power of the AP may not be the same as the maximum power of the clients. Most smartphones are limited to an effective isotropic radiated power (EIRP) of 14–15 dBm in

5 GHz and 6 GHz (for any channel width). This limitation comes from the phone form factor,[8] but also reflects regulatory limitations on the maximum amount of energy that the phone, which is intended to be operated near your body (in particular, your head), can emit.[9] Larger devices (e.g., laptops) can operate at higher power. A client operating at 14 dBm and positioned at the edge of the cell of an AP operating at 23 dBm (for example) may not be able to use the same MCS as the AP, resulting in exchange asymmetries.

■ As the BSA increases, the number of clients to the AP also increases, including the clients at the edge of the cell using low MCSs—and, therefore, consuming a lot of airtime for each transmission. The risks of contention and collisions increase with the number of clients. Controlling the client count is one method to control the efficiency of the cell operations.

To account for these two considerations, a common design is to set the AP power at about the same power as the weakest key target client. For example, if the key clients are laptops (22 dBm maximum EIRP) and smartphones (14 dBm maximum EIRP), setting the AP to 14 or 17 dBm[10] EIRP is a reasonable approach.

Channel Reuse

At this point, you have also noted that the minimum data rate at the edge of the cell increases with the channel width. It may be tempting to use wide channels to achieve the largest possible cell-edge data rate. This intent is counterbalanced by the limited number of channels. Recall that the term *cell* comes from cellular designs, where the equivalent of the BSA is represented by a hexagon, as the simplified representation of a circle. This approximation was useful because hexagons contain almost the same area as a circle with similar diameter, but can be arranged as a repeating geometric pattern (Figure 8-2). The same logic can be applied to other wireless deployments, including 802.11. However, given that most 802.11 installations are located in indoor buildings with rather rectangular or square-like dimensions, it is often easier to design and deploy while using square cells as the best representation of the intended coverage area. However, the square design is not a very good representation of a circle area. The overlap between cells may be large along the square's edges, and small (or nonexistent) around the corners.

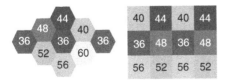

Hexagon cells   Square cells

FIGURE 8-2   Cell Types

8 Battery limitations, but also the difficulty of inserting multiple efficient antennas while respecting the aesthetics constraints of the phone.
9 These limitations include all the energy emitted by the phone—typically Wi-Fi, Bluetooth/BLE, and cellular. The cellular radio continues operating when Wi-Fi is active.
10 The AP likely has a better receive sensitivity than the client, allowing for some difference between the AP and the client EIRP.

To avoid overlapping BSS interferences, neighboring cells are set to different channels. The primary question that this type of motif attempts to answer is that of frequency reuse, or channel reuse: If the AP in a cell is set to some channel (e.g., channel 36), how far is the nearest cell on the same channel? Answering this question is equivalent to asking this one: Can you create a group of cells (a cluster), all on different channels, that would form a pattern that you can reuse over and over again? As you can see in Figure 8-3, some patterns "work," but others do not. On the left side of the figure, you can form a cluster of cells on three different channels, which pattern can be repeated infinitely. A cell on a given channel (e.g., A) has other cells on the same channel at equal distances in all directions. Because each pattern has three cells, the reuse factor is $\frac{1}{3}$.

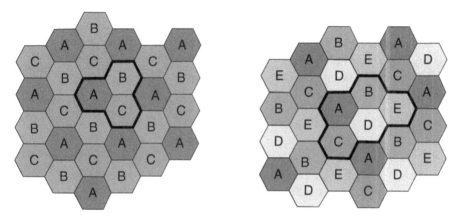

FIGURE 8-3   Valid Reuse Factor (Left) and Invalid Reuse Factor (Right)

On the right side of Figure 8-3, you can see an invalid reuse pattern. You can form a cluster with five channels and repeat the cluster pattern infinitely (this is relatively easy with hexagonal shapes), deciding on a reuse factor of $\frac{1}{5}$. However, a cell on a given channel does not have other cells on the same channel at equal distances in all directions. Take the center cell on channel A, for example. You can find other cells on the same channels three cells away in some directions, but two cells away in other directions.

It may seem challenging to find valid reuse factors. With hexagons, reuse factors formulae borrow from the geometry of hexagons. If i is the first channel (thus $i = 0, 1, 2$, etc.) and j the next channel ($j > i$), the number of channels you can group in a cluster is defined by the equation: $N = i^2 + ij + j^2$. When substituting i and j with 0, 1, 2, and so on, you find that the possible values of N are 1, 3, 4, 7, 9, 12, 13, 16, 19, and so on. This result means that you can form valid clusters of cells on 3 channels (as you saw in Figure 8-3), 4 channels, 7 channels, and so on, with a channel reuse factor of $\frac{1}{3}$, $\frac{1}{4}$, $\frac{1}{7}$, and so on.

The logic is somewhat similar when the cells are represented as squares. However, the geometry of squares is usually more familiar to the human mind than the geometry of hexagons. As soon as you can form a square with cells on different channels, you have a pattern that can be duplicated infinitely. You can see two valid examples in Figure 8-4: on the left, a cluster of 4 (2 × 2) with a reuse factor of $\frac{1}{4}$, then on the right a cluster of 9 (3 × 3) with a reuse factor of $\frac{1}{9}$. You can extend this logic to form a pattern of 16 cells (4 × 4), 25 cells (5 × 5), and so on.

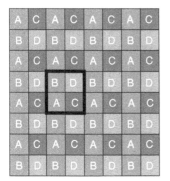 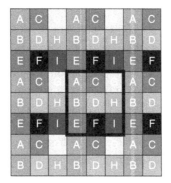

FIGURE 8-4   Valid Reuse Factors for Square Cell Representation with a Reuse Factor of ¼ (Left) and ⅑ (Right)

"Invalid" does not mean that you are not allowed to use that particular reuse factor. Each deployment and its constraints tend to lead to deployment decisions that are best in that particular context. However, "invalid" means that your design is unbalanced, with reused channels that are closer in some directions than in some others, and with the likelihood that you could use a smaller reuse factor (and thus larger channels) with the same co-channel interference outcome.

Regardless of the cell shape approximation (hexagon or square), larger clusters tend to be preferred. Such clusters increase the separation between cells on the same channel, and thus reduce the overlapping BSS (OBSS) interferences. Keep in mind that this interference is caused not only by the APs in each BSS, but also by the STAs. You can see this effect in Figure 8-5. The APs in BSS 1 and BSS 3 are on the same channel, and the signal from one AP may be only marginally detected in the BSS. Below the minimum receive sensitivity threshold of –82 dBm in Table 8-2, that would allow one side to detect co-channel interference (CCI) from the other. However, when a client at the edge of one cell transmits, its signal is detected above the CCI threshold by the other—in this example, at –65 dBm instead of below –82 dBm. In this scenario, both cells overlap.

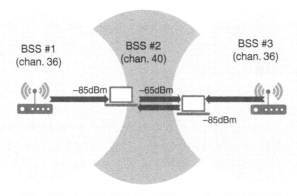

FIGURE 8-5   The Effect of Cell-Edge STAs on OBSS STA CCI

As the number of cells between APs on the same channel increases, this CCI risk declines. However, the number of channels you can use on a given band is limited. As the width of the channel in each cell increases, the number of available channels is reduced, limiting the reuse factor to a smaller range.

At this stage of the design, the main task is arbitration between an increased channel width (higher CCI risk and smaller reuse factor, but higher cell-edge minimum data rate) and increased reuse factor (lower CCI risk, but reduced channel width and lower cell-edge minimum data rate). The environment in which the APs are deployed plays a big role in this decision. Open spaces tend to constrain the system to narrower channels (and larger reuse factors) with high AP density (small cells with up to 1 AP per 625 sq ft area), whereas buildings with many and/or thick walls allow for larger channels and smaller reuse factors without a negative effect on the CCI (and larger cells of 1 AP up to 5000 sq ft). Table 8-8 shows some common deployment densities for indoor WLANs designed for 5 GHz/6 GHz coverage supporting time-sensitive applications (e.g., real-time voice, AR/VR, video-conferencing).

TABLE 8-8   Typical Cell Areas in 5 GHz/6 GHZ BSSs

Cell Area	Typical Distance Between APs	Cell Edge to AP Distance
1000 sq ft / 92 m^2	36 ft / 11 m	18 ft / 5.5 m
1200 sq ft / 111 m^2	40 ft / 12 m	20 ft / 6 m
1500 sq ft / 140 m^2	44 ft / 13.5 m	22 ft / 6.75 m
2000 sq ft / 185 m^2	50 ft / 15.2 m	25 ft / 7.6 m
2800 sq ft / 260 m^2	60 ft / 18.2 m	30 ft / 9.1 m

Designing for Wi-Fi 7

Wi-Fi 7 does not change the transmit power or the channels of 802.11ax (Wi-Fi 6) and earlier radios. Wi-Fi 7 introduces 4096-QAM in all bands. As we have seen in Chapter 5, "EHT Physical Layer Enhancements," and in this chapter, this modulation has a short range. The main two features that 802.11be introduces and that have an impact on coverage and capacity are access to 320 MHz channels in the 6 GHz band on the one hand, and multi-link operations (MLO) on the other hand. You can choose to include or exclude these new features based on the future device mix (and spectrum availability).

Brownfield Versus Greenfield

The first key consideration for Wi-Fi 7 site planning is that it is much like Wi-Fi 6E (for 6 GHz) or Wi-Fi 5/6 (for 5 GHz). In other words, the Wi-Fi 7 coverage area will be similar, if not identical, to that of those previous generations (for the same minimum data rate, channel bandwidth, and so on). So, unless you are engineering a new high-density (HD) venue, a high-speed Wi-Fi direct link, or a long-range metropolitan point-to-point (pt-to-pt) link exploiting the highest or lowest 802.11 data rates available, you can be reasonably confident in placing similar Wi-Fi 7 APs in the same locations as their Wi-Fi 6, Wi-Fi 6E, or Wi-Fi 5 counterparts (i.e., those with broadly similar antenna gains and pattern). This equivalence means that a Wi-Fi 6–oriented site survey is a reasonably good predictor of a Wi-Fi

7 deployment. Therefore, whether you are integrating Wi-Fi 7 APs into a pre–Wi-Fi 7 existing WLAN (brownfield deployment) or designing a new deployment (greenfield deployment), the design considerations are approximatively the same. For many users, the primary objective when choosing Wi-Fi 7 is to exploit the new 6 GHz spectrum that was first supported by the 802.11ax-based Wi-Fi 6E standard.

Wi-Fi deployments can range from small personal networks, to city-wide metro networks, to everything in between. For simplicity, this chapter considers five typical deployment scenarios:

- **Hotspot:** An AP covers a small area (e.g., automobile, wireless laptop peripherals).

- **Home:** A few APs cover a residence or small office with Internet and consumer IoT access.

- **Commercial:** Tens of APs cover a building with Internet access and commercial IoT access.

- **Enterprise:** Hundreds of APs provide mission-critical access (e.g., interactive media, industrial IoT).

- **Outdoor metropolitan:** A set of APs forms a long-range pt-to-pt or pt-to-multipoint link.

Like previous Wi-Fi generations, Wi-Fi 7 increases the peak rates through wider channels (320 MHz versus 160 MHz)) and higher-complexity modulation and coding schemes (4096-QAM versus 1024-QAM). This results in a peak spectral efficiency of 12 bps/Hz—that is, 2.882 Gbps per spatial stream in a 320 MHz channel. These capabilities lead to impressive theoretical performances, as you can see in Table 8-9.

TABLE 8-9   Peak 802.11be Data Rates per Device

Channel Bandwidth (MHz)	1SS (IoT)	2SS (Mobile)	4SS (Indoor AP)
20	172 Mbps	344 Mbps	688 Mbps
40	344 Mbps	688 Mbps	1376 Mbps
80	721 Mbps	1442 Mbps	2884 Mbps
160	1441 Mbps	2882 Mbps	5764 Mbps
320	2882 Mbps	5764 Mbps	11,528 Mbps

The goal of a WLAN design is to achieve the right data rate and latency (and performance requirements) for all STAs, not to ensure that each STA reaches the latest 802.11 standard peak data rate at all times. Reaching the design goal typically means that the peak data rate may or may not be reached, but that all STAs can successfully operate in the BSS, and between BSSs (either neighboring AP devices, or co-located AP radios on different bands).

Designing 5 GHz and 6 GHz Coverage

The number of channels available per AP cluster (i.e., a set of APs with unique channels) is based on the selected channel bandwidth in the band of operation. In 2.4 GHz, you typically have only three or four 20 MHz channels, so 20 MHz is the maximum recommended bandwidth in that band. For 5 GHz and 6 GHz, more channels of various sizes are available. The Wi-Fi 7 design starts from the 6 GHz band, because it is the band where the most performance can be achieved. This band has the largest number of channels, as indicated in Table 8-9, and the least effects from legacy devices.

TABLE 8-9   5 GHz and 6 GHz Channels in the FCC and ETSI Domains

Band	Channel Width	Number of FCC Channels	Number of ETSI Channels
5 GHz	20	24	23
	40	11	10
	80	7	6
	160	3	2
6 GHz	20	59	24
	40	29	12
	80	14	6
	160	7	3
	320	3	1

Recall from Chapter 3, "Building on the Wi-Fi 6 Revolution," that the number of channels available in the 6 GHz band depends on the regulatory domain. Rules may be changing, and you should check the local regulations before deciding on a 6 GHz channel plan. In many regulatory domains, you can set your APs to low power indoor (LPI) mode. In some regulatory domains, you can use standard power (SP). You should keep in mind the rules for LPI usage:

- The AP, and its WLAN, must be indoors. The AP cannot be deployed outdoors or cover an outdoors area, and the AP cannot be weatherized (i.e., set in a weather-protective enclosure with the intention of providing outdoors coverage).

- The AP must have a permanently attached antenna. This requirement means that the coverage area depends on the antenna radiation pattern attached to the AP. You cannot use an external antenna to provide a different coverage (e.g., direction).

- The AP cannot be battery powered (this regulation is meant to prohibit APs that are "temporarily" deployed outdoors). This requirement has important consequences in the site survey phase of Wi-Fi system design. In most cases, site surveys use APs on tripods, positioned temporarily to evaluate a possible coverage area. In many cases, the AP is battery powered so that it can be easily moved around. When surveying for 6 GHz coverage, you may need to clearly label the AP (e.g., "for survey only"), and add personnel to monitor the AP.

- The maximum power is 5 dBm/MHz (see Chapter 3), and the STA maximum EIRP is 6 dB below the AP maximum EIRP. Therefore, you likely do not want to set the AP to its maximum power (see the "Cell Size" section in this chapter).

Check whether your local regulations allow 6 GHz deployments outdoors. If such deployment is allowed, you must follow the AFC rules. The AFC is accessible by those commercial entities (e.g., an enterprise, a commercial or service provider operator) willing and able to register themselves and their individual AP locations (accurately) with the regulatory agency. For example, in the FCC domain, this entity must be an authorized AFC provider that is connected to the FCC database. The location and physical details of the AP antenna's installation (e.g., the down tilt) are used to determine whether allowing the AP access to the controlled 6 GHz spectrum would cause incumbent interference. Once permission to use the channel has been granted by the AFC, the AP location and possible interference with the incumbents must be reverified each time the AP reboots and every 24 hours.

These elements are important, because you should start your design from the 6 GHz band, if it is allowed in your regulatory domain. The 6 GHz band offers less interference and does not suffer from the legacy compatibility requirements as the 5 GHz and 2.4 GHz bands do.

The range and coverage of a 6 GHz AP (Wi-Fi 6E or Wi-Fi 7) is largely the same as that of a 5 GHz AP. You can use the standard free space path loss (FSPL) propagation model to predict the range of the Wi-Fi radio at the bottom of the 5 GHz band (channel 36, set to 5180 MHz) to the top of the FCC 6 GHz band (channel 233, set to 7115 MHz). For reference, the FSPL, or ratio of receive power and transmit power, is calculated as follows:

$$\frac{P_r}{P_t} = G_{TX} G_{RX} \left(\lambda / (4\pi d) \right)^2$$

The fundamental propagation difference (removing the effect of antenna gains, channel bandwidth, and transmit power) comes from the relationship between the FSPL and the channel center frequency contribution, represented as the wavelength squared (λ^2). By taking the ratio of ($\lambda_1^2 / \lambda_2^2$) for the two center frequencies f_1 and f_2, you can calculate the pathloss difference between these two frequencies: $(c/f_1)^2/(c/f_2)^2$ or $(f_2/f_1)^2 = (7115/5180)^2 = 1.88$, or in dB simply 2.7 dB. The conclusion from this calculation is that, if you deploy an AP device with one AP radio set to the lowest 5 GHz channel (channel 36) and another AP radio set to the highest 6 GHz channel, the 5 GHz signal is about 2.7 dB stronger than the 6 GHz signal at any point of the BSS. In an ideal (perfect and noise-free) environment, this difference represents a 5 GHz cell that is 19% larger than a 6 GHz cell. Measurements on many 5 GHz/6 GHz clients show a reported RSSI difference of 1.5–2 dB. Practically, and given that the 6 GHz band is less affected by legacy interference than the 5 GHz band, this difference can be ignored. In most cases, you would design the 5 GHz and 6 GHz bands as covering the same area.

The AP power in 6 GHz (using LPI) may be different from the AP power in 5 GHz, as you can see in Table 8-10. This table uses the FCC domain as an illustrative example. Refer to Table 3-5 (in Chapter 3) for more values.

TABLE 8-10 5 GHz U-NII-2-C and 6 GHz U-NII-5 EIRP Versus Bandwidth

Device	20 MHz	40 MHz	80 MHz	160 MHz	320 MHz
U-NII-5 (6 GHz LPI)	18 dBm	21 dBm	24 dBm	27 dBm	30 dBm
U-NII-2-C (5 GHz)	23 dBm	23 dBm	23 dBm	23 dBm	23 dBm

If you set your APs to the maximum power (e.g., because all of your clients are laptops), then the 5 GHz BSA is larger than the 6 GHz BSA. In that case, you are designing the WLAN for 5 GHz, with the property that cells will have a smaller inner zone where 6 GHz will be available. The difference between the 5 GHz and the 6 GHz BSAs may be considerable. In the free pathloss equation, a 6 dB difference represents twice the distance. Thus, if your channel is 20 MHz wide, and your useful BSA for 5 GHz stops *2d* feet (or meters) away from the AP, then you need to walk halfway to the AP (to *d* feet/meters) for your 6 GHz BSA to be available, as illustrated in Figure 8-6.

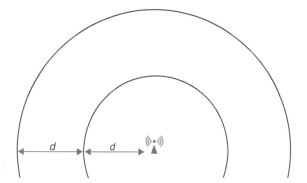

FIGURE 8-6   6 dB Difference Between an Inner 6 GHz BSA and a Larger and Stronger 6 dB, 5 GHz BSA

As the channel becomes wider, the difference has an inverse relationship, because of the different ways that power is calculated in 5 GHz and in 6 GHz systems. However, the AP beacons (and most probe responses) are sent over 20 MHz. Therefore, your clients discover the BSSIDs over 20 MHz. Notably, most vendor algorithms account for the fact that the 6 GHz EIRP depends on the channel width, and that the beacon contains less power than the entire channel. In short, this difference is important if your APs are set to maximum power and your channel is either narrow (20 MHz) or very wide (160 or 320 MHz). For all intermediary widths, the difference between 5 GHz and 6 GHz LPI is marginal. In most deployments, you would not set your AP to maximum power, especially if you expect smartphone clients. Instead, you would set the 6 GHz radio power slightly higher than the 5 GHz radio (e.g., 6 GHz radio set to 17 dBm and 5 GHz radio set to 15 dBm), resulting in BSAs of equivalent sizes.

Designing for Wide (320 MHz) Channels

The primary challenge when designing a WLAN for wide channels (320 MHz, but also 160 MHz) is channel reuse. For example, there is only one 320 MHz channel in the ETSI domain, and three 320 MHz channels in the FCC domain. A channel reuse factor of ⅓ (as shown on the left part of Figure 8-3) is possible, and a reuse factor of ¼ (all neighboring cells on the same channel) is certainly valid. However, in both cases (very probably with the ⅓ scheme, and almost certainly with the ¼ scheme), you will have to manage overlapping BSS interferences. If your design is intended for an environment with a lot of obstacles (walls), then that design may work without too much CCI. When you are designing a small network (e.g., home, hotspot), you may not need to install more than one (or three) APs, and the CCI considerations do not apply—at least if you do not have, or do not worry about, neighboring WLANs. In most cases where your network includes more than three APs, the 320 MHz design is difficult to realize in practice. The interferences from neighboring cells on the same channel will reduce the throughput of the 320 MHz cell to a point that will negate the wide channel benefit.

If you set your AP to LPI, then the EIRP will limit the AP power. The regulatory limits in the FCC domain (5 dBm/MHz and 30 dBm EIRP) mean that a 320 MHz signal will exceed the maximum EIRP of 30 dBm if set at the PSD maximum of 5 dBm/MHz. When the AP is set to 320 MHz at maximum LPI power (30 dBm), each 20 MHz segment must operate well below 5 dBm/MHz, which obviously limits the size of the BSA. Adaptive APs sent the narrower frames (including beacons and probe responses) at optimal power (typically 18 dBm for 20 MHz, as shown in Table 8-8), but the full 320 MHz channel will be available only at low power. The ETSI regulations also recognize LPI, but allow this mode only in U-N-II-5 and U-N-II-6, and with a limit of 10 dBm/MHz and 23 dBm as the AP maximum EIRP. Therefore, the AP should also be set at 23 dBm maximum for a 40 MHz, 80 MHz, or 160 MHz channel. Practically, this means that the PSD rule finds its limit in the ETSI domain at 23 dBm EIRP. As a consequence, if you design 320 MHz channels, high data rates (high RSSI, high SNR) will be achieved only near the AP, regardless of the domain.

Standard power allows for an increase of power. However, a 320 MHz scheme supposes that almost the entire band is free from incumbent interference. There are only a few locations where these conditions are expected to be realized. With 320 MHz channels, when an incumbent is discovered, there is no other channel for the AP to move to in a ⅓ or ¼ scheme. If the AP is indoors, it can switch to LPI. Outdoors, the AP must either switch to vacate the affected channel, which may mean switching to a narrower channel, or apply puncturing.

For all these reasons, a channel width of 320 MHz may be more commonly used in small networks. In enterprise deployments, 80 MHz is likely to be the common width in 6 GHz, which is also the assumption behind the preferred scanning channel (PSC) design.

Designing for MLO

The dual connection allowed by MLO improves the STA operations—in particular, intra AP-device roaming, where the process of moving between AP radios is as simple as selecting one or more radios to transmit at each TXOP (e.g., based on RSSI). From an AP configuration standpoint, the WLAN should have MLO enabled and each AP radio (2.4 GHz, 5 Hz, 6 GHz) should be configured to advertise the availability of the others' links (e.g., with the reduced neighbor report, detailed in Chapter 3), as is routine for 6 GHz deployments. You can also deploy the APs with dual 6 GHz radios. In that case, you need to plan the same band twice. To reduce adjacent channel interference (ACI), you should attempt to space each 6 GHz link as far as possible from the others (e.g., one AP radio on channel 7 and the other on channel 215). Keep in mind that the reuse factor is affected by this design.

In some cases, it may be advantageous to restrict certain STAs or certain latency-sensitive traffic types (e.g., TID 5–7) to one link (usually the 6 GHz link) that has the best performance. To implement this kind of policy, the AP must support the Wi-Fi 7 TID-to-link mapping (T2LM) capability, allowing the designer to steer any STA's link without critical traffic away from the reserved 6 GHz premium link. This type of design yields the best outcome for business-critical applications such as AR/VR and industrial IoT (IIoT).

When MLO is enabled, the MLO-capable STA may be using capacity on two (or more) AP radios. This link consumption affects the coverage and capacity plans significantly. In some cases, the STA has the same volume of traffic to send as it would have in a single-radio case. The STA then splits its traffic between the links, appearing as a lower-bandwidth device on each band. In other cases, the STA applications are limited by the single-link bandwidth, and the multi-link availability allows the STA to send more traffic. In this situation, the applications might possibly consume up to the peak bandwidth designed for each STA on each band. A single client device then appears as two STAs, each consuming the expected peak bandwidth. This phenomenon may double the number of STAs in the WLAN.

To complicate matters, there are several types of MLO to consider:

- **STR:** Capable of 2.4 GHz + 5 GHz, 2.4 GHz + 6 GHz, 5 GHz + 6 GHz, or in some cases dual 6 GHz simultaneously.

- **NSTR:** Like STR, except throughput is limited in 5 GHz + 6 GHz.

- **eMLSR:** Capable of association to multiple bands, but capable of using only one at a time.

The STR case is the simplest to plan for. You simply double (or triple) count each MLO-STR–capable device, counting it once in each of the bands it can operate under for both coverage (minimum data rate) and capacity purposes. Because the links are expected to be active simultaneously, you can design the cells with a minimum data rate on each band suitable for this device type. In theory, if all STAs were STR in 5 GHz and 6 GHz, you could treat 5 GHz as a backup link and set the 5 GHz radio to a narrow channel (e.g., 40 MHz), making it more reliable (due to reduced CCI) albeit with lower capacity. However, in most deployments, not all Wi-Fi 7 STAs are expected to be STR, and legacy 5 GHz devices may need to be accommodated. STR does have an advantage in that either link can be used at any time. Both 5 GHz and 6 GHz links could be deployed with the same reuse factor. This design typically results in the 6 GHz channel being twice the channel width of the 5 GHz channel due to the relative amount of spectrum in each band.

The NSTR case is similar in effect to STR, with the only notable performance difference being that a 5 GHz + 6 GHz NSTR device will experience reduced throughput and possibly higher latency due to the need for the STA and AP to coordinate the transmission across both bands so that both frames end at the same time on both links (see Chapter 6, "EHT MAC Enhancements for Multi-Link Operations," for details). This may mean either padding one transmission to force end-of-frame alignment (reducing throughput) or deferring transmission until the frames are the same size (increasing latency). However, the RF planning considerations are generally the same as for STR with one caveat: The NSTR STA can also be an STR STA if there is sufficient frequency separation between the 5 GHz and 6 GHz operation channels (e.g., 200 MHz). Given this issue, it may be advantageous to explicitly plan this separation, and to override RRM if necessary.

The eMLSR case is more complex, as the STA can receive control on any of its two connected links, but will transmit/receive on only one at any given time. These exchanges are controlled by the AP in the downlink and by the STA in the uplink. The factors for link selection are not specified by Wi-Fi 7 or 802.11be, but the motivation for eMLSR is to achieve power savings on the STA. Thus, it is advantageous for the STA to use the least amount of airtime to transmit a frame and to use the fastest link. The STA would use the slower link only if the faster link became unavailable. This behavior assumes that the coverage is available on each band. The eMLSR STA will likely "camp out" on the higher-speed link (usually 6 GHz for indoor) and most of the time will not leverage the other bands. When designing your WLAN, you could consider the eMLSR STA as a 6 GHz-only STA, which occasionally happens to use the 5 GHz link. But if the 6 GHz coverage is sufficient, the 5 GHz exchanges will be rare and can be ignored in the design.

Irrespective of the mode that the clients are expected to use, MLO is useful only if all three bands (2.4, 5, and 6 GHz) have sufficient link RSSI to allow proper STA operations. This requirement dictates a 6 GHz-driven cell size selection, as the 6 GHz link is likely to be the weakest in the SP and LPI modes. Although you may be able to design different coverages for the 5 GHz and 6 GHz bands, with different cell-edge MCS/data rate selections, such a mismatch is not recommended for MLOs. When this design is used for MLO, the link selection may be biased on a semi-permanent basis, resulting in low utilization of one of the bands.

The fate of 2.4 GHz may also be called into question with Wi-Fi 7 MLO. Historically, this band was used as either a coverage channel (e.g., for low-power IoT) or a backup channel for 5 GHz. Now, however, with 6 GHz as the likely primary channel, 5 GHz may serve as a fast-backup channel; that is, switching back and forth will occur automatically and transparently to the user applications. In this context, the utility of 2.4 GHz is diminished. If your WLAN has no legacy IoT devices, you might consider eventually decommissioning the 2.4 GHz radios. You might also decide to deploy the 2.4 GHz channels only at the edge of the network for onboarding. For example, they could be deployed in the lobby of a building, from where the 2.4 GHz signal may reach far into the parking area, and allow for auto-association before users enter the building.

Designing for P2P Coexistence

An 802.11/Wi-Fi device that connects (or tethers) directly to another 802.11/Wi-Fi device and does not attach to an infrastructure AP (infrastructure BSS) is referred to as a peer-to-peer (P2P) device. In the 6 GHz band, this may also be referred to as the client-to-client (C2C) mode, based on regulatory designations. From a practical standpoint, the P2P and C2C mechanisms are similar. Because P2P devices do not associate with the infrastructure BSS (AP), their operating channel may or may not be the same as the AP's channels. If the channels used by the P2P devices are the same as, or overlap with, the AP's, then interference may occur and cause degradation of both the WLAN and P2P link.

The protocols used by P2P are based on 802.11, but some unique P2P group discovery and topology formation protocols have also been standardized by the Wi-Fi Alliance's Wi-Fi Direct program that are required for interoperability. A P2P network consists of two or more devices a user has near their person, such as a smartwatch, a smartphone, or AR glasses. These devices connect to each other and share updates, synchronize state, and may possibly stream live video/audio as needed. An emerging use case is as a replacement for a wired HDMI cable (i.e., wireless HDMI) where high-rate raw or semi-compressed video/audio is exchanged continuously between a video/audio source (e.g., laptop, PC) and a display device (e.g., smart glasses or AV/VR headset). Unlike a smartwatch, the wireless HDMI link has the potential to consume significant airtime and, therefore, may degrade the infrastructure BSS performance, if both the BSS and the P2P link are on the same channel. The P2P group leader (e.g., a smartphone) may also be associated with the infrastructure AP via another link/radio or the same link/radio (e.g., alternating between them) and perform an associated function such as cloud data access.

P2P networks may appear in the WLAN. If so, you need to decide how to best accommodate them. If the P2P link is used for critical real-time services such as high-rate video/audio transport (e.g., wireless HDMI, VR), then you should explicitly consider their throughput and latency targets in your design. Given the distance between the P2P devices (e.g., a few feet, one or a few meters), it is reasonable to assume the higher Wi-Fi 7 modulation schemes (specifically Wi-Fi 7's 4096-QAM) are achievable at least for one or two spatial streams. This would lead to a 1.4–2.8 Gbps peak data rate on 80 MHz channels. In this scenario, the challenge is predicting how much bandwidth the P2P link can acquire, free of CCI from the infrastructure AP, if operating on the same channel. If the anticipated load (channel utilization) on the AP is low (e.g., 10%), then this coexistence likely does not present a problem. However, if the anticipated AP load is high (e.g., 50%), then the P2P exchanges may cause noticeable interference to the AP BSS, and vice versa. To avoid the excessive interference issue, Wi-Fi 7 offers two specific mechanisms:

- Channel-based segregation (channel usage)
- Time-based (TXOP) sharing

In the channel usage approach, an AP, with the assistance of some radio resource management (RRM) software, selects a preferred channel and bandwidth on which the P2P groups should operate. It then advertises this channel to all the STAs via the beacon. The P2P group leader (whether associated to the AP or not) can then receive the beacon and instruct its group members to consider changing to the recommended channel (e.g., after scanning). For this channel-usage method to work best, the AP should act as follows:

1. Select for the P2P suggested exchanges low-utilization channels and/or channels where other APs are received at low-RSSIs (e.g., outside AP coverage).

2. Coordinate with other APs via RRM to create available P2P channels, and keep them free from AP BSS traffic.

3. Select the channel bandwidth based on the envisioned P2P application needs (e.g., 1 GBps needs 80 MHz at MCS 10 with 2SS).

The amount of spectrum RRM may allocate to P2P depends on the amount of spectrum available and the channel reuse factor in effect—that is, the number of free channels between the allocated infrastructure channels. One challenge of P2P groups using channels adjacent to the infrastructure channels is that the P2P device itself may be close to the AP and induce adjacent-channel interference (ACI). This challenge is especially relevant for medium-power VLP devices that may have EIRP of similar levels to the AP itself (e.g., 10 dBm). To mitigate this issue, a spectrum analysis of the P2P group operation is advised to help determine the ideal channel separation.

In the time-based TXOP sharing scheme, the AP and P2P group operate together on the same channel, and the AP grants channel access to the associated P2P group leader for transmissions between group members. This methodology is more complex because the AP must account for the specific needs of the P2P application. Fortunately, the SCS QoS Characteristics IE (QC) allows the P2P group leader to express the needs of the group in terms of periodicity (e.g., every 35 ms a video frame is sent from the iPhone to the AR glass) and time duration (e.g., 100 µs for a 10KB frame at MCS11). The real-time duration is difficult to estimate given that the actual data rate between the P2P STA and P2P group leader is dynamic. Therefore, a very conservative estimate is usually needed, which can be wasteful in terms of airtime.

Owing to this waste risk and the scheme's complexity, the time-based TXOP sharing approach may not see wide adoption. However, you should be aware of this capability and potentially disable the feature if it is enabled and the expected airtime consumed by the P2P group is excessive.

Designing High-Density Environments

In many WLANs, the APs need to be placed in positions where propagation is favorable (e.g., ceiling level). That can lead to very large cell edges and either very low frequency reuse for a given area or high CCI. In some cases, directional antennas can help reduce CCI by sectorizing the cell or reducing its azimuth plane projection (i.e., the area on the ground the signal is received on). In general, though, CCI remains an issue in these scenarios, as illustrated in Figure 8-5.

One approach to solve the CCI problem is to reduce the AP power, which also decreases the useful part of the BSA. However, STAs at the edge of the cell may transmit at high power—possibly at the same power as the AP. AP power control is efficient only if it is coupled with techniques to ensure that the STAs stay close to the AP, and then, when getting to the edge of the useful part of the BSA, roam to another AP.

The higher MCSs introduced by 802.11be (MCS 12 and MCS 13, with 4K QAM) allow for the possibility of very small cells with high data rates. Suppose a design is based on 6 GHz/80 MHz channels, an AP EIRP of 24 dBm, and 18 dBm for STAs in U-NII-5. In this context, MCS 12 and 13 correspond to 647 and 720 Mbps, respectively, with 45 dB and 48 dB SNR required, respectively (see Table 8-2). A design goal could be to ensure at least MCS 12 everywhere in the useful part of the BSA. Suppose that the AP includes a 4 dBi antenna with a noise figure of 6 dB, that the typical STA transmit power is 12 dBm, and that the system comes with an antenna of –2 dBi and a noise figure of 10 dB. Then the link budget equation (see the "Link Budget and Cell-Edge Data Rate" section earlier in this chapter) allows you to envision different possible power settings on the AP. You can also use this equation to evaluate the distance at which a STA in a neighboring BSS can start detecting the AP (based on the signal at that distance, and the minimum sensitivity requirements in Table 8-2), the distance at which

MCS 12 can be achieved, and the distance at which a STA in a neighboring cell can be detected. In these open-space environments, the AP-to-AP (i.e., ceiling to ceiling) and AP-to-STA (i.e., ceiling to ground) links can be approximated as LOS, with only the STA-to-OBSS STA link experiencing NLOS (i.e., ground level to ground level). Therefore, the pathloss equation can be used directly, without incorporating the effect of obstacles. Table 8-11 displays these different ranges for MCS 12 and different power settings on the AP.

TABLE 8-11 HD Range (m)

AP EIRP	8 dBm	11 dBm	14 dBm	17 dBm	20 dBm	23 dBm
AP CCA	8 m	11 m	15 m	21 m	28 m	39 m
MCS range	1.2 m	1.7 m	2.3 m	3.1 m	4.3 m	5.9 m
OBSS MCS range	8.2 m	8.7 m	9.3 m	10.1 m	11.3 m	12.9 m

You can see that the target MCS (12) can be delivered 5.9 m (19 ft) from the AP when the AP is set to 23 dBm (Table 8-11 assumes an 80 MHz-wide channel), The OBSS interference from a cell-edge STA to an OBSS STA is 12.9 m. Any OBSS STA beyond that distance will detect the STA 23 dBm transmission below the –43 dBm required level (see Table 8-2) and will not demodulate the transmission. With BSS coloring, the STA may be able to transmit at the same time as the local MCS 12 transmitting STA. However, a challenge is that the AP creates CCA-level interference on an OBSS for 39 m to achieve the desirable cell-edge rate. Therefore, OBSS STAs may be able to transmit at the same time as the MCS 2 local STA, but their MCS is likely to be low (MCS 0 at 12.9 m, and increasing with the distance). Smaller power levels result in smaller cells that may still be acceptable in a home, commercial, or enterprise environment.

If you design your 6 GHz band with 80 MHz-wide channels, you can plan for a reuse factor of 13, and therefore 4 entire cells of separation (each of width "twice the radius of one cell" plus the radius of the originating cell) for $4 \times 2 + 1 = 9$ radii of separation. If each cell has a radius of 19 ft, then the distance from each AP to the nearest STA in the OBSS is $9 \times 19 = 171$ ft, which is enough to avoid CCA-based contention from the AP (calculated in Table 8-11 as 39 m/128 ft). Figure 8-7 illustrates this scenario.

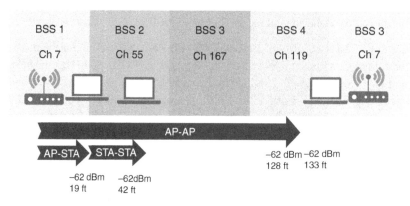

FIGURE 8-7   High-Density Cell Separation Example

In many professional WLAN systems, the AP can use features such as 802.11k/v to encourage a STA to roam to a neighboring AP when the system believes the quality will be better. Recall from Chapter 2, "Reaching the Limits of Traditional Wi-Fi," that the 802.11k protocol allows the STA to share RSSI measurements of its own and other BSSs (beacon reports) to the AP; in return, the AP can suggest the best APs for the STA to roam to (neighbor report). The 802.11v protocol supports the BSS transition management (BTM) commands from the AP to encourage the STA to move to a suggested nearby AP based on the AP's view of the signal quality to and from the STA. Together, these protocols, when configured on the AP, can help the STA move to the neighboring BSS when it reaches what you decide is the edge of the useful BSA.

Designing for Industrial IoT

In complex environments, such as industrial IoT, time-sensitive network (TSN) equipment, such as motion controllers, is used to monitor and command automated ground vehicles (AGV) or autonomous mobile robots (AMR) using tight control loops. These loops require deterministic delivery of periodic reports (e.g., positions every 100 ms) from the AGV/AMR. In response, the controller issues a command (e.g., continue, stop), which must be delivered before a deadline (e.g., 10 ms). This scenario is illustrated in Figure 8-8.

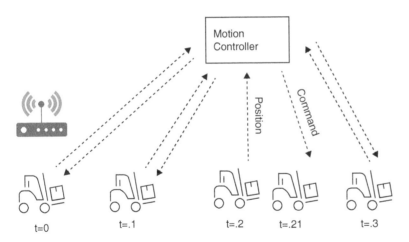

FIGURE 8-8   IIoT Controller

To enable low bounded latency (determinism) for these control loops, you can use two Wi-Fi 7 capabilities designed for such operations. The first capability, the stream classification service (SCS) with QoS characteristics (QC), allows the endpoint (STA) to describe the time-critical flows, such as periodic position reports followed by a command. The second capability, restricted target wake time (R-TWT),[11] allows the STA to specify a series of time-slots during which only it is allowed to communicate with the AP; other STAs in the BSS may not transmit during these timespans. Taken together, these features allow the timely AP scheduled delivery of the position and command messages and ensure that other STAs in the same BSS do not send at the same time.

11 For more details on SCS and r-TWT, see Chapter 7, "EHT MAC Operation and Key Features."

In this IIoT use case, the STA could request, for example, two SCS flows from the AP as follows:

- **Report:** Uplink direction, maximum and minimum scheduling interval (SI) = 100 ms, delay = 10 ms, MSDU size = 100B, start-time = 0 ms

- **Command:** Downlink direction, maximum and minimum scheduling interval (SI) = 100 ms, delay = 10 ms, MSDU size = 10B, start-time = 0 ms

The key is that the application packet generation times and the SCS service schedule are synchronized via the start-time. The alternative method (not explicitly defined in 802.11be/Wi-Fi 7) is that the R-TWT session is synchronized to the traffic generation times (via the TWT start-time and periodicity), which are used by the AP for uplink and downlink STA scheduling.

Time synchronization is required for optimal operation. The system clock can be synchronized to an external 802.1AS/gPTP clock or to the Wi-Fi clock (TSF). The standards do not define how this synchronization happens. However, without such synchronization, the packets arriving at the Wi-Fi interface from the source will encounter undesirable queuing delays and will not meet their delivery deadlines.

We can envision more IIoT or wireless TSN (WYSN) use cases being supported in this way—for example, wireless PLCs, safety controllers, and similar deterministic applications. The ability to schedule time-sensitive traffic is unique to Wi-Fi 7 and is not supported in previous 802.11 and Wi-Fi generations.

Summary

You now have many of the building blocks required to design Wi-Fi 7 networks. All 802.11 designs start from the same foundation: collecting the clients' characteristics and the WLAN requirements. With that information, you can design your first cell, making sure that each device can get the minimum data rate needed, right from the edge of the cell. Fulfilling requirements usually means choosing the channel width for each BSS. Then, the power at which each AP is set drives the size of the BSA. Next, you need to extend the design to the multi-cell scenario, and choose a channel reuse factor that combines channels as wide as needed with the requirement to limit co-channel interference. These considerations are valid for any 802.11 design, but Wi-Fi 7 design commonly starts from the 6 GHz band and expands to 5 GHz, and possibly 2.4 GHz. In most cases, it is difficult to design for 320 MHz, and most multi-AP networks will end with narrower channels (e.g., 80 MHz). Designing for MLO may compel you to adjust your AP power and BSA size. You may also need to account for special cases, such as incorporating P2P exchanges in your WLAN, designing small WLANs that require high data rates, or accommodating time-sensitive networking traffic.

Once deployed, the network will always be evolving. As more Wi-Fi 7 devices enter the market, your WLAN requirements may also change and push you to perform updates and develop new designs. In the future, when Wi-Fi 8 enters the market, it will undoubtedly change the way 802.11 networks are deployed yet again.

Chapter 9

Future Directions

future channel guidance: future channel guidance communicates likely future channel information so that stations (STAs) can efficiently move their activity when the absence of Beacon frames is noticed.

802.11-2020, Clause 3

As 802.11be is being published and the Wi-Fi Alliance completes its Wi-Fi 7 release 2 program, the industry continues thinking about the future of Wi-Fi. The technology has proved to be immensely successful, and new use cases, new scenarios, and their cohort of challenges are being brought to the IEEE 802.11 Working Group every day. Multiple groups are constantly in operation—some that started when 802.11be was active, and others that follow 802.11be to continue what the HE group started. This concluding chapter aims to give you an insider's view of these different streams that will define what ends up being called Wi-Fi 8.

Wi-Fi 8 Directions

This section reviews the likely (and some less likely) directions for 802.11 as the industry stretches toward Wi-Fi 8: distributed resource units, better roaming, multi-AP coordination, and a security evolution. Other innovations are likely to mature at the same time and may well be part of the Wi-Fi 8 products, such as client privacy, client location, and client sensing.

Actors and Timelines

Wi-Fi 8 is the Wi-Fi Alliance's presumed brand name for the generation of Wi-Fi after Wi-Fi 7. At time of writing, the effort to build it is well under way at IEEE, with a great flowering of ideas and scores of topics and subtopics being explored. Based on past experience, many of these will be whittled down during the standardization process, with the strongest and most implementable ideas being preserved—and perhaps and some others, too.

At IEEE, the process began in earnest with the Ultra High Reliability (UHR) Study Group, which paved the way for the UHR Task Group, 802.11bn. The UHR Study Group recognized the limited opportunities for a PHY speed increase (discussed later in this section) and so identified reliability as the key theme for Wi-Fi 8. The Project Authorization Request (PAR) and Criteria for Standards Development (CSD) explicitly mentioned the following use cases as targets for the next major generation of 802.11:

- Highly reliable and consistent Wi-Fi connection with predictable timing and delivery of data packets for AR/VR immersive experiences and the metaverse

- Deterministic and reliable Wi-Fi connectivity for manufacturing (e.g., digital twins) and industrial automation (sensors, robot/drone motion control)

- Near-seamless mobility providing reliable and consistent user experiences for clients roaming in residential settings, enterprise environments, public venues, manufacturing scenarios involving automatic guided vehicles, and smart agriculture applications for autonomous control of machines

- Lower power consumption to reduce energy bills, address the growing concerns about global warming, and extend the battery life of untethered devices

The timeline as of this book's time of writing is compressed: A first draft of the amendment is expected in May 2025, followed by a second (and more complete) draft in May 2026. Once the 802.11 Working Group is well satisfied with the maturity of the draft (typically around drafts 4–6 and targeted for May 2027), the draft will be opened to a larger group of reviewers, the Sponsor Ballot pool, and further refinements. This will continue until final IEEE approval, which is anticipated in May 2028. In parallel, the Wi-Fi Alliance typically initiates Marketing and then Technical Task Groups to drive interoperability certification. Much of the industry values timeline predictability in this process, so there are already proposals for Wi-Fi 8 to follow the evolving Wi-Fi 7 timeline, but with four years added:

- Wi-Fi 7 (release 1): launched in January 2024

- Wi-Fi 7 (release 2): launching in January 2026(?)

- Wi-Fi 8 (release 1): launching in January 2028(??)

- Wi-Fi 8 (release 2): launching in January 2030(???)

The sections that follow describe some of the problems under investigation and the features proposed as solutions for them. These are speculative discussions because change is absolutely to be anticipated: Upon deeper investigation, sometimes the performance gains are not there, or the implementation complexity is too high, or the competing interests cannot find a suitable compromise in time.

Directions for the PHY Layer

The PHY layer is often the first to be considered when designing the next Wi-Fi generation. Almost any improvement to the PHY is likely to require chipset architecture changes, and those can happen only if they are deemed possible at scale. Therefore, the pushes in the PHY domain are usually a mix of moderate improvements, which most interested parties can agree with, and more exotic ideas, which need to be pondered for a long time before they become mainstream.

No Speed Increase

The idea of 480 or 640 MHz PPDU bandwidths was canvassed early in the UHR discussions. As described in Chapter 4, "The Main Ideas in 802.11be and Wi-Fi 7," widening the bandwidth is the best way to achieve greater single-link throughput. However, China has no allocated spectrum, such as 6 GHz, that can support these bandwidths, and Europe's 6 GHz spectrum allocation can support only a *single* 480 MHz channel. Even the 1200 MHz available in countries such as the United States restricts this to one 640 MHz channel and one 480 MHz channel. Usage of this bandwidth by one AP makes high reliability much harder to achieve by other nearby Wi-Fi networks and wireless systems. Given these challenges, the 802.11 members backed away from this proposal.

Other approaches to increase PHY data rates are even less appealing. More than eight RF transceivers in a band, for up to three bands, is hard to justify given the cost, power, and small number of devices capable of handling eight spatial streams in the market today. The investment required to deliver 16384-QAM (16K QAM) outweighs the limited customer benefits from such a short-range technology.

Accordingly, no basic PHY speed increases are anticipated to be defined in 802.11bn. Instead, extra spectrum seems to be the most promising path (see the "Integrated Millimeter Wave" section later in this chapter), but it will not be part of release 1 of Wi-Fi 7.

Distributed RUs

Recall from Chapter 3, "Building on the Wi-Fi 6 Revolution," that 6 GHz is a shared spectrum with incumbents at certain locations that need protection. In consequence, the regulations allow either a lower power (such as Low Power Indoor [LPI] operation) or a controlled power (such as Standard Power [SP] operation). The AP's automatically determined location, albeit with uncertainty, is sent to an automated frequency coordinator that determines which powers on which channels should cause minimal interference to incumbents.

Standard power operation is a vital technology to unlock the full capabilities of the 6 GHz band and is available today. However, it requires some engineering investment in AP localization, such as GNSS receivers, indoor location determination, fees to an AFC provider, and so forth. Not all vendors are capable of making such investments. Even for those that are ready to invest, it can be difficult to obtain an accurate location in challenging environments, such as at basement levels and in buildings with highly RF-absorbing glass windows.

The Distributed Resource Unit (DRU) is a technology that follows the LPI regulations as issued, yet is still able deliver higher power. The LPI regulations limit the transmit power spectral density to, say, –1 dBm EIRP per 1 MHz for clients. Wi-Fi 6 and Wi-Fi 7 have 12.8 subcarriers per 1 MHz, on average, and the resource units are (almost exclusively) made up of contiguous subcarriers. In such cases, the –1 dBm transmit power allowance during the UHR-modulation portion of a PPDU (including the Data field) is shared across the 12 to 13 subcarriers of a client in any 1 MHz. DRUs adopt an alternative way to define RUs: Each STA's tones are semi-regularly distributed across the PPDU bandwidth (or a portion of it, in the case of very wideband PPDUs) such that each STA is energizing many fewer than 12 to 13 subcarriers per MHz. You can see the difference between traditional RUs and distributed RUs in Figure 9-1.

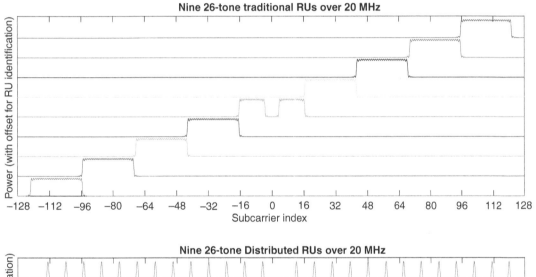

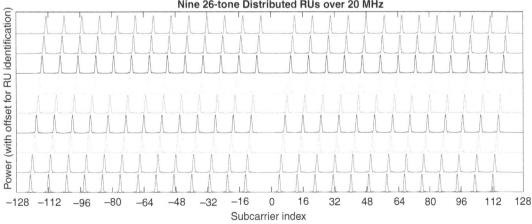

FIGURE 9-1   Comparison of Traditional RUs (Upper) and Distributed RUs (Lower) for Nine 26-Tone RUs over 20 MHz

DRUs increase the power available to the transmitter. For instance, 2 STAs transmitting 484-tone RUs distributed over 80 MHz can each transmit 3 dB more under the current regulations; 4 STAs transmitting 242-tone RUs have 6 dB more, and 8 STAs transmitting 106-tone RUs have 9 dB more. However, 16 STAs transmitting 52-tone RUs do not quite get 12 dB more because, in each 1 MHz, 802.11 is limited to 12 to 13 subcarriers.

Meanwhile, the preamble is unchanged, due to the requirements for CCA and receiver acquisition, and in many scenarios is not as robust as DRUs. As a result, the DRU technology is best suited to triggered UL-PPDUs where the preamble is "simulcast" by each user.

Integrated Millimeter Wave

802.11 has defined two generations of PHY at 60 GHz and thereabouts: 802.11ad-2012 and 802.11ay-2021. Although communications in this band can be stunningly fast, the deployment of these products has tended to be limited to point-to-point communications between buildings and the like due to the high power consumption required to support the 2160 MHz bandwidth (in 802.11ad) and the bandwidth up to 8640 MHz (in 802.11ay). These bandwidths require analog-to-digital converters (ADCs), digital-to-analog converters (DACs), and baseband processing to operate at very high speed and have commensurately high power draws. Also, propagation at 60 GHz is more difficult: Signals transmitted through obstacles are heavily attenuated, and tend not to diffract around those obstacles.

Accordingly, we can think of the integrated millimeter wave approach as a revisiting of these wideband design choices. Instead of aiming for the highest data rate, it aims for a power consumption (and complexity level and then data rate) consistent with smartphones and laptops. Here are the main ideas:

- Operate at familiar bandwidths like 160 or 320 MHz.

- Take advantage of Wi-Fi 7's MLO. If and when the 60 GHz link is blocked by obstacles, switch over to another link at 2.4/5/6 GHz.

- Take advantage of the lower bands (like 2.4/5/6 GHz) to assist with the discovery of APs operating at 60 GHz.

- Reduce the implementation barrier to entry by taking a familiar PHY design, such as 802.11ac (Wi-Fi 5), and overclocking it by a straightforward factor, such as 8. In this way, a traditional 20 or 40 MHz bandwidth becomes a 160 or 320 MHz bandwidth at millimeter wave, and the traditional data rates are increased by a factor 8.

However, these ideas were not accepted as part of the 802.11bn scope. Accordingly, the Integrated MilliMeter Wave (IMMW) Study Group is pursuing a separate track at 802.11, with a separate Task Group, and with the goal of being ready in time for Wi-Fi 8 revision 2. The integrated millimeter wave approach will still have cost and power implications for products, and is liable to be initially confined to higher-end devices.

Exotic Ideas

Many other ideas are under consideration and are publicly available on the IEEE 802.11 document server under UHR SG or TGbn.[1] A few of these topics are highlighted here:

- Renewed support for high-mobility STAs.

- Beamforming for UL OFDMA transmissions, to improve user orthogonality.

- Improvements to LDPC coding rules, to ensure more consistent decoding reliability versus PSDU length.

- Better support for a mix of older and new clients via a frequency-domain aggregate of PPDU formats.

- Better reliability via hierarchical modulation. The sparser gray-coded bits in a constellation can be assigned to high-priority streams and the denser gray-coded bits in a constellation can be assigned to low-priority streams.

- Improved reception in multipath by requiring a tighter EVM specification so that approximate maximum likelihood receiver processing, if applied, performs better.

The early phase of a Task Group is usually focused on translating the use cases surfaced in the Topic Interest Group and the Study Group (UHR, in the case 802.11bn) into requirements. However, it is also a great time to present new solutions or possibilities for the next generation of Wi-Fi. Some of these ideas are very new, and will require a lot of discussion before they are considered feasible, interesting, and in scope for the 802.11bn TG. No doubt, some of these ideas will make it to Wi-Fi 8.

Directions for the MAC Layer

MAC improvements might appear to be easier to design than PHY features, because MAC functions are about new elements, new frames, and new fields, all of which can be coded in software. However, they aim to solve complex problems, and require a careful design. It is therefore common to see many ideas proposed, of which only a few will prove efficient enough at solving critical issues to warrant integration into the next generation of the 802.11 standard.

Better Roaming

Wi-Fi roaming has always been important in larger deployments: carpeted enterprises, hospitals, college campuses, theme park rides, oil and gas fields, and so forth. Good solutions such as 802.11r fast transition (FT) are available. Still, more reliable roaming has traditionally received less attention as a key Wi-Fi metric given that many familiar Wi-Fi deployments, such as in the home or the coffee-shop hotspot, often consist of a single AP. However, given the advent of very-short-range MCSs; mesh systems for larger homes such as Wi-Fi Easy-Mesh; the availability of "fiber to the room" in housing

---

1 802.11 document server: https://mentor.ieee.org/802.11/documents.

developments in countries such as China, Japan, and South Korea; and the renewed attention given to integrated millimeter wave, it has become clear that multi-AP deployments are becoming even more prevalent and improved roaming can provide important user benefits. Even a fixed device such as an IP camera bolted to a wall can benefit from improved roaming if its closest AP is taken out of service (e.g., for firmware updates).

Accordingly, multiple proposals are being made to 802.11bn to improve roaming efficiency. Other proposals seek to enable something much closer to a make-before-break architecture—that is, establishing a connection to the next AP without terminating the connection on the current AP—instead of the current break-before-make design.

One approach under discussion is most suited to the home, where the traditional AP would be disaggregated into a single "upper MAC" device connected to multiple "lower MAC, baseband, and RF" devices distributed in different rooms around the home. To a client, the system would appear to be very similar to a single AP MLD with many links per band. Only the "upper MAC" device would have subnet connectivity.

A more scalable approach would create a new entity, the Seamless Mobility Domain (SMD), which is logically somewhere between the AP MLD and the 802.11r mobility domain. The SMD is a management plane entity only, distributed across multiple AP devices. It ensures continued secure connectivity by assisting with moving a client's connectivity context from an old AP MLD to a new AP MLD. The client connectivity context includes things like security keys, the state of the packet number counter, the state of the sequence number counters, and agreements on issues such as Block Ack and target wake time (as well as an unchanged IP address). An important goal is for the client's supplicant to authorize a *network* of APs rather than an individual AP. In general, the data path with SMD is somewhat similar to Wi-Fi 7. Specifically, the data at the top of the AP MLD is locally switched or tunneled back to a wireless LAN controller, with two upgrades during the roaming period:

- The client can continue to retrieve buffered downlink frames for a short period from the old AP even after it starts to retrieve frames from the new AP, all the while preserving the traditional 802 requirements for in-order delivery and de-duplication.

- As an advanced option, the architecture can support moving buffered MSDUs from the old AP MLD to the new AP MLD.

Better Quality of Service

Given that the Wi-Fi PHY already supports multi-gigabit per second speeds, achieving higher quality of service (QoS) for flows is the next natural goal. In general, achieving this goal requires lower and more predictable latency and jitter.

Wi-Fi 7 has already made major QoS strides, as described in previous chapters. Nevertheless, some potential gaps remain, which the features described in the sections that follow attempt to address.

Multi-AP Coordination

To a very large extent, when scheduling and attempting to deliver traffic to clients on time, each AP on a channel acts independently of other APs. This independence makes good sense for an unlicensed spectrum where anything can happen at any time. It also makes Wi-Fi enormously resilient and robust—more so (to the authors' best knowledge) than any other wireless system. At the same time, providing an outstanding solution for the worst-case scenario can leave some opportunities on the table for better-case scenarios. For example, one could imagine the scenario in which all devices in a neighborhood are Wi-Fi devices and are prepared to help each other to ensure that the most urgent traffic is delivered with the highest priority, regardless of which AP is conveying the traffic.

This mutuality is relatively straightforward when all APs are part of the same administrative domain, but comes with some caveats in other circumstances. Fortunately, the former same-domain case is a common occurrence, and solutions for it can be used judiciously with the latter cross-domain case.

Multi-AP coordination (MAPC) is an enormous topic, with many possible directions—ranging from low complexity with good gains to ultra-high complexity with massive gains (Table 9-1). It may be that 802.11bn starts at the simpler end of the complexity spectrum as a first, pragmatic step. Importantly, MAPC can deliver gains when nearby APs are on the same or overlapping channels, so its importance increases with higher AP density and/or wider AP channel operation.

TABLE 9-1 Flavors of Multi-AP Coordination

Flavor	Complexity	Mechanism	Advantages
Optimal Wireless Networking Architecture (OWNA)	Highest	IQ samples are fronthauled between multiple AP "radioheads" and a centralized lower MAC and baseband in the wiring closet, which enables: ■ On the downlink, high-scale distributed (i.e., multi-radiohead) beamforming and MU-MIMO with collision avoidance ■ On the uplink, high-scale distributed diversity and UL-MU-MIMO with collision resolution ■ Dynamically and locally picking one of these modes, or something simpler, such as pathloss-based spatial reuse Radioheads are almost perfectly synchronized in time, carrier frequency, and carrier phase.	Highest possible uplink and downlink throughput, rate-versus-range, and reliability

Flavor	Complexity	Mechanism	Advantages
Joint Transmission (JT)	Very high	Distributed (i.e., multi-AP) DL-MU-MIMO. Data for all APs' clients is replicated across the APs, and all APs participate as if they were a single virtual AP with an antenna count equal to the total number of AP antennas. For each TXOP, the TXOP-holding AP is the leader AP; other follower APs almost perfectly synchronize in time, carrier frequency, and carrier phase to the leader. Clients are sounded from all APs at once; then beamforming matrices are calculated in a distributed manner. The fronthaul (which might be a wireless mesh) carries replicated MSDUs and coordination control rather than IQ samples.	Highest possible downlink throughput and rate-versus-range, with potential for collision avoidance (i.e., the downlink part of OWNA and with relaxed fronthaul requirements)
Coordinated Beamforming (C-BF)	Very high	Distributed DL-MU-MIMO, except each AP directs data to only its own clients and steers nulls toward the clients of other APs. For each TXOP, the TXOP-holding AP is the leader AP; other follower APs almost perfectly synchronize in time and carrier frequency to the leader. Same sounding requirements as for JT. The encoding, modulation, and beamforming matrices are applied just to the AP's own clients. The fronthaul carries MSDUs for the AP's own clients, plus coordination control.	Highest possible downlink rate-versus-range, with potential for collision avoidance
Coordinated Orthogonal Frequency Division Multiple Access (C-OFDMA)	High	Nearby APs coordinate to share their channel bandwidth, with fine-grained time and carrier frequency synchronization. For instance, a leader AP might contend for 320 MHz and consume 240 MHz itself and grant 80 MHz to another AP that reports it also has (urgent) traffic.	Opportunities for greater efficiency and determinism
Coordinated Spatial Reuse (C-SR)	Medium	For each TXOP, a leader AP serves its own clients. It then enables one, or perhaps more, nearby follower APs that report they have (urgent) traffic to use a portion of the remaining TXOP at the same time as the leader AP or among themselves.	Opportunities for greater efficiency and determinism
Real-Time Dynamic Bandwidth System (RT-DBS)	Low–medium	In dense enterprise networks, AP channel widths are typically much less than 320 MHz to minimize inter-BSS contention and avoid client unfairness. However, when neighboring APs report light traffic, a coordinated AP with a high traffic load might temporarily switch itself and its clients to a much wider channel width to deliver the traffic more quickly.	Opportunities for greater throughput whenever the traffic is spatially unbalanced

Flavor	Complexity	Mechanism	Advantages
Coordinated Restricted Target Wake Times (C-R-TWT)	Low–medium	With C-R-TWT, each AP reports its R-TWT agreements to its neighboring APs, and they all mutually respect each other's R-TWT SPs. To avoid SP collisions, C-R-TWT works best when a common service interval is agreed; it should be an integer submultiple of the beacon interval. However, the natural service intervals of QoS flows vary, and are rarely integer (sub) multiples of the beacon interval. Thus, SPs commonly overlap, and then behavior becomes variable.	Considerably greater determinism
Coordinated Service Periods (C-SP)	Low–medium	C-SP is similar to C-R-TWT in the sense that the requirements of periodic QoS flows [e.g., via the SCS(QC) protocol] are shared between APs and a time-sliced SP approach to channel access is adopted. However, C-SP includes specific mechanisms so SPs are gracefully and efficiently rescheduled even when SPs would nominally overlap.	Considerably greater determinism
Coordinated Time Division Multiple Access (C-TDMA)	Low	For each TXOP, a leader AP serves its own clients. It then enables another nearby follower AP that reports it has (urgent) traffic to exclusively use a portion of the remaining TXOP. The leader AP might perform this process multiple times in the same TXOP.	Potential for greater determinism

BSR and Flow-Level Upgrades

The buffer status report first defined in Wi-Fi 6 reports how much traffic per AC each client has, in units of octets. Wi-Fi 8 is evaluating upgrading this information to report how urgent the traffic is (i.e., how imminent its delivery deadline is). Making this information available to the AP and thence to neighboring APs that support per-TXOP MAPC technologies such as C-OFDMA, C-SR, and C-TDMA can significantly help with overall scheduling and ensure that the most urgent traffic is scheduled most quickly.

A key question for priority is which applications are key to the deployment and its users. The QoS features in Wi-Fi 7 are somewhat pragmatically defined:

- The resolution of the Trigger frame is limited to a preferred AC.

- The resolution of the SCS (QC) protocol for the uplink is limited to a TID.

- Buffering at the client might be per AC rather than per TID or per SCS flow.

To ensure QoS policies are achieved with greater precision (and so free up wireless resources for other flows), protocols can be enhanced with TID-level or flow-level (i.e., SCS ID level) singalongs and behaviors.

L4S

Low-latency, low-loss, and scalable throughput (L4S) has been defined by the Internet Engineering Task Force (IETF) under RFCs 9330, 9331, and 9332 to reduce the transit delay and jitter of packets across the Internet. The L4S designers observed that most TCP connections are greedy—that is, TCP attempts to use all the available bandwidth, until the weakest link becomes oversubscribed and packets start being lost. At that point, the TCP window suddenly shrinks, before slowly building up again. At that point, the cycle repeats again.

Multiple flavors of TCP have been invented to attempt to mitigate this seesaw effect. L4S attempts to solve this issue in a creative way, with an approach that consists of two parts:

- Each network node along the way observes its transmission buffer. If the buffer (toward the next node) is congested, the node marks the passing L4S packet with a flag that indicates congestion. In its acknowledgment back to the packet source, the end consumer (e.g., your laptop or phone) indicates that congestion was signaled, causing the server to reduce its flow a little bit. This way the flow is always tailored to the near real-time capacity of the entire network, including its weakest link.

- Given the previous function, L4S packets should not accumulate in any node, waiting for other L4S packets ahead of them to be transmitted. Because of this property (and because non-L4S packets do not display this property), L4S does not build queues. Instead, L4S-enabled nodes implement a queue that is parallel to the regular queue and entirely dedicated to L4S traffic. This queue is relatively shallow—only a few packets in the queue are sufficient to trigger the "congestion" flag. However, because of the near real-time adaptation of L4S to the network conditions, the queue is usually almost empty, ensuring that traversing L4S packets do not get delayed.

In combination, these two properties make L4S well suited for delay-sensitive traffic whose volume can be modulated, such as traffic in AR/VR/XR applications. However, L4S is a Layer 3 technology. When the last (or the first) nodes in the chain are an AP and a STA, the L3 principles may need to be extended to Layer 2. For uplink packets, the STA needs to signal to the AP the L4S nature of a flow, so that the AP schedules the STA at undelayed intervals. The AP, for uplink and downlink flows, needs to know (at Layer 2) that they are L4S, so that the AP can direct the packets accordingly in an Ethernet (uplink) or 802.11 (downlink) shallow queue. Unlike an Internet router, the AP is not a L3 object. Thus, when congestion occurs, the AP needs to be able to signal congestion in an L2 or L2 signal.

L4S is an interesting technology for 802.11 and matches the 802.11bn amendment's asserted goals. Its implementation may just be a matter of adding new messages to the AP and STA exchanges.

In-Device Coexistence

Modern smartphones and laptops are often multiple-radio devices. Triband Wi-Fi, multiband cellular (for smartphones), and Bluetooth in 2.4 GHz and GNSS are most common, but UWB (for ranging) and Bluetooth in other unlicensed bands are seeing increased attention, too. Furthermore, the device's Wi-Fi resources might be virtualized to support multiple purposes—for example, Internet connectivity through the client's infrastructure AP, plus wireless docking and/or peer-to-peer data transfer and/or screen sharing and more.

All of these activities make the design of clients challenging. Greater ecosystem value might be realized if their implementation burden is transferred to APs in a way that can lead to better system QoS and/or better Wi-Fi performance. For instance, when the client is busy transmitting (say) audio to earbuds, the client might be unable to receive Wi-Fi signals at the same time. The in-device coexistence (IDC) features under discussion for 802.11bn would enable the client to indicate a periodic or upcoming absence to the Wi-Fi AP; the AP could then schedule its transmissions to the client at an earlier or later time instead of fruitlessly transmitting to the client while it is unavailable.

Although this kind of signaling can require the AP to reschedule its traffic "on a dime," it is possible with minimal protocol changes because Wi-Fi is inherently resilient to unexpected events. However, oftentimes the Wi-Fi traffic is more urgent than the client has planned for. To deal with this situation, there are complementary proposals to enable each side to report the priority of planned traffic as well as absences, with the possibility that a reported absence could be canceled to ensure that the higher-priority traffic is transferred in a timely manner.

P2P Coordination

Clients that support multiple radios operating in unlicensed spectrum, as well as suffering from IDC issues, may also introduce interference to the infrastructure network as a side effect of connecting directly to other personal-area client devices (i.e., laptops, wireless docks, tablets, phones, watches, HMDs, and other wearables). Conversely, the infrastructure traffic poses the risk of interference to these personal-area peer-to-peer (P2P) links. In many cases, this interference is tolerable and does not need further management. However, consider advanced use cases such as automated manufacturing, high-density wireless docking, infrastructure AR/VR (i.e., an HMD using remote rendering from the local/cloud compute accessed through an infrastructure AP), and P2P AR/VR (i.e., an HMD using remote rendering at a P2P client device). In those scenarios, it is desirable if both systems can operate at the same time without causing harmful interference.

Ideas to resolve this dilemma have been developed in Wi-Fi 7 and the achievable features are seeing deployment. It is anticipated that refinements will continue to happen in 802.11bn, and that broader adoption will become possible now that the protocol implications are understood in greater detail.

Dynamic Sub-band Operation

Not all clients can operate at 320 MHz, and some clients with sporadic traffic (e.g., IoT devices) may operate at lower bandwidths, such as 20 MHz. As a result, the primary 20/40/80 MHz of a wideband

BSS is often more occupied than the secondary 80/160 MHz. Rather than the AP scheduler leaving such secondary channels idle, the dynamic sub-band operation (DSO) proposal is intended to mitigate this scenario by providing the AP with the means to direct clients, on a per-TXOP basis, to the secondary channels to receive their intended RUs/MRUs/DRUs.

Non-Primary Channel Access

A second concern arises with wideband BSSs: If an overlapping BSS (OBSS) begins a narrow (e.g., 20/40 MHz) TXOP on the primary channel of the wideband BSS, then the wideband BSS is stalled. In that scenario, no AP or client is permitted to transmit for the duration of the OBSS TXOP. Non-primary channel access (NPCA) is a proposal to enable APs and clients to agree to independently jump to an "alternate" primary channel; there, they can contend and, if the channel is free, exchange data up to the duration of the OBSS TXOP. Once the OBSS TXOP completes, the AP and clients return to their normal primary channel and continue channel access there.

AP Power Save

In the 802.11 architecture, APs are always awake and ready to serve clients. This state consumes power even in the absence of clients or when there is little client traffic, such as late at night. In the Wi-Fi Alliance's Wi-Fi Direct specification, an AP would have the ability to sleep periodically if the attached client(s) are also Wi-Fi Direct capable.

Two distinct problems are associated with this behavior:

- The traditional AP of the home and enterprise (and the smartphone-based hotspot AP) offers Internet connectivity to any device that is nearby. In essence, the AP must support the billions of existing clients, and any change can be enabled only over a multi-year period as legacy devices fall out of use (or when a new band is enabled).

- The virtualized Wi-Fi interface of a client device may be intended as a cable replacement where the peers are individually and specifically connected.

The solution to both problems is similar, but the adoption cycle for them in the traditional AP case will take longer. Proposals for AP power save include the following options:

- The AP and the client(s) negotiate a mutual sleep cycle with regular on and off periods.

- Like a Wi-Fi 7 EMLSR client, the AP operates an always-on (or mostly on) low-power, low-capability radio (e.g., 20 MHz, 1SS, low MCS) that is able to receive RTS, Probe Request, and other basic frames, and to transmit Beacon frames. It keeps its high-capability radio (capable of supporting multiple wideband spatial streams and high-MCS PPDUs) in a light sleep mode until awakened by the low-capability radio or when there is a lot of downlink traffic to transmit.

Artificial Intelligence and Machine Learning

In recent years, artificial intelligence and machine learning (AIML) has become the central topic of almost every IT conversation, and 802.11 has not been spared. An AIML Topic Interest Group was formed in 2022, with the initial goal of forming a Task Group focusing on the introduction of AIML techniques into 802.11. Ultimately, though, the group concluded that AIML is not a set of features, but rather a set of tools that can be used to improve 802.11 operations.

For example, AIML techniques can be used to improve sounding operations. Each time it needs to perform a sounding exchange with a STA, the AP needs to send an NDP frame, possibly over multiple radio chains. The STA needs to respond with a sounding matrix, which can be very large if, for example, there is a 320 MHz-wide channel and 8 spatial streams. Machine learning techniques could be used to shrink this matrix to a smaller form, using probabilistic mathematics to decompress the matrix, or they could be used to rebuild the elements that were not explicitly mentioned in the report.

Similarly, efficient roaming is always at the heart of any multi-AP deployment. AIML could be used within the infrastructure to predict the next-best AP for a STA. Such a prediction would obviously be based on the STA movement, but also on the load on the neighboring APs, the prediction of other STAs' movements, and their expected traffic volume and type; in addition, it would account for the time at which the STA should roam to maintain optimal goodput performances. In a world with multiple APs, AIML could help these APs coordinate their channels and their transmission, so as to optimize the use of the RF resources.

The AIML TIG found many examples where AIML could help 802.11, but concluded that the AIML effort would likely not entail the development of an amendment, where use cases are found and the protocol is improved to solve these use cases. With AIML, new problems that AIML can help solve are expected to surface on an ongoing basis. For these reasons, the TIG became a Standing Committee—that is, a group focused on a topic (AIML) and operating for many years (until the 802.11 chair declares their work complete or no longer useful). Problems and ideas related to machine learning in 802.11 can be brought to this committee, before being suggested to one or several active Task Groups for further development.

Security

Major updates in Wi-Fi security are often discussed both in the 802.11 forums and in the Wi-Fi Alliance. Therefore, it is common that improvements are proposed at the start and during the development of each new 802.11 effort.

Extended MAC Header Protection

Several fields in the MAC header are masked out when the frame body protection is calculated, and therefore are not protected. This design was because these fields change (or may change) during retries, but the packet number (PN) of a frame should not change.

However, there are some security concerns related to the following masked-out fields:

- The Power Management bit
- The Sequence Number field
- The HT Control field

Accordingly, there are proposals to add a separate message integrity check (MIC) for these fields that is recalculated even for retries.

Control Frame Protection

With strong data and management frame protection in place, attention turns to other possible attacks, such as those targeting control frames. These cannot be more than denial-of-service (DoS) attacks, and mitigations for the most severe DoS attacks are already available, but it seems advisable to prevent even them. There are proposals to add MICs to certain control frames that might create a magnified DoS effect, such as the following:

- Trigger frames, because an attacker could perform a battery attack by repeatedly sending them to a victim client to cause it to transmit at a high-duty cycle.
- Block Ack Request and Block Ack frames, because (in the absence of client mitigations or the protected Block Ack feature) an attacker could cause the victim to advance its starting sequence number beyond the current operating range. This would make retries ineffective and/or cause a gap in accepted frames at the receiver.

Mandatory Support for 802.1X

Given the sustained resilience of 802.1X security (i.e., WPA3-Enterprise), it is proposed to mandate support for 802.1X in 802.11bn APs and clients, and to enable all security modes to be performed via EAP methods.

Post Quantum

The advances in quantum computing have been steady, and quantum computers are likely to start appearing on the market within the next decade. At that time, their computing power will exceed the power of current machines by such a large margin that the entire security world will be shaken. In 802.11 protocols (as in many other protocols), authentication and encryption rely on mathematical operations that use large numbers that are known to the AP and the STA, but kept secret. An observer could guess the operation being used, but would have immense difficulties performing that same operation without knowing the secret numbers. Working backward to the secret numbers from the result of the mathematical operations (which may be inferred from the encrypted frames exchanged by the AP–STA pair) would also be immensely difficult. However, the level of this difficulty becomes more questionable if the attacker uses a quantum computer. Reverse engineering of the encryption exchange may still be very difficult, or it might suddenly become accessible and possibly trivial.

To prepare for this development, security researchers and mathematicians have been working on new mathematical schemes—for example, those involving elliptic curves—that are so computationally expensive to reverse engineer that quantum computers would still require immense computation times to break encryption or authentication keys. If elliptic curves start appearing in 802.11, most of the 802.11 security schemes would still rely on pre-quantum structures. One possible charter of 802.11bn is to design a fully updated, quantum-resistant set of authentication and encryption exchanges, so that 802.11 will be ready when the first quantum computers become available to the general public.

Localization and Sensing

802.11 has been such a successful technology that most buildings come with their sets of AP devices and STA devices, ranging from laptops to phones, IoT, and more. As the number of devices grows, it becomes useful to keep an inventory of these objects, and to know their individual location. In turn, the presence of Wi-Fi in many buildings has inspired many protocol designers to use Wi-Fi for navigation as well.[2] Both functions—asset localization and navigation—are on the verge of being realized, and new applications are emerging that use the raw 802.11 signals to track non-802.11 objects.

Localization and Sensing Before 802.11be

The idea of using 802.11 signals to locate[3] a STA has been within the scope of Wi-Fi since its very beginning. Since the early days of 802.11, infrastructure companies have proposed services in which a set of APs, using the RSSI and modified free pathloss equations, would convert the signal of a STA probe request into an estimated distance to the queried AP. The proposed method was crude, because any environmental factor (e.g., walls, but also reflections) would affect the received signal strength and lead to a misleading distance estimation. For example, a simple plaster wall between the STA and the AP could be enough to reduce the signal by 6 dB, but in the free pathloss equation, that 6 dB reduction of the signal translates to twice the distance. This issue could somewhat be mitigated with multiple APs and multilateration techniques, but accuracy stayed in the 10–20 meter range in most cases. The fact that the STA location could be computed as a side effect from the standard 802.11 SSID discovery mechanism (i.e., without the STA willful cooperation) also caused privacy concerns (see the "A Privacy-Respecting 802.11" section later in this chapter). Additionally, this location possibility was just a clever use of the 802.11 mechanisms and was not intended or described in the 802.11 standard.

The first standardization of the process of ranging into 802.11 came in the 802.11-2012 revision, albeit as a side effect of another mechanism. At that time, the IEEE 802.1 Working Group (defining IEEE 802 architectures) had chartered a Time Sensitive Network (TSN) Task Group, whose goal was

---

2 Navigation, or "blue dot," is the ability of an end device, such as a phone, to display on its screen its location on a map, and by extension, to propose steps to reach another location.

3 In the language of RF technologies, *localization* is the action of finding the location of an object.

to design protocols that would allow time-sensitive traffic to traverse network nodes without being delayed behind other queued traffic. One possible mechanism for achieving this goal is to know in advance the time at which each TSN packet is expected to enter a given node, and to make sure that the node has an empty queue ready for that traffic traversal. Knowing the time of arrival implies synchronizing times between nodes, and also knowing the travel time between one node and the next. When this process is extended to 802.11, and one node is the AP and the next node a STA, there needs to be a process to estimate the travel time from the AP to the STA.

Introduced in 802.11v-2011, and then into 802.11-2012 with the 802.11v amendment integration, the Timing Measurement (TM) protocol[4] allows a STA, associated to an AP, to ask that AP for timing measurements. The AP would then send a measurement frame (carrying primarily an exchange token number, to track individual frames in multiple exchanges, in addition to the TSF field[5] that indicates the local time on the AP) at time t1, that would be received at time t2 by the STA. The STA would then send back an acknowledgment frame, which would leave the STA at time t3 and reach the AP at time t4. At this point, the AP would know t1 and t4, and the STA would know t2 and t3. In a subsequent timing measurement message, the AP would send the {t1, t4} value pair. Using the time indicated in the AP message (with the TSF), the STA could compute its time offset t_o to the AP with the simple formula $t_o = [(t2 - t1) - (t4 - t3)]/2$ and thus know the travel time from the AP to the STA. You can see this mechanism on the left side of Figure 9-2. When a frame is scheduled to travel to the STA, the STA knows how long after the AP transmission the frame will arrive. It is then easy to ensure that the STA is ready to receive the frame at the right time. As the STA moves, this request can be made again to update the AP–STA travel time.

As the 802.11v amendment was integrated into 802.11-2012, it became obvious to the 802.11 designers that the protocol could be used to compute the distance instead of the time offset, simply because the 802.11 signals, like light and any other radio wave in the spectrum, travel at the "speed of light" (c). Therefore, the travel time can be converted to distance by just multiplying t_o by c. You can see the general mechanism on the right part of Figure 9-2.

A few other improvements were needed to make this process usable for location purposes. For example, timing measurement is limited to a STA and its associated AP. Location requires measurements of the distance to multiple APs to allow for multilateration[6] calculation. Therefore, the improved mechanism needed to allow for ranging to APs to which the STA has no association. The AP also needs to transmit its location to the STA, so the STA can convert position (relative to the AP) to location (latitude/longitude coordinates). The 802.11 designers then refined the TM protocol and introduced[7] into 802.11-2016[8] its Fine Timing Measurement (FTM) variant.[9]

---

4 See 802.11-2020, clause 11.21.5.
5 See 802.11-2020, clause 11.1.3 for how a STA uses the TSF in general to maintain time synchronization with the AP.
6 The determination of a position from the distance to different reference objects (APs in this case). Typically, three distances are needed to obtain a position in two dimensions—hence the term *trilateration*.
7 As a direct insertion into the revised 802.11 standard, without being the result of a dedicated Task Group work.
8 To help you relate this introduction to the timeline of the 802.11az discussed next, note that FTM was discussed and inserted into the 802.11-2016 draft in the years 2014 and 2015.
9 See 802.11-2020, clause 11.21.6.

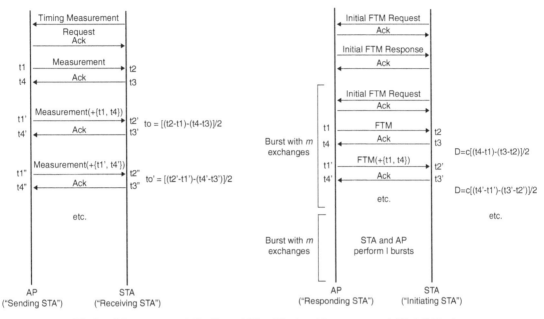

FIGURE 9-2   Timing Measurement (Left) and Fine Timing Measurement (Right) Exchanges

In both TM and FTM, the STA has the initiative of starting the dialog, and is also the only site to obtain all four timers (t1, t2, t3, t4). The AP is merely responding. As a consequence, the infrastructure has no direct ranging-related benefit to the transaction. This key factor limited the adoption of FTM in the early years of its availability. In addition, the time of arrival (t2 or t4) is computed by the receiver by observing the start of the frame preamble on the circuit behind the antenna. The exact time of its arrival is not always accurate. Recall from Chapter 1, "Wi-Fi Fundamentals," that a STA should be able to detect an incoming preamble within 4 µs of the signal true arrival time. But in the span of 4 µs, light has traveled 300 meters (close to 1000 feet). This type of accuracy is, of course, unacceptable for ranging, and many 802.11 chipsets could do much better (often with a ranging accuracy down to 2 to 4 meters).

FTM also introduced, in the initial FTM request from the STA, the idea that the STA could negotiate with the AP multiple bursts of multiple measurements, to maximize the number of samples the STA would work from. However, the ranging accuracy of FTM was limited by the laws of physics and the ability of a receiver to detect a preamble fast enough.

802.11az, 802.11bk, and Beyond

To improve and modernize FTM operations, the 802.11az Task Group was created in 2015 and was in full operation at the time 802.11be was designed (the 802.11az amendment was published in 2022).

The 802.11az amendment introduced many improvements, of which four are of particular interest for our purposes:

- Several new modes were introduced. For example, a trigger-based mode allows the AP to range with multiple STAs at the same time—following 802.11ax MU MIMO operations, and thus saving airtime. In the new passive mode, the APs exchange frames that the STAs detect. The STAs are silent, which increases the STA privacy, but also allows an unlimited number of STAs to perform location operations at the same time.

- The 802.11az ranging frame, with an 802.11ax format, repeats the High Efficiency Long Training Field (HE-LTF) segment of the PHY header. This gives the receiver multiple opportunities to detect and measure the preamble. Knowing where the HE-LTF is positioned in the header, the receiver can then deduce when the beginning of the preamble is supposed to have reached the receiver antenna. This mechanism results in improved ranging accuracy. Specifically, where 802.11-2016 FTM achieved about 2- to 4-meter ranging accuracy for an 80 MHz channel exchange, 802.11az offers 1- to 2-meter accuracy under the same conditions.

- 802.11az allows for larger bandwidth. 802.11-2016 enabled FTM over 20, 40, and 80 MHz. 802.11az also allows exchanges over 160 MHz. This change is important, because accuracy is also a function of the channel width. Larger channels result into higher ranging accuracy.[10]

- With 802.11az, the STA can share back to the AP its timers or its computed range (with the Location Measurement Report [LMR]). In some cases (e.g., enterprise managed assets), the AP can mandate this sharing. In (most) other cases, sharing is optional, meaning it is under the control of the STA and its user. This mechanism allows the infrastructure to benefit from the exchange—for example, enabling a store to offer private navigation to its customers, but also track the store's assets (e.g., barcode scanners).

Whereas 802.11az designed the main features of FTM, 802.11be introduced the idea of 320 MHz channels in 6 GHz. This topic was of great interest for protocol designers focusing on location services, because larger channels result in higher ranging accuracy. However, it was too late to insert this possibility into 802.11az, because the 802.11az boundary was 802.11ax and no later amendment. The 802.11bk Task Group was then created in 2022 specifically to extend 802.11az to EHT 320 MHz wide channels in 6 GHz.[11]

802.11bk enables all the modes that 802.11az allows: passive mode in high-density public areas, such as conference centers, airports, and alike; and LMR for corporate assets. Meanwhile, 320 MHz offers the possibility of sub-meter ranging accuracy, advancing the location services in 802.11 from general zone accuracy (10–20 meters, with RSSI), to visual accuracy (2–4 meters, with 802.11-2016 FTM), to body accuracy (1–2 meters, with 802.11az), to sub-body accuracy (less than 1 meter, with 802.11bk).

---

10 The reasoning behind this principle is complex, but a short and simple explanation is that the STA samples a 40 MHz channel twice as fast as it samples a 20 MHz channel. Therefore, the STA needs half the time to detect that a new signal is incoming. The same factor of 2 repeats for 40 MHz to 80 MHz channels, and for 80 MHz to 160 MHz channels.

11 The publication date for the 802.11bk amendment is the end of 2024.

At that scale, the location engine can tell if the barcode scanner is in your left hand or right hand, and in a hardware store, your navigation application can tell you that the box holding the 1-inch nails that you are looking for is on the shelf 30 cm (one foot) to your right.

Is there a need to go further? Certainly, location accuracy enables new use cases as it improves, and new use cases call for even higher accuracy. Protocol designers are dreaming of using future protocols, like those enabled by the Integrated MilliMeter Wave (IMMW) Study Group in the 41–70 MHz bands,[12] to detect the movements of your phone in your hand, or the movements of your head while wearing smart glasses. Such fine accuracy would, in turn, enable new use cases, where—for example—small movements of your head or your hand are sufficient to translate into large strides in virtual universes.

The Hybrid AP

The idea of locating a device using 802.11 signals has always involved a hybrid scenario. In some cases, the goal is really to locate the 802.11 transmitter, but in most scenarios, the 802.11 device is a proxy for something else. Perhaps the 802.11 card is in a personal device (e.g., a smartphone), and the location of the card is the same as the location of the phone, and both are proxies for the location of the person. Even when 802.11 radio frequency identifier (RFID) tags are used that emit specific 802.11 messages for asset tracking, the tag is merely a proxy for the object to which it is attached.

This idea of locating something or someone with 802.11 signals has many implications, because it suggests that an 802.11 AP device can be used for more than just connecting STAs to a network. After more than 25 years of existence, Wi-Fi has become so ubiquitous that AP devices are expected in most indoor venues at intervals of a few tens of meters (50–100 feet at most). The AP device has become an element of the indoor landscape, an expected element of furniture, with the unique property of being connected to a network and to a source of power. This presence empowers the AP to perform new functions, and ranging for location purposes is one of them.

The ranging function is by no means limited to STAs. One key burden of Wi-Fi deployment and management is to locate the AP itself. The AP device is often mounted on the ceiling—and its barcode, or its MAC address, printed on the AP body, is usually on the side of the AP facing the ceiling, and thus invisible as soon as the AP is mounted. In other places, the AP is hidden from view for aesthetic reasons. Once the deployment is completed, documenting the position of each AP is laborious, but necessary for many practical reasons—troubleshooting, coverage gap analysis, and more. After a few years and multiple cycles of building maintenance tasks, during which APs may be moved, updated, or removed, maintaining an up-to-date record of each AP's location can prove challenging. The issue is even more pressing for APs supporting 6 GHz, as those APs need to report their location to be allowed to use standard power.

These needs have led AP vendors to leverage the FTM technology for the benefit of the infrastructure itself. By making the APs perform FTM ranging between each other, infrastructure management systems can learn the relative distances of AP pairs. If the positions of a few of these APs are known,

12 Higher frequencies, with shorter wavelengths, allow the detection of smaller movements.

the management platform can build a graph with the likely locations of all detected APs in the building. When an AP is removed, it stops responding to FTM messages, and can be identified as such. Any new AP can be reported by the others and, if it supports FTM, positioned accurately on the graph. If the new AP does not support FTM, the knowledge of the other APs' positions helps put the new AP's rough location on the map.

In some cases, however, relying on the known positions of a few APs is too much to ask. To facilitate the location process, GNSS receivers can be installed inside AP device enclosures. Depending on the AP device's location in the building, it may receive enough GNSS signal to learn its own location.[13] As the GNSS receiver is connected to the AP device operating system, the AP can then report its location to the network management platform. Once the locations for a few of these AP devices are known, FTM enables the management system to learn the locations of all the other APs on the floor.

In some scenarios, FTM may not be accurate enough to establish the distance between an AP pair. FTM is generally oblivious to obstacles. It is true that RF signals slow down in dense materials like walls, but this effect is negligible. Therefore, the time of flight of an FTM frame is about the same in open space as it is through one or a few walls. As long as the FTM frame can be detected, it can be used. This property is a great advantage of FTM over the RSSI method. However, FTM suffers in canyon scenarios, where the direct line-of-sight signal is blocked by an obstacle, and the FTM signal reaches the receiver after reflection (Figure 9-3). In that case, the reported distance can be much larger than the real distance, to a point where the FTM measurement becomes unusable.

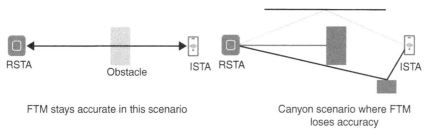

RSTA   Obstacle   ISTA   RSTA   ISTA

FTM stays accurate in this scenario   Canyon scenario where FTM
                                                    loses accuracy

FIGURE 9-3   Canyon Scenario for FTM

To complement FTM, some infrastructure vendors have started to equip AP devices with other radio technologies, which can be used to provide a "second opinion" on the AP device-to-AP device distance. One such technology is ultra-wide band (UWB), based on other IEEE technologies (primarily IEEE 802.15.4a and 802.15.4z). As the UWB frame uses an ultra-wide channel, it is very accurate (down to a few centimeters or inches). However, UWB emits at lower power than 802.11, so it is more limited by the presence of walls. Even so, UWB can be used to provide very accurate distance measurements

---

13 Contrary to an urban legend, GNSS signals are often partially available inside. AP devices located deep inside the building may not receive any GNSS satellite signal, but APs closer to the edge of the building may receive partial but usable, or sometimes excellent, signal. The AP device may need to collect many GNSS signals for a long time before reaching a conclusion on its geo-position. But unlike a mobile device, the AP device does not move. Therefore, the GNSS signals are cumulative.

between AP devices that support the technology and are in (UWB) range of each other, improving the quality of the position graph.

Another technology is Bluetooth Low Energy (BLE). The 802.11 standard has supported RSSI-based ranging for a long time. Bluetooth version 5.1 (released in 2019) also supports angle of arrival (AoA) and angle of departure (AoD) functions, that can help detect reflection.[14] RSSI-based ranging is not very accurate, but it provides the "second opinion" (especially in combination with AoA/AoD) that may be enough to identify and mitigate FTM inaccuracy scenarios.

At this point, the AP device potentially includes a GNSS receiver, plus UWB and BLE transceivers. Such an AP device is now well beyond the 802.11 AP functions. These other radio technologies can, of course, be used for their own merit. UWB is primarily used to ensure accurate ranging, but the BLE radio can also be used to manage BLE tags.

One phenomenon that ranging technologies have difficulty assessing is the notion of a floor. It is possible to determine the distance between two APs, but it is often difficult to know if these two APs are on the same or different floors. Many GNSS receivers incorporate an altimeter; AP devices (or their GNSS module) may do the same. Some altimeters are very accurate pressure sensors. When connected to the network, they can use weather information at the calculated location to determine the expected air pressure at each height, and then deduce the exact altitude of the receiver, often with an accuracy of 1 or 2 meters. When the sensor is within an AP device (or module), it may not have weather information, but it can report the accurate pressure value to the network management platform through the AP device host. AP devices that report the same pressure are likely at the same height on the same ceiling. In this way, the management system can determine which AP is on which floor.

The pressure sensor is an environmental sensor. Many such sensors come in bundles where the same bundle can measure pressure, humidity, and more. Given these capabilities, it is tempting, when equipping an AP device with a pressure sensor, to add other environmental functions. For example, the AP could become a smoke detector and a sort of weather station. Its goal may not be to predict the weather, but elements like humidity can be used to detect the presence of humans in the room, and to help in office buildings with functions like booking meeting rooms and "free room" detection. Through these extensions, the AP device, having become a normal object of the indoors environment, is slowly becoming a hybrid object, capable of performing multiple functions, or capable of hosting modules that just need power and network connectivity.

802.11bf and Sensing

A clear sign of this trend of turning the AP device into a multifunction object is the idea of using the AP, in cooperation with other static 802.11 objects, to sense the presence of humans or objects. This trend may change deeply the way people interact with 802.11 objects.

---

14 Bluetooth 6.0, targeted for 2025, is set to support an accurate ranging system similar in principle to FTM. This system is called Bluetooth Channel Sounding (formerly known as High Accuracy Distance Measurement [HADM]).

RF Sensing

The idea of using RF signals to detect objects is as old as radars. Radars send RF signals, which reflect off objects (e.g., a plane in the sky) before coming back to the radar antenna. The echo is caused by the presence of an obstacle, and provides proof that an object is in the signal path. In the case of an airport radar pointing to the sky, that can only mean a flying object. By repeating the radar signal several times and using clever mathematics, the radar system can also detect the direction and speed of the object, along with some of its physical characteristics, such as its size.

Radar Sensing Versus Wi-Fi Sensing

Radars are well known, and for that reason are the first point of comparison for 802.11 practitioners learning about 802.11 sensing. However, the mechanisms used in Wi-Fi sensing and in most radars (e.g., airport radars) have major differences.

For example, airport radars measure the time difference between the emitted signal and the reflected echo (the round-trip time of the radar signal) to estimate the aircraft's distance from the airport. The radar signal itself does not carry any timestamp. If a pulse is sent at t0 and comes back 0.5 ms later, the reflecting object must be (300,000 × 0.5) about 150 km (93 miles) away. Measuring a change in this round-trip time allows the system to calculate the aircraft speed. If the next pulse comes back in 0.45 ms, then the object is now 134 km (83 miles) away, and thus 16 km (10 miles) closer. The time between pulses allows the system to calculate the aircraft's speed.

Airport radars are also very directional. You may have seen these antennas, which look like large satellite dish antennas, continuously rotating above airport installations. If you were able to slow their movement down and see the RF signals, you would observe that the antenna sends a signal in a narrow direction (with a width of just a few degrees), receives the possible echo from that direction, and then rotates slightly and repeats the operation. When an airplane is detected, the progressive change of direction at which the plane is detected from one rotation to the next provides to the system the plane's angular speed. The angular change is then combined with the round-trip time change to calculate the true speed of the aircraft.

In contrast, 802.11 antennas usually do not rotate, so the angular speed cannot be determined in a normal 802.11 setting. Additionally, the round-trip time is difficult to use, because 802.11 systems are half duplex (explained further in this section). The 802.11 sensing operation follows the same general principle as underlying radars, in that a signal is sent. That is, reflection of the signal from an object is used to understand the presence (and possibly position and movement) of the object. However, 802.11 sensing technology is very different from radar technologies.

Systems based on 802.11 also use RF signals, which are affected by the environment. Recall that the effect of the environment on the 802.11 signals is one reason why transmitters, at a given distance from the receiver, may dynamically change their data rate (or MCS) over time. Radars, for example, send very short pulses separated by silences. This mechanism allows the receiving system to be silent when the reflected signal returns, switching very rapidly between the Tx and Rx functions. Sometimes radars include one transmitter and one receiver in the same unit to avoid this Tx/Rx switching. Such a system, measuring the echo of its own signal, is called monostatic, because a single (mono) static (nonmobile) system performs both transmission and echo measurement. By contrast, 802.11 systems send long frames instead of short pulses, and are half duplex (no dual Tx/Rx functions). It is therefore difficult for an 802.11 device to send a signal and measure its reflection on objects of the environment. By the time the frame transmission is completed, the Wi-Fi transceiver needs to switch to Rx mode, which takes some time. As a result, the Wi-Fi system cannot properly measure the reflection of the end part of the frame. Circuits that switch very rapidly between Tx and Rx are certainly feasible, but are much more costly—more than most Wi-Fi users are ready to pay for their AP or STA device.

Sensing with 802.11, in most cases, requires bistatic or multistatic systems. With bistatic systems, one device sends a signal (e.g., a static printer, a Wi-Fi thermostat on the wall), and a second system (e.g., the AP) receives and measures the signal. With multistatic systems, two or more devices receive the same signal, and then a central system compares the reported signals to conclude something about the environment.

With a bistatic (or multistatic) system, you cannot really measure the time between the transmitted signal and its echo—because there is no echo. You also cannot reliably measure the time of flight between the transmitter and the receiver, unless the frame carries a very precise timestamp (802.11 frames do not) and unless the receiver and the transmitter times are very carefully synchronized (802.11 systems are usually not carefully time-synchronized). The only measurement that can be reliably made in such bistatic/multistatic systems is the differences between one received signal and the next one, which reflect general changes in the environment itself. The main conclusion that 802.11 sensing can make about the environment is therefore movement, including the presence of an object that was not there before or the absence of an object that was there earlier.

To understand the underlying mechanism behind this type of conclusion, recall the structure of the OFDM signal (discussed in Chapter 1). With a modulation like QAM, the 802.11 receiver measures each subcarrier's amplitude and angle. Now suppose that an 802.11 transmitter sends the same signal over and over again. This idea is interesting in the sensing context, because all 802.11 frames start with the PHY header and its training field, whose structure is known and always the same from one frame to the next. In a perfectly quiet open space, the receiver would measure the same amplitude and the same angle from one training field in one frame to the same training field in the next frame.

If you were to plot these representations (for each tone, the amplitude and angle of the OFDM transmission), you would see that the plot for one frame training field overlaps almost perfectly with the plot for the next frame training field. The word "almost" applies here, because subtle power and therefore amplitude variations may occur during the transmission. For example, such variations may be evident when the power amplifiers (PAs) at the transmitter incrementally warm up during a transmission. You can see this near-perfect overlap on the left side of Figure 9-4. Each of the four plots represents the amplitude of the received signal. The horizontal axis is the frequency: The channel center frequency at index 0 and 40 MHz on each side combine to form an 80 MHz channel. The vertical axis represents the amplitude of the signal at the given frequency.[15] Each of the four plots represents the signal of a STA as it was measured on each of the 4 radio chains of an AP. Each line across the plot is one frame, and the first 200 frames are shown. As you can see, the lines for each frame almost perfectly overlap.

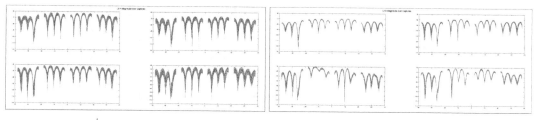

FIGURE 9-4   Amplitude Changes on an 802.11 Sensor

If an object is now introduced into the environment, especially a moving object (e.g., a walking human body), it will alter the transmitted signal, usually via a mix of absorption and reflection effects. The amplitude (and angle deviation) of each subcarrier will change slightly when the object is introduced into the room, even if it is not mobile, and even if it is not exactly on the path between the transmitter and the receiver. Its mere presence in the zone is sufficient to change the reflection pattern of the environment. As the object moves, the effect on the environment also changes, so the amplitude and the angle of each subcarrier on the receiver end may be slightly different. You can see this effect on the right side of Figure 9-4, the scenario where a human was walking in the room where the STA and the AP were positioned. You can see that the amplitude of each frequency changes from one frame to the next, causing the plots to be blurry.

In this way, 802.11 signals can be used to detect presence (or absence) and movement. Sudden presence (or absence) is easy to measure. The amplitude of the movement that can be detected is another area where 802.11 differs from other technologies. All RF signals obey the same laws of physics, which limit the resolution (the amplitude of the movement that can be detected) to a ratio of the signal bandwidth, similar in principle to the idea of a faster sampling rate on larger channels (see the discussion of 802.11az in this chapter). For sensing, the size of the movement that can be detected follows the equation $s = c/2BW$, where s is the smallest movement size that can be measured, c is the speed of light, and BW is the bandwidth of the signal. You can see that a 20 MHz signal can measure movements of 7.5 meters and larger. So, if you introduce a human body into the environment, the amplitude graph will immediately reflect the presence of the new object. If the human body stays at the same position for a while, the amplitude graph will stabilize, producing in the end a pattern similar to the left side of Figure 9-4. If the human moves a bit, the graph will still stay stable. The human needs to move by 7.5 meters or more for the graph to reflect a change that will expose the movement with certainty. Even large channels, like 80 MHz, allow only measurements of movements close to 2 meters. Therefore, traditional 802.11 can detect movements (like a human walking), but not much more. Some complex

---

15 The measurement is normalized, and the vertical axis units are irrelevant.

systems can be set to measure the movement of an arm, but such granularity approaches the limit of what is possible with such channel information and requires a complex setup (e.g., multiple antennas). As soon as more than a few objects are present and moving, the movement is lost.

Larger channels allow for a better resolution. For example, 6 GHz offers 320 MHz channels (see Chapter 5, "EHT Physical Layer Enhancements"), with a possible resolution of about 50 cm (less than 2 feet). 802.11 was also defined in the 57–71 GHz range (the millimeter wave range of the spectrum), with channels of 1.76 GHz width that can be bonded to even larger channels. At a 1.76 GHz bandwidth, the resolution reaches 17 cm (about 5 inches), albeit at the cost of a shorter range than is possible with a 2.4/5/6 GHz signal. When the channels are bonded, smaller movements become detectable, such as the movements of your fingers in the air (over a phone screen, for example), or even small movements like those of your chest when your heart beats or when you breathe. New use cases become possible: You can control applications on your 802.11 device by just moving your fingers (e.g., swiping in the air left or right), your head, or some muscles of your face (e.g., frown or smile to lock or unlock your computer).

This type of scenario supposes a monostatic system, or rather a pair of co-located monostatic systems, forming a bistatic system, where one radio sends and the other receives the echoes of the first radio transmission. Such co-location is necessary because with a true bistatic system, someone walking behind you might interfere with the movements of your fingers and cause a false response. Notably, 60 GHz transmissions have a very small wavelength, and can be properly exchanged only if the antennas are directional. The phone, or the laptop, would therefore have directional antennas pointed toward you. Your body would then form enough of an obstacle that someone behind you would not introduce significant interference, if the receiving system is also in front of you.

As yet, most APs and most client devices do not incorporate these kinds of millimeter wave radios. Until such radios become common, the main use case of 802.11 will be in the 2.4/5/6 GHz bands, for presence and movement detection. This use case is still very beneficial. When multiple receivers are present, comparison of changes in the signal on the receivers, processed with clever algorithms (e.g., machine learning), can be used to recognize patterns. The system can, for example, recognize that someone is entering your home (even if the 802.11 systems are not at the entrance, but rather in your living room and/or bedroom), climbing a stair, or entering or leaving a room. The 802.11 system can also differentiate one person from two or more people. In a professional setting, this mechanism allows users to know whether a meeting room in an office building is occupied. In schools, this presence detection can be coupled with HVAC systems to turn the heating (or air conditioning) off when a classroom is empty. In factories, the movement of robots or workers can be followed and alarms triggered in case of an anomaly (e.g., "man down," presence in hazardous areas).

802.11bf

As the sensing technology continues to evolve, better accuracy and new use cases will appear. Meanwhile, the 802.11bf amendment was created to enable standard 802.11 devices to perform sensing functions. As 802.11 sensing relies on bistatic or multistatic systems, at least two 802.11 devices need to work together to enable proper sensing capabilities. A sort of universal protocol is needed to drive these exchanges, if both devices are not from the same vendor.

To that aim, 802.11bf defines two roles:

- The **sensing initiator** is the device (STA or AP) that starts the RF sensing measurement exchange. In most cases, it is also the device that needs the sensing measurement report. For example, an AP is connected to some presence detection application, and wants to sense the environment to report to that application.

- The **sensing responder** is the device (STA or AP) that responds to the sensing initiator messages. In other words, the responder is a system that implements 802.11bf, and can therefore be triggered. Before 802.11bf, sensing 802.11 systems needed to be hacked and artificially put in a sort of monitor mode where they would consider only the 802.11 preamble and its characteristics. By default, an 802.11 system consumes the preamble to validate and interpret the frame; it then reports only the payload of the frame to the upper layers, as long as the receiver address matches the address configured on the receiving system.

The 802.11bf amendment makes a distinction between the lower band (2.4/5/6 GHz) systems and the 57–71 GHz systems (called directional multi-gigabit [DMG] in 802.11 parlance), because the operations in the millimeter band are technically different from those of the lower bands. In the context of this chapter, these differences do not matter and are ignored.

The sensing initiator and sensing responder define who is starting the sensing dialog, and who is responding. However, 802.11 is not just about "wanting to perform sensing," but also about "performing sensing." This second action also requires at least two participants:

- The **sensing transmitter** is the entity (STA or AP) that sends a PPDU that will be used for sensing measurements.

- The **sensing receiver** is the other entity (STA or AP). It receives the sensing PPDU sent by the transmitter, and measures the amplitude and angle of each subcarrier. In this way, it evaluates the effect of the environment on the preamble sent by the sensing transmitter.

As the receiver cares about only the preamble, the PPDU is usually a null data PPDU. Such a PPDU carries a PHY header but no Data field.

This dual system, which includes both the initiator/responder and transmitter/receiver roles, creates three different scenarios:

- The sensing initiator is also the sensing transmitter. It sends an NDP to the other device, which acts as both the sensing responder and the sensing receiver. The sending responder/receiver processes the PHY header to analyze the channel (performs sensing). In some situations, the sensing initiator/transmitter might be operating under the control of some external entity (an app somewhere in the network), and the sensing responder/receiver may be configured to send the result of its sensing operation to that entity. However, in the vast majority of cases, the entity needing the result is likely to be the sensing initiator. The sensing responder inserts the sensing result in a frame (called a sensing measurement report), and sends that frame to the sensing initiator. This choreography consumes two frames—the NDP and the report—and supposes that the sensing receiver is always ready to sense (i.e., expects the NDP and is ready to process it).

■ In a simpler scenario, the sensing initiator is not the sensing transmitter. The sensing initiator sends to the sensing responder a trigger (called a "request" for the DMG case), asking the responder to send the NDP frame for sensing purposes. The sensing responder thus becomes the sensing transmitter, and sends the NDP frame. The sensing initiator acts as the sensing receiver; it receives and processes the NDP frame, thereby performing the sensing operation.

■ Both of the previous scenarios can be combined. The sensing initiator first operates as the sensing transmitter, and sends an NDP frame to the sensing responder/receiver. The sensing receiver immediately responds with another NDP frame, meaning that the sensing receiver becomes the sensing transmitter. This NDP frame is followed by a report frame on the sensing performed for the first NDP frame. Thus, the sensing initiator gets both a report (telling the initiator how the responder evaluated the environment) and an NDP frame (allowing the initiator to perform the sensing function for the NDP signal sent back from the responder). As these exchanges occur at short intervals, they give a good symmetrical picture of the environment at a given point in time, and can also be used to track small movements.

Figure 9-5 illustrates these three modes.

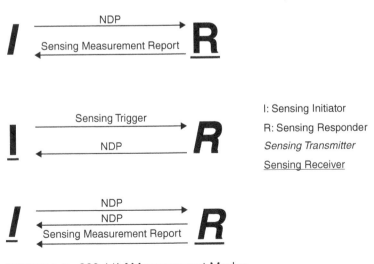

FIGURE 9-5   802.11bf Measurement Modes

These different roles for the sensing function are defined for both the AP and the STA. This flexibility allows the sensing functions to be activated in multiple ways. As you might expect, at association time, the STA and the AP exchange capabilities that signal their ability to perform sensing, and to fulfill one or more of the four roles. Once this exchange is complete, some application somewhere needs some sensing function. Note that 802.11bf does not define how this communication happens, as it is usually proprietary and, if the application is on a server behind the AP, probably occurs over a wired connection anyway. There can be more than one application in need of sensing—for example, presence detection in each meeting room of an enterprise building during the day, and movement at the entrance of the building and the windows during the night.

Each application receives the list of STAs and APs supporting sensing and each role, and selects the actors that are useful for the application goals. In practice, you may help the application by positioning the devices on a map and configuring the sensing operations. At that point, a first set of action frames allows a sensing initiator and a sensing responder to establish a sensing session. This exchange may be somewhat complicated if both are STAs, as their communication needs to transit through the AP. The exchange will be simpler if one of them is the AP. In all cases, the initiator sends an action frame to the responder, inviting the responder to a sensing session (with its associated ID). It also suggests session parameters (called measurement setup parameters), such as the mode (e.g., VHT or HE), the number of radio streams that should be involved in the transmission, and the number of participating antennas for the receiving side, among others. A given STA (or AP) that supports the responder role might be invited by one or more initiators to one or more sessions, depending on the underlying requesting applications, which is why the session ID is useful. The responder can accept the parameters or suggest new ones.

At this point, the actors are selected and ready to operate. Once sensing is needed, the initiator sends a new action frame establishing a new session instance. The instance includes an instance ID, but also mentions the session ID. Now, the exchange depicted in Figure 9-5 occurs. There can be multiple instances in a given session, with a given setup. Once the instances complete, the session can be either terminated or maintained as enabled, if more instances are expected.

These exchanges are almost fully integrated with 802.11, as the 802.11bf amendment considered the principles of 802.11be.[16] This means that MLO and MU-MIMO procedures are part of the 802.11bf choreography. You can imagine an 802.11bf AP using the DL-MU-MIMO or OFDMA procedure, or sending a trigger (e.g., NDPA) to a group of STAs to get these STAs to send NDPU sensing messages in an UL-MU-MIMO frame. The integration is partial, because at the time this book was written, MLO sensing still needed to happen on a per-link basis (pretty much like non-MLO). Yet, it could be useful to fully integrate the idea of the MLD, so as to range on two bands at the same time, with each band providing its own accuracy and view of the environment.

---

16 Each 802.11 Task Group is provided with an order number, based on its expected publication date. As 802.11bf is expected to be published after 802.11be, it is the responsibility of the 802.11bf participants to consider each new draft version of 802.11be, and to integrate its mechanisms (if applicable) into the 802.11bf draft. This way, a new amendment always considers the other amendments that were published earlier, avoiding parallel developments in a vacuum (with the risk of gaps or contradictions between amendments that are developed in parallel).

Even without full MLO support, the integration of 802.11 sensing into APs and STAs is likely to change our interactions with 802.11 objects. In particular, 802.11 objects can become more efficient. For example, your phone can know if you are nearby and it should be in light sleep mode, or if it is in a drawer and can enter a deep sleep mode that will keep the battery at full charge for a very long time. If your home or office 802.11 AP can know when you are here, the whole notion of a smart home or smart building is transformed. The IoT sensors (e.g., door, window) can be deactivated automatically when the AP knows that you are there, and activated when you leave, without the need for presence sensors in all rooms. Your laptop, knowing that you are not around, can start an update that is likely to take 30 minutes, but save that task for later if you are around. In other words, the 802.11 objects around us can become assistants of human life, beyond their strict 802.11 functions.

AMP and 802.11bp

The idea that the energy of 802.11 frames could be used to sense the environment also led to the idea that this energy could be harvested. In the world of IoT, many sensors require battery power, which also defines the lifetime of the IoT object. The designers of IoT protocols are always looking for ways to reduce the energy consumption of battery-equipped objects, but also realize that, in a world where Wi-Fi is omnipresent, the 802.11 energy could potentially be harvested to recharge the batteries. After all, an AP sends a beacon about every 100 ms, day and night, even if there are no STAs to receive it. Most of the time, this energy is wasted. Similarly, any frame is usually an energy wave received by many devices (all STAs in the BSA). The receiver consumes the frame data, and the others simply ignore it. None of these frames (beacons or others) represent a lot of energy, but they could still be collected and used by harvesting objects.

In 2022, the 802.11 WG formed an Ambient Power (AMP) Topic Interest Group to study whether 802.11 could bring useful contributions to the growing world of energy harvesting. The idea of IoT object harvesting energy is not new, and the group soon found that the energy transmitted with Wi-Fi is very small, especially compared to well-studied sources like light. However, Wi-Fi operates day and night.

The group also determined that there are two types of AMP objects:

- Some objects are **entirely passive**. They typically use a coil that can be configured before deployment, similar in concept to the anti-theft tags that are often attached to books. When the object receives enough energy, the coil radiates back that energy. In some cases, the radiation is always the same message. As an example, imagine an AMP tag attached to a fire extinguisher and radiating the date of its last maintenance check. In other cases, a passive mechanism allows the object to radiate information that depends on environmental conditions. As an example, imagine that the coil is partially obfuscated by a piece of fabric that shrinks or expands depending on the temperature or the humidity; the sequence radiated back reflects the length of the coil that is exposed to the RF signal, and thus can be made to reflect the current temperature or humidity level.

■ Other objects are **semi-passive**. They do not really use a battery, but rather rely on a capacitor. As the object harvests the energy from the environment, the capacitor charges, until it has enough energy to generate a full frame on its own.

The passive AMP devices only radiate energy. Practically, this limitation means that their signal happens only while another object is also sending energy (in the case of 802.11, a STA or AP is sending a frame). One interesting property of these objects is that the radiation can be offset. In other words, the 802.11 transmission may occur on one channel, whereas the reflected energy utilizes a slightly different frequency, thereby avoiding direct interference with the 802.11 transmission. The semi-passive AMP objects can generate their own frames, and so can be set to transmit at a time when no other 802.11 object is transmitting. Practically, however, the requirement to perform CCA consumes energy, and using a frequency offset may be an interesting way to remove this requirement.

Because these objects are very simple, they have no or very limited computing capability. In consequence, the modulation they send must be simplified to the maximum extent possible. There is no hope that these objects would be able to transmit with OFDM in the near future, but they could send a DSSS signal (i.e., use the legacy 802.11 transmission structure). Their transmission would not be very powerful, and would depend on the amount of energy collected. The AMP TIG found that lower frequencies carry more energy than higher frequencies.[17] Proof-of-concepts showed AMP devices could receive energy from an AP (10 meters away) and radiate it back (or emit their own signal), and that a receiver on a smartphone (placed 10 meters away from the object) could read and understand the radiated message.

In the face of these strong results, the AMP TIG became a SG, then the 802.11bp TG in early 2024. The goal of the group is to define protocols and mechanisms for operations of ambient powered (i.e., energy harvesting) devices in the 2.4 GHz and sub-1 GHz bands. The group is targeted to complete its work in 2028. By that time, you may start seeing IoT objects that can be installed anywhere (e.g., door sensors, or sensors you attach to any object in your environment), that will never need to be recharged, and whose information you can read directly on your smartphone.

A Privacy-Respecting 802.11

The rapid expansion of smartphones in the 2010s was paralleled by a growing concern about privacy. A laptop in an enterprise environment might possibly be a shared device, but a smartphone is very likely a personal device with a single user. If the phone transmission can be used to identify this person (personally identifiable information), then any analysis of the phone traffic carries information about the activity of the person using the device (Personally Correlated Information [PCI]). A similar concern had led the 802.11 designers to architect different levels of authentication and encryption security (see the "802.11 Security" section in Chapter 2, "Reaching the Limits of Traditional Wi-Fi"), with the aim

---

17 See Chapter 1. In essence, the size of the antenna of an RF device should be half of the intended wave length. It is possible to cheat a bit with this rule, but it is generally true. Lower frequencies mean longer wavelengths (because one is the inverse of the other), and therefore longer antennas and a larger energy collection surface, and in the end more energy collected at equivalent signals.

of limiting an eavesdropper's ability to read user information carried within 802.11 frames. However, with the development of artificial intelligence and machine learning (AIML) techniques, a concern emerged in the mid-2010s that the STA MAC address would provide a stable reference point that an attacker could use to analyze all traffic to and from a given STA, especially in combination with the STA location. The 802.11be Task Group was not chartered to solve this issue, but three other groups around 802.11be worked on the problem.

802 MAC Addresses

Vendors did not wait for the designers of 802.11 to address this issue. By 2014 and 2015, Apple iOS 8 and Microsoft Windows 10 had implemented the idea that the STA would use randomized MAC addresses for network discovery (iOS and Windows) and could also change its MAC address at intervals, while associated (Windows).

> **Note**
>
> This practice was labeled with different names, especially as vendors tried to make the process appear to be a vendor-specific innovation, giving the feature names like "private address" and the like. The privacy community—in the IETF, then in the 802E Recommended Practices for Privacy, and then in the 802.11 Working Group and its Task Groups—ended up calling the feature "randomized and changing MAC address" (RCM), with the understanding that the RCM address is always a local address, as explained further in this section.

This initiative caused quite a stir, because it was done without considering the possible effects of such changes on the device experience in 802.11 networks. In fact, the rotation of the associated MAC address immediately broke the principles of many hotels' guest Wi-Fi networks, which charge customers for Wi-Fi and use the MAC address to ensure that the same user is not charged multiple times for the same device. The operating system vendors argued that the random MAC address was chosen within the scope allowed by the 802 umbrella standard.[18] Indeed, the standard defines the formats of a MAC address, and indicates that the MAC address can be universal or local. A universal MAC address is unique. A company receives from the IEEE a company ID for part of the MAC address—for example, the 24, 28, or 36 most significant bits of the 48-bit address, depending on whether the company asks for a MAC address large (MA-L), medium (MA-M), or small (MA-S) block.[19] The company then has exclusive use of the remaining 24, 20, or 12 bits of the block, and can allocate a unique individual value for each individual device it produces. The advantage is that the device can appear in any network on the planet, while remaining confident that its Layer 2 identifier is unique and that no collision with other addresses will occur.

18 The 802 standard applies to all underlying standards (802.11, among others). It defines the general principles underlying the PHY, MAC, and LLC layers, as well as the services offered by each layer, but also the key principles for each layer, including, for the MAC layer, addressing.

19 See 802-2014, clause 8.2.2, table 1.

All of these blocks are known to map to universal addresses because of their structure. In an 802 MAC address, the least significant (last) two bits of the first octet have a specific meaning. The last (least significant bit), called the individual/group (I/G) bit, indicates whether the MAC address is an individual address (bit unset) or a group address (bit set). The second-to-last least significant bit, called the universal/local (U/L) bit, indicates whether the MAC address is a universal address (bit unset) or a local address (bit set). This structure is illustrated in Figure 9-6.

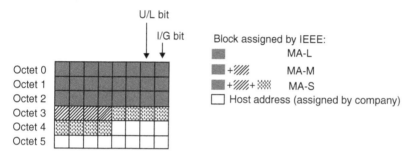

FIGURE 9-6   Structure of the 802 MAC Address

The local address is defined (in the 802 standard) based on the idea that some devices may never leave their native network. As such, there is no risk of collision with another device, if the administrator plans their MAC addressing structure properly (e.g., assigning local addresses to virtual machines and virtual servers). To eliminate the need for an organization to purchase an address block from the IEEE in that case, the 802 standard allows an address to have only a local value, which means that two different networks might potentially have overlapping local MAC addresses. As the matching devices are supposed to never leave their native network, there is no risk of collision. Therefore, an administrator can set the U/L bit, and then unset the I/G bit to associate the address to a single device. The rest of the address can be set to any value of the administrator's choice[20] (even a random value), as the address is only locally significant.

Vendors used this local address structure when assigning RCM addresses. For iOS devices, a key idea was that as long as the STA did not require any service from the AP (i.e., the RCM is used for probing, not for association) beyond basic discovery, collisions did not matter. For Windows devices, a key idea was that the risk of collision in any given network was minuscule, and could therefore be ignored.

However, none of these schemes was designed in collaboration with the IEEE. Thus, there was no guarantee that the random local MAC address structure would not break basic 802.11 operations.

---

20 The 802C-2017 amendment proposes an organization of the local address, called the Structured Local Address Plan (SLAP), that would allow administrators to set other bits to indicate if the address is local, but within an IEEE-assigned block (Extended Local Identifier [ELI]); entirely local but assigned by an administrator, likely in local blocks (Administratively Assigned Identifier [AAI]); or completely random (Standard Assigned Identifier [SAI]). One difficulty of 802C practical implementation is that administrators implementing random MAC addresses without reference to the 802C structure may randomly assign the ELI, AAI, and/or SAI bits, making it difficult for an observer to determine whether the random MAC uses an 802C structure.

802.11aq

The 802.11aq group was the first to look into the RCM address problem. The 802.11aq amendment (published in 2018) focused on pre-association discovery of services,[21] so this was a key group to look at the effect of using a random local MAC address (before association occurs) for such discovery. Combing through the 3500 pages of the 802.11-2016 standard was out of scope (and a difficult endeavor), so the 802.11aq group focused on only pre-association scenarios.

The group concluded that there was no harm in using RCM addresses, pre-association, provided some conditions were met. Specifically, the group concluded that a STA could use a randomized local MAC address[22] (built from 802C or not, and with the U/L bit set), but should keep the same address when a transaction is started on the AP, and when a state is set on the AP for the STA. For example, a STA could use one RCM address for a first probe request, and then another RCM address for the next probe request. However, for each probe request, the STA creates a transaction on the AP: The STA asks a question to the AP, to which the STA expects a response. During that time, the STA must keep the address it started the transaction with. Practically, this means that if the STA sends a probe request using address A, the STA should be able to properly receive a response from the AP sent to address A.

When the STA associates, it creates a state on the AP. It starts a transaction with the authentication request, continues the transaction with the association request, and receives an AID. From that point onward, the AP has a live record of the MAC and its AID. The STA cannot change its MAC address without breaking the stability of that state. Therefore, the STA cannot change its MAC address once associated.

When the 802.11aq group examined the 802.11-2016 standard a bit further, it observed that many roaming scenarios assume a backend exchange between the APs, to carry the context of the STA (and its state) from one AP to the other. Therefore, the 802.11aq TG concluded that the state is present not just on the AP, but on the entire ESS.[23] A STA that associates to a first AP must keep the same MAC address for the entirety of its session on the AP, and also as the STA roams from one AP to the other within the ESS. It is only when the STA disconnects (or gets disconnected by the AP) that it can, upon making a brand-new association (but not a reassociation, which is the restoration of an existing state rather than a new state), use a new RCM address.

802.11bh

The 802.11-2018 amendment brought much needed clarity to the RCM practice. However, the amendment also underlined a limitation of the 802.11 standard. In some scenarios, a STA may want to change its MAC address for the next association. A typical example is a public guest network, like that found in an airport or a hotel, where you may not want observers to notice that you are a returning user.

---

21 Your laptop could, for example, check whether a printer or other key service that an application on the laptop is seeking is available on this network, before deciding to associate.

22 See 802.11-2020, clause 12.2.10.

23 Chapter 1 explains why the ESS carries the STA state between APs.

> **Note**
>
> This concern came to be known as the J-Lo effect, although the name is a misnomer. The general idea is that paparazzi spend time at airports to track trips taken by celebrities such as Jennifer Lopez, J-Lo. However, maintaining such a presence is time-consuming. It would be much easier to eavesdrop on the celebrity's home and catch their phone's MAC address. After installing a small and hidden AP at the airport, and observing the MAC addresses attempting to discover the AP's SSID (or attempting to connect), a paparazzo can place a small script that causes the AP to send a message when key MAC addresses are detected. This way, the paparazzo gets automatic messages when known celebrities (and their phones) are detected as moving through the airport.
>
> However, due to the adoption of RCM, this mechanism has not worked for years, especially since Android 10 (in 2019) adopted the pre-association RCM scheme. Therefore, paparazzi have changed their approaches. Their preferred practice today is to hide an AP that advertises the same SSID as the celebrity's home SSID. Any MAC that attempts to connect to that particular SSID is obviously configured to connect to the celebrity home's SSID and must be the celebrity or someone living in the same home. This detection is called the J-Lo effect. RCM does not solve this problem (which is why the name is incorrect): The personal device will try to connect automatically, even when changing the MAC address, causing the celebrity's returning presence to be detected even when RCM is implemented.
>
> Celebrities are special cases. For most non-celebrity people, RCM and a new MAC address are sufficient to expect that an observer will not be able to identify the returning user.

Even at home, in an apartment building where your Wi-Fi signal may be observable by neighbors, you may not want observers to know whether you (i.e., your phone MAC address) are home. If your home has multiple 802.11 devices, RCM can help make individual user detection more difficult.

However, there are many scenarios in which some elements of the network (the AP or services behind the AP) may need to recognize a returning device. In enterprise networks, for example, each device is associated with a series of privileges and access rights, which may need to be reassigned as soon as the device returns. In enterprise and home networks, connection or performance issues may need troubleshooting. But if the STA MAC address has changed (because the STA disconnected or could never fully connect), troubleshooting becomes difficult or impossible if the support personnel have no idea what the MAC address of the STA was when the issue occurred, which could be multiple "sessions" ago.

In public venues, users may want to be recognized. The case of hotels that charge guests for Wi-Fi access based on the device MAC address is obvious. Suppose a guest connects to the hotel Wi-Fi and pays for the access or for premium service that will be valid for the entire stay. In the morning, the guest leaves the hotel to go to work or meetings. When they return in the evening, their device picks a new RCM for a new association—and then the guest needs to pay again for the same service.

In stores, some customers have fidelity apps that display coupons or special offers based on customer identification. If the customer's device changes its MAC address at each visit, the customer may not be recognized and may lose their fidelity advantages. There are many other examples where device identification is useful, from parental control (stopping a young child's device access to the Internet after a given hour in the evening, to encourage sleeping over browsing all night) to smart homes (where devices and their types are often recognized based on the MAC address) and more.[24]

With 802.11-2020 (i.e., as amended by 802.11aq), there is no mechanism for the STA to be recognized when it returns to some networks, but not recognized when it returns to some others. The operating system may allow the user to configure RCM or no-RCM for each SSID, but asking for the user's input also presupposes teaching the entire issue of RCM and tracking, which are difficult concepts for many non-expert users to grasp. Additionally, the user may want the infrastructure, but not an eavesdropper, to recognize the device. So, the MAC may still need to change, even in scenarios where the returning device should be recognized by the AP.

To deal with this issue, the 802.11bh group was created in 2021. Its goal is to find solutions for these scenarios where a STA, using RCM, needs to be recognized by the AP at its next (new) association. The 802.11bh designed two mechanisms, which can be used independently or jointly: one mechanism where the STA chooses its identifier, and another mechanism where the AP allocates an identifier to the STA.

The first mechanism is called Identifiable Random MAC address (IRM). The IRM is an RCM address, and thus a local address. At the beginning of the procedure, a STA associates to the AP with some MAC address. As you would expect, both the STA and the AP express support for IRM with bits set in the Capabilities field of their initial exchanges. The expectation is that the network will implement one of the five forms of security (see the "802.11 Security" section in Chapter 2). As an example, we will consider the open with encryption form here. At the end of the association, the STA engages in a four-way handshake with the AP. In the last message of the four-way handshake (sent from the STA to the AP, at a stage where encryption key exchange is complete), the STA includes an IRM Key Data Encapsulation element (KDE) that includes an (encrypted) IRM address. By doing so, the STA indicates to the AP "this is the MAC address I will use next time I come to you." The AP can reject the IRM if that IRM is already known by the AP—perhaps because by coincidence (although the collision space is small) another station has already indicated because another STA has already indicated its intention to use this IRM, or because the other STA is currently using that MAC. The STA then generates a new IRM and sends it in a new attempt of the same message.

The AP stores the IRM, along with the current STA MAC address. After the session completes, when it returns to the same AP, and if it wants to be recognized, the STA uses as its MAC address the IRM it indicated in the previous association. This IRM could even be used before association, such as for probing, FTM ranging, or other exchanges where the STA may want to be recognized by the AP. Using that IRM as its address, the STA then sends a new IRM in the fourth message of the four-way handshake, indicating its address for the next session. These IRM exchange principles are illustrated on the left side of Figure 9-7.

24 There are, of course, many other cases where a STA may not want to be recognized upon returning to the same network.

The IRM scheme is flexible, giving all the power to the STA to choose its next MAC address, and to decide whether it wants to be recognized. The risk of collisions exists, but is usually small in most settings. That risk increases in large settings, however. Suppose, for example, on a large university campus with 60,000 students,[25] each student carries at least a phone and a laptop (possibly other 80.11 devices such as tablets, game consoles, and the like). Such a campus connects 120,000 to 150,000 devices. If each device changes its MAC address daily, and if the network management system keeps track of each MAC address for a week (for troubleshooting, liability, or statistical purposes), the risk of collision at the scale of the campus is 1:35.[26] That means that several collisions will be reported each week. Each event does not necessarily mean that a STA (IRM-generated) MAC address will collide with the active MAC address of another device in the same BSS—but it does mean that the management system will receive the declaration of an IRM that is the known MAC address of another STA (that MAC address may currently be active or not), causing the management system to confuse two STAs.

To avoid this issue, and also accommodate enterprise scenarios where the infrastructure controls the behavior of the enterprise assets, 802.11bh allows another mechanism, called device ID. The logic of device ID is similar to that of IRM, but the process is reversed. To illustrate this mechanism, we will again take the open with encryption scheme as an example. The STA first associates using some MAC address. Then, in the third message of the four-way handshake (recall from Chapter 2 that at this point the AP has installed the keys), the AP sends an encrypted Device ID KDE as an identifier for the STA. Unlike in IRM, where the IRM is a MAC address, the device ID can be any type of value—a MAC address, a string, or any other structure that the local network wants to use to identify STAs. The STA continues its session with the MAC address it had started using at the authentication phase. Here again, just as with IRM, it follows the 802.11-2020 requirement (derived from 802.11aq-2018) that a STA must not change its MAC address as soon as it has caused a state to be created on the AP.

At some point, the STA ends its session. The next time it comes back to the AP, probably using a new MAC address, in the second message of the four-way handshake, the STA (if it wants to be recognized) indicates in a Device ID KDE the device ID it had received in the previous session. This allows the AP to recognize the STA and map the new MAC back to the previous MAC address. In the third (encrypted) message, the AP assigns to the STA a new device ID, for the next session. The AP can also just reassign the same device ID if it so desires. These mechanisms are illustrated on the right side of Figure 9-7. The second message of the four-way handshake is not encrypted, so any observer can see the Device ID KDE. However, seeing this element does not help an eavesdropper. The observer does not know which MAC address the STA used in the previous session. The only conclusion that can be drawn from the Device ID KDE observation is that the STA is returning, and was assigned this device ID before.

If it does not recognize the device ID provided in the second message of the four-way handshake, the AP indicates this failure when it assigns a new device ID to the STA in the third message of the four-way handshake.

---

25 Such a large number of students is not unusual in "university towns" in the United States, China, and many other countries.
26 See contribution 11-23/2062.

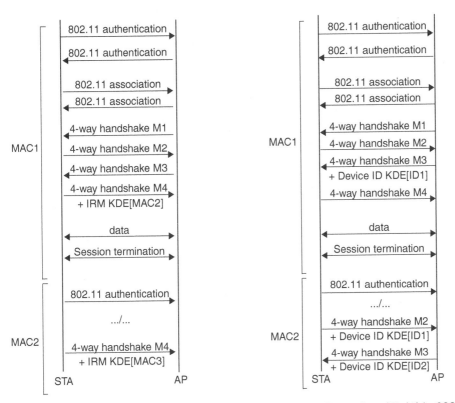

FIGURE 9-7   The IRM Procedure (Left) and the Device ID Procedure (Right) in 802.11bh

One major difference between IRM and the device ID approach is that the device ID value is exchanged after association in most cases.[27] Therefore, the returning STA cannot be recognized when it sends messages before association. 802.11 includes an unauthenticated mechanism for a STA to create a secured tunnel to an AP before association, known as the Pre-Association Security Negotiation (PASN) protocol.[28] With PASN, the STA and the AP exchange public keys, which are then used to encrypt the traffic they exchange. That is, the STA uses the AP public key to encrypt traffic sent to the AP, and the AP uses the STA public key to encrypt traffic sent to the STA. If the AP and the STA are not part of the same key infrastructure, the procedure can be unauthenticated, in the sense that the traffic is encrypted, but the STA has no guarantee that the AP is genuinely part of the infrastructure. When the returning STA and the AP are part of the same organization, this uncertainty is eliminated. In any case, the device ID mechanism allows the STA to indicate its device ID value in the first PASN frame it sends to the AP, allowing the pair to exchange (e.g., FTM ranging messages) while the AP knows the identity of the STA.

---

27 One exception is FILS, because the four-way handshake occurs along with the association. The device ID values are exchanged in the association messages.

28 First defined in 802.11az.

802.11bi

802.11bh addresses the issue of identifying the returning STA in an RCM context, but the Task Group is not charged with improving STA privacy. (The 802.11bh charter does state that the mechanisms that the group defines should not worsen the STA privacy problem.) Given that frames can be observed by almost anyone in range of the RF signal, the privacy issue with Wi-Fi goes beyond mere MAC address changes between sessions. The MAC address may need to change during sessions. The traffic pattern of a STA (e.g., the sizes of frames and the intervals between them) may be sufficient to identify the STA if it is the only one in the BSS sending that traffic. Additionally, frames contain identifiers beyond the MAC address, such as sequence numbers that are easy to follow from one frame to the next.

To address the issue of 802.11 privacy, the IEEE 802.11 Working Group also created the 802.11bi Task Group in 2021. It is working in parallel to the 802.11bh group, but has the broader and longer goal of improving the privacy of 802.11 networks.

As this book was being written, the 802.11bi group was still in its early phases, but had already made important progress. One finding is that "privacy" may refer to the exchanges of the STA, but also to the frames received or sent by the AP (e.g., in your home, or for the virtual AP on your phone when set to tethering mode). The AP problem is very different from the STA problem. The AP sets the tempo, and acts as the stable point of the BSS. Hiding the AP identity implies changing its MAC address often, which creates all sort of issues for STAs. Therefore, the 802.11bi group has defined two concepts: (1) client privacy enhancements (CPE), sets of features that aim to improve the non-AP STA privacy, and (2) BSS privacy enhancements (BPE), sets of features that aim to improve the privacy of both the AP and the non-AP STA. BPE is more difficult than CPE to accomplish, so the group first focused on the latter.

The primary goal of CPE is to prevent an observer from easily identifying a STA and its traffic. This requirement implies that the STA should be difficult to uniquely identify as it associates, and difficult to uniquely identify once it is associated and exchanging data frames. The association and reassociation phases need a more careful choreography. In many cases, the STA or the AP mentions identifiers that were created earlier, in an attempt to circumvent the need to go through phases that were completed before. For example, in the reassociation phase, a STA or an AP can mention a PMKID that was used in the previous association, making obvious the link between this session and the previous one, even if the STA MAC address has changed. The same issue exists with SAE: The STA can mention the SAE credential identifiers it intends to use, allowing an observer to know when a STA is returning to the same BSS.

From probe requests to final association and key generation completion, the 802.11bi group has identified more than 30 elements that could be used to recognize a STA. Obfuscating these elements is possible. If the element assumes the existence of previous keys, then it may be possible to encrypt the element in some way. When such encryption is not possible, forcing the reestablishment of new parameters (instead of simply reminding the other side of previous parameters) may be a better approach. In both cases, backward compatibility is broken, as older devices will not understand the AP's insistence of running again through a phase that was completed before, or will not understand an encrypted element that was supposed to be provided in the clear before. Therefore, 802.11bi networks are unlikely to be backward compatible with legacy networks.

The same compatibility issue exists for post-association frames. Recall from Chapter 1 that the 802.11 header contains multiple fields, providing information about the STA or about the transported flow (e.g., the three or four MAC addresses, the QoS field, the SN and PN values). The 802.11bi group concluded that the best course of action was to encrypt all fields in the header, leaving potentially the RA field as the only one visible. The RA and the TA are called over-the-air MAC addresses (OTA MAC). At association time (after the four-way handshake), the AP and the STA would agree on an OTA MAC rotation scheme such that each side would know the OTA MAC that the other side will recognize as its own, without the need to formally send the next MAC address value in advance. An over-the-air observer would see frames with just the TA field in the clear. Some frames would mention the same TA—likely a STA sending to the AP, or the AP sending to a STA, if the receiver has not rotated its MAC address yet. Other frames would show a different TA that would stop being used at random intervals. The observer would also not know if the visible TA is a STA or the AP, as the AP could have a different rotation scheme for its own address with each STA, thus appearing as different TAs.

One key difficulty remains in regard to the pace at which the MAC address changes. If all STAs change their OTA MAC at a fast pace (e.g., at each TXOP), then the collision risk mentioned earlier in this chapter becomes a collision certainty. In the university campus example, a rotation per TXOP would cause the management system to see more than 200 million new MAC addresses every hour, making collisions unavoidable. At the same time, if there are only a few STAs in the BSS, changing the OTA MAC at a fast pace is the best way to provide an impression of mass (a BSS with many STAs) and make the STA tracking task more difficult. In contrast, changing the OTA MAC requires computation on both the AP and the STA. Some STAs may have limited computational ability, such as an IoT device with a very basic CPU and the requirement to conserve battery energy. The AP may also be limited in the number of its own OTA MACs and the STAs' OTA MACs that it can compute per unit of time. Lastly, some STAs may have strong privacy requirements (e.g., your smartphone), whereas others may have more relaxed privacy requirements (e.g., a smart light).

To accommodate all these needs, the 802.11bi protocol allows for three rotation schemes:

- An automated rotation scheme is the default. At association time (at the end of the four-way handshake), the AP agrees with the STA on the rotation scheme to use, and informs the STA of the rotation pace—that is, the duration for which an OTA MAC can be used before being changed. This duration, called an epoch, can last anywhere from one TBTT (beacon interval time) to many hours. In most cases, the epoch duration will depend on the network type, and will be configured by the local network administrator. For example, in a public network (airport, coffee shop, or other hotspot), where eavesdroppers are possible, the rotation pace may be fast, with the epoch lasting only a few seconds. In an enterprise network, where eavesdropping risks are limited, the epoch may last a few minutes or a few hours. The STA acknowledges the epoch structure and starts following the OTA MAC rotation pace.

- Sometimes the default pace might not match the STA's requirements: It might be too slow for privacy-conscious STAs, or too fast for CPU-constrained devices. In these cases, the AP can also advertise the existence of groups that apply a different rotation pace. In most public networks, it is likely that there will be a privacy-conscious group with a faster pace as well as a more relaxed-pace group. Of course, there can be many groups—as many as the administrator wants to create. The STA can learn the groups' individual epoch durations, and select one group to join instead of applying the default epoch value. For better privacy, the AP advertises (in protected frames) a count of each group's members. This information can help the STA make its decision. If the group contains many members, then the STA has a better chance of hiding in the crowd, and a slower pace may be acceptable. When all the STAs, AP and non-AP, change the OTA MAC at the same time, it is more difficult for an eavesdropper to track a STA if 100 MACs change at the same time, than if only 2 MACs change (the AP and one STA) at a given point in time.

■ If the STA does not like the default value and does not find a group whose rotation pace it likes, it can ask for an individual epoch structure. The AP can refuse the proposal if the structure contradicts the AP's policy or causes too much computation complexity for the AP. If the AP accepts, then the STA gets its own epoch structure. If two or more STAs happen to request the same individual structure, then the individual structure can become a new group.

These three schemes provide a good mix of flexibility and efficiency. However, their (intended) consequence is that the STA stops being trackable over the air, which makes troubleshooting more difficult. Notably, the infrastructure is required to store many OTA MACs for each STA, and a more advanced algorithm is required to track a STA across multiple epochs from an air capture. These schemes also mean that the STA's MAC address is very unstable. In most cases, the AP will convert the STA OTA MAC into a more stable MAC address that the AP will use to represent the STA to the wired network and its services (DHCP and others).

Summary

As the developments highlighted in this chapter suggest, the future of Wi-Fi is bright. The 802.11bn group is designing what will one day become Wi-Fi 8. Though it may not have a lot of new PHY features, it will certainly offer great improvements in terms of QoS, roaming, and the ability to move an association from one AP to the next one with literally zero delays and zero losses. Meanwhile, the improvements in the field of location and sensing open the door to new possibilities. Not only could indoor navigation become a solved problem, but your 802.11 device could have the capability to know when you are there and how you are moving in front of its screen, let you pilot functions with gestures, or simply go into deep sleep when you are not around. Moreover, now that privacy has become a central concern, your 802.11 device may be able to change its visible parameters at a fast pace, making it more challenging for an eavesdropper to know what you are doing, and complementing the 802.11 security picture.

Index

Numerics

Register Your Product at informit.com/register

Access additional benefits and save up to 65%* on your next purchase

- Automatically receive a coupon for 35% off books, eBooks, and web editions and 65% off video courses, valid for 30 days. Look for your code in your InformIT cart or the Manage Codes section of your account page.
- Download available product updates.
- Access bonus material if available.**
- Check the box to hear from us and receive exclusive offers on new editions and related products.

InformIT—The Trusted Technology Learning Source

InformIT is the online home of information technology brands at Pearson, the world's leading learning company. At informit.com, you can

- Shop our books, eBooks, and video training. Most eBooks are DRM-Free and include PDF and EPUB files.
- Take advantage of our special offers and promotions (informit.com/promotions).
- Sign up for special offers and content newsletter (informit.com/newsletters).
- Access thousands of free chapters and video lessons.
- Enjoy free ground shipping on U.S. orders.*

* Offers subject to change.

** Registration benefits vary by product. Benefits will be listed on your account page under Registered Products.

Connect with InformIT—Visit informit.com/community

 Pearson   **inform**IT*